Proceedings in Technology Transfer

Series Editors

The book series "Proceedings in Technology Transfer" provides a valuable platform for researchers, practitioners, and policymakers to share their experiences and insights related to technology transfer. Our aim is to facilitate knowledge exchange between academic researchers and industry professionals, accelerating the introduction of new technologies and driving economic growth.

Our series publishes books and conference proceedings that focus on original research findings, best practice recommendations, emerging trends, and technology evaluations. The series covers all topics which explore the transfer of technology in various industry sectors that require innovations and economic growth.

Join us in advancing the technology transfer and its impact on innovation. Publish your proceedings with us and offer insights and expertise to professionals in your field.

Manjula Siriwardhana · Sivanappan Kumar ·
Madhusanka Liyanage ·
Jayathu Gunathilake Samarawickrama ·
Iresha Udayangani Atthanayake ·
Sumudu Samarakoon
Editors

Selected Proceedings of the 3rd International Engineering Research Symposium, IERS 2025, 30 October, Colombo, Sri Lanka

Engineering Solutions for Climate Actions, Digital Transformations and Green Industries

Ministry of Education
Ministry of Technology

NERDC

Springer

Editors
Manjula Siriwardhana
National Engineering Research and Development Centre of Sri Lanka
Ekala, Sri Lanka

Sivanappan Kumar
Faculty of Engineering
Naresuan University
Phitsanulok, Thailand

Madhusanka Liyanage
School of Computer Science
University College Dublin
Dublin, Ireland

Jayathu Gunathilake Samarawickrama
Faculty of Engineering
University of Moratuwa
Moratuwa, Sri Lanka

Iresha Udayangani Atthanayake
Faculty of Engineering
University of Moratuwa
Nugegoda, Sri Lanka

Sumudu Samarakoon
Centre for Wireless Communications
University of Oulu
Oulu, Finland

ISSN 2948-2321 ISSN 2948-233X (electronic)
Proceedings in Technology Transfer
ISBN 978-981-95-7983-9 ISBN 978-981-95-7984-6 (eBook)
https://doi.org/10.1007/978-981-95-7984-6

This Springer imprint is published by the registered company Springer Nature Singapore Pte Ltd.
The registered company address is: 152 Beach Road, #21-01/04 Gateway East, Singapore 189721, Singapore

Preface

It is with great pleasure that we present the proceedings of the 3rd International Engineering Research Symposium 2025 (IERS2025), organized by the National Engineering Research and Development Centre (NERDC) of Sri Lanka. IERS2025 focuses on cutting-edge engineering research and innovations under the theme of "Engineering Solutions for Climate Actions, Digitalization and Green Industries". IERS2025 was held in Galadari Hotel, Colombo, Sri Lanka, on 30 October 2025.

This symposium is the annual engineering and research conference conducted by NERDC, and this year, the 3rd International Engineering Research Symposium 2025 (IERS 2025) was organized in partnership with several international and national universities and research institutes, namely, the 6G Flagship, University of Oulu, Finland, Netslab, University College Dublin, Ireland, The Australian National University and the Center for National Online Distance Education Service of the Open University of Sri Lanka.

IERS2025 brought the leading industrial experts, academics, researchers, and students to a common platform to explore the application of technological advancements and engineering principles to drive the digital transformation and green industrial revolution. It serves as a platform for academics, researchers, industry professionals, and aspiring students to share groundbreaking research ideas, foster collaborations, and explore the latest innovations in the field of engineering and technology while creating an environment for influential policy decisions towards sustainable development and digital transformation.

IERS2025 accepted full proceedings in multi-disciplinary engineering themes such as Chemical and Process Engineering, Climate Change Mitigation and Adaptation, Climate Smart Agriculture, Computer Science, IoT, AI, Cyber Security, Block Chain and Wireless Communication, Electrical, Electronic and Telecom. Engineering, Energy and Environmental Management, Green Infrastructure and Civil Engineering, Industrial Digitalization and Manufacturing Innovations, Mechanical and Mechatronics Engineering and Water Management and Hydrology. This book contains manuscripts of research work from many different engineering disciplines. All the manuscripts were pre-reviewed and presented in parallel sessions at IERS 2025 on 30 October 2025 at the Galadari Hotel, Colombo, Sri Lanka.

Prof. Chrishantha Abeysena, the Minister of Science and Technology, Sri Lanka, joined the symposium as the chief guest along with several other executives from the ministry. The keynote speeches were delivered by Prof. Rahula Attalage, Pro Vice-Chancellor (Academic) of SLIIT Malabe Campus, Sri Lanka, and Prof. Saman Halgamuge, FIEEE, from the University of Melbourne, Australia. Dr. Madhusanka liyanage from the University College Dublin, Ireland, and Prof. P.M.C. Thilalakerathne, Vice Chancellor of the Open University of Sri Lanka, delivered two insightful guest speeches. We wish to express our sincere gratitude to all the speakers.

Each oral presentation that was presented by the researchers was evaluated by a panel of experts in three parallel sessions under the following sub-themes.

- Engineering for Digitalization.
- Engineering Product Development.
- Engineering Solutions for Climate Actions and Green Industries.

Each session was started with an insightful speech by the session chairs, Dr. Chitra Lakshmikanthan, Dr. Iresha atthanayake, and Mr. Eranda Adikary. We appreciate the services rendered by all the panel members and moderators for their invaluable contributions in establishing an outstanding platform for knowledge exchange and fostering meaningful discussions on Engineering Innovations for sustainable development and digital Transformation.

We are equally thankful to the authors for submitting high-quality research papers. Each manuscript in this proceedings book was carefully reviewed by a panel of distinguished academic and professional experts from around the world. We deeply appreciate the significant efforts these reviewers have devoted to the process.

We are also pleased to acknowledge the guidance and support provided by the members of the advisory and editorial committees, as well as the many volunteers whose contributions were instrumental in making this event a success. Additionally, we extend our sincere thanks to the academic collaborators and partners of this symposium and the other organizations whose generous financial sponsorship played a crucial role in the success of this international conference.

These proceedings serve as a valuable resource for the advancement of engineering research and contributing to global green and digital transformations.

October 2025

Manjula Siriwardhana
Sivanappan Kumar
Madhusanka Liyanage
Jayathu Gunathilake Samarawickrama
Iresha Udayangani Atthanayake
Sumudu Samarakoon

Organization

National Engineering Research and Development Centre of Sri Lanka

Organizing Committee

Conference Chair

Manjula Siriwardhana

Conference Secretary

D. R. S. K. Wimalarathna

Programme Committee

S. A. P. S. Silva
J. K. I. P. Jayasiri
S. M. V. P. D. Senanayake
I. D. M. H. Jayantha
S. P. K. Wanasinghe

Advisory Committee

Daniel Macdonald
S. Kumar
Abdul Salam
H. C. Ambawatte

Collaborators and Academic Partners in IERS2025

6G Flagship Program, University of Oulu, Finland
NetsLab, University College Dublin, Ireland
The Australian National University, Canberra ACT, Australia
The Center for National Online Distance Education Service (C-Nodes), Open University of Sri Lanka

Contents

Engineering Solutions for Digitalization

An Assistive Vision And-Audio System for the Visually Impaired: Region-Based Object Detection, Voice Interaction, and Stereo Vision 3
S. K. Abeyweera and P. S. Abeyweera

Comparative Analysis of Convolutional Neural Networks for Automated Classification of Cucumber Leaf Diseases 14
S. T. K. Peiris and H. K. I. S. Lakmal

Design and Friction-Optimized Modeling of a Tendon-Driven Continuum Robot Arm for Confined Space Maintenance 26
W. D. Navodya and H. K. I. S. Lakmal

Design of a Self-Oscillating Flyback-Based High-Voltage Pulse Elephant Deterring System 36
Y. S. P. Weerasinghe, S. M. S. S. Senarathna, D. R. S. K. Wimalarathna, T. A. S. Anuruddha, G. D. K. Samaranayake, and J. G. Samarawickrama

Development of an Aquabot for Coral Monitoring and Marine Data Collection Using Image Processing Techniques 47
W. A. M. P. L. Adhipaththu, A. M. M. H. B. Abeysinghe, N. Y. Ranasinghe, U. V. H. Sameera, and B. L. Sanjaya Thilakarathne

Development of Defect Reducing Mechanisms for Paper Bag Manufacturing Using Machine Vision System with YOLOv8 Model 58
Himasha Shanuka and Roshan Thilakarathne

Dynamic Modeling and Simulation of R-Hex Hexapod Robot for Efficient Locomotion in Challenging Terrains 73
W. C. Nirmal, K. J. P. Fernando, H. K. I. S. Lakmal, M. W. P. Maduranga, P. N. M. Kumara, and S. L. Kulatunga

Multi Directional Stabilizing Active Wing 89
W. A. D. D. Ransika Weerakkody and Geethal Siriwardana

Speaker-Consistent Speech Enhancement with Auxiliary Conditioning 103
Thumula Rajakaruna and H. K. I. S. Lakmal

Engineering Product Development

Design and Development of Magneto Rheological (MR) Fluid Based Soft Gripper 113
R. Gangadevi, B. C. Mohana Gokul, A. Nikhil, Kalluru Snehitand, and S. Senthil Raja

Design and Implementation of a C_{10} & C_{20} Capacity Tester for Lead Acid Battery Discharge Evaluation 121
T. Chandana Peiris, L. N. Rathnayake, S. P. R. Jayasinghe, and J. P. D. S. Athuraliya

Design and Simulation of a Corrugated Tube Shell and Tube Heat Exchanger for an Ethanol Production Line 131
W. K. P. G. N. Charuni and K. T. K. M. De Silva

Development and Evaluation of an Experimental Dual Chamber Hot Air Dryer for Drying Turmeric 140
J. M. H. B. Dewanthi, P. W. A. W. Pathiraja, A. L. K. Tharaka, U. G. A. S. Kumara, T. K. K. S. Pathmasiri, and K. T. K. M. De Silva

Development of a Point Absorber Type Wave Energy Converter for Southern Sri Lanka Wave Conditions 150
K. G. S. M. Rathnapriya, R. G. P. D. Rankothge, R. R. Gamage, W. A. D. Sampath, I. H. D. Senevirathne, and H. C. P. Karunasena

Development of a Theoretical Model to Predict Filtration Efficiency of Electret Based Masks 161
W. M. I. S. Jayaweera and D. Attygalle

Development of Multi Level Transformerless Inverter for Grid Integration of Solar Photovoltaic Power Plant 171
Rattan Kumar Venkatesan and Chitra Lakshmikanthan

Evaluation of Rice Husk Ash as a Reinforcement Material for the Development of Soil-Based Roofing Sheet 186
M. S. H. Karunanayake, S. M. V. P. D. Senanayake, T. N. Fernando, A. J. Fernando, and N. P. A. K. G. Pramuditha

Root Cause Analysis and Mitigation of Defects on Handle Attachment in Porcelain Products Using 6σ-DMAIC Methodology 196
Vihanga Chandrasekara and Roshan Thilakarathne

Smart Bike Guardian: Accident Notification and Theft Detection System 208
S. D. Marasinghe, P. K. A. P. Fernando, P. T. S. Silva, and R. M. H. K. Rathnayaka

Engineering Solutions Towards Climate Actions and Green Industries

Anaerobic Co-Digestion of Cheese Whey with Garden Waste: Effect of Mass Ratio and Garden Waste Particle Sizes 225
S. M. U. S. T. J. A. Subasinghe, S. P. H. E. Dharmawickrama, and S. M. W. T. P. K. Ariyarathna

Analysis of Emissions from Diesel and Electric Buses: Towards a Sustainable Public Transport System 235
T. Thilakshan, A. G. T. Sugathapala, and J. M. S. J. Bandara

Concepts of Circular Economy to the Sustainable Development of the Sri Lankan Construction Industry 250
U. M. Samararathne, I. U. Atthanayake, and W. N. Kawmudi

Design Methodology for Shoreline Protection in Sarawak Shoreline Using Rubble-Mound Breakwater 264
P. A. Ratnasinghe and K. K. Kuok

Developing a National Safety Standard in Using Flammable/Toxic Refrigerants in HVAC/R Applications in Sri Lanka 280
C. H. Kannangara, J. B. Senarath, and W. P. D. Fernando

Development of an Oil-in-Water Emulsion Model for Ocean Oil Spills in Tropical Conditions 290
Lalani Kaluarachchi, Ajith de Alwis, and Indrajith D. Nissanka

Investigate the Volumetric Heat Transfer Coefficient for the Rotary Drying of Coir Pith 300
S. A. P. S. Silva, H. K. A. Mahela, S. Sarmilan, A. D. U. S. Amarasinghe, and M. Narayana

Modeling of Thermal Zones and Heat Transfer in BFB/CFB Incineration of Sludge 311
K. A. C. G. Siriwardhana, K. R. R. Mahanama, I. U. Atthanayake, and D. H. G. S. R. Somasundara

Numerical Investigation of $MA_3Bi_2I_9$-Based Lead-Free Perovskite Solar Cells: Impact of ETL Selection and Interface Band Alignment 322
W. G. A. Pabasara and G. A. Sewvandi

Numerical Simulation and Analytical Calculation of Heat Transfer in a Roof Assembly with Coir-Fiber Reinforced Natural Rubber Composite Insulation 332
W. S. A. Sudarshana, L. K. T. Srimal, and R. Gallage

Urban Air Quality Mapping Using GIS: A Case Study of Gampaha District, Sri Lanka 341
K. P. D. D. Jayasekara and H. I. D. Liyanage

Author Index 353

About the Editors

Dr. Manjula Siriwardhana is attached to the National Engineering Research and Development Centre of Sri Lanka as a Principal Research Engineer. She earned her PhD from the Australian National University (ANU), Canberra, ACT, Australia. Her research interests include renewable energy, climate change, energy management, energy project evaluation, GHG quantification, climate resilient green growth and Carbon neutrality.

Prof. Sivanappan Kumar is with the Faculty of Engineering, Naresuan University, Thailand. He is also an Emeritus Professor at the Asian Institute of Technology (AIT), Thailand. With over three decades of academic experience, he has been involved in teaching, research, training, and outreach in the fields of renewable energy, climate change, greenhouse gas mitigation and sustainable development. He has held several senior academic and administrative positions. His research interests include renewable energy resource assessment, solar thermal and photovoltaic technologies, energy efficiency in buildings and industries, greenhouse gas mitigation technologies, low carbon and smart cities, and Sustainable Development Goals.

Dr. Madhusanka Liyanage is an Associate Professor/Ad Astra Fellow at University College Dublin, Ireland, and an Adjunct Professor at the University of Oulu, Finland. He has held visiting positions at leading institutions, including the University of Oxford, Lancaster University, UK, Sorbonne University, France, University of New South Wales (UNSW), University of Sydney, and CSIRO, Australia. His research interests are in the areas of cybersecurity, 6G networks, and privacy-preserving AI.

Dr. Jayathu G. Samarawickrama is a Senior Lecturer in the Department of Electronic & Telecommunication Engineering, Faculty of Engineering, University of Moratuwa, Sri Lanka. He obtained his Ph.D. degree in the field of humanoid robotics from the University of Genova and the Italian Institute of Technology in Italy. He worked as a Postdoctoral Fellow at the University of Genova, Italy, for one year on a European Union-funded project. Currently, he is a Director General at the at National Engineering Research and Development Centre of Sri Lanka, located in the Industrial Estate, Ekala, Ja Ela, Sri Lanka. His research interests are in the areas of Humanoids/Robotics, Electronic Systems, FPGA/DSP/GPU based systems, Embedded Systems, Computational Vision, Stereo Vision, Vision Heads.

Dr. Iresha Udayangani Atthanayake is a Senior Lecturer in Mechanical Engineering at the Open University of Sri Lanka. She holds a PhD in Fluid Mechanics from the University of Warwick, specializing in rotating flow and particle image velocimetry. With extensive research experience, her work focuses on computational fluid dynamics, thermodynamics, and sustainable energy solutions.

Dr. Sumudu Samarakoon received his Ph.D. in communications engineering from the University of Oulu, Finland. He is currently an Assistant Professor with the Centre for Wireless Communications, University of Oulu. His main research interests are in heterogeneous networks, radio resource management, machine learning at the wireless edge, and game theory.

Engineering Solutions for Digitalization

An Assistive Vision And-Audio System for the Visually Impaired: Region-Based Object Detection, Voice Interaction, and Stereo Vision

S. K. Abeyweera and P. S. Abeyweera(✉)

Department of Mechanical Engineering, Faculty of Engineering, University of Peradeniya, Kandy, Sri Lanka

pamoda.abeyweera@eng.pdn.ac.lk

Abstract. This study presents a novel assistive system for individuals with visual impairments, designed to enhance independent mobility by describing the surrounding environment. The system integrates real time voice guidance, stereo vision-based distance estimation and object recognition. Objects are localized using a 3×3 grid camera segmentation. The distance estimation is done through a calibrated stereo vision setup. YOLOv8 is used for fast and accurate object detection and Vosk is incorporated into offline voice recognition to interpret user commands. Depth estimation was achieved using two iPhone 7 cameras, with a fixed baseline, and a polynomial regression model correlating pixel disparity with real-world distances. The system demonstrated a root mean square error of 122 mm, with 85% of estimations falling within a 5% tolerance margin. On the experimental hardware, the system achieved an average response time of 368.5 ms, indicating further optimization is needed for real-time use. While indoor testing yielded reliable results, outdoor performance was affected by dataset limitations. The results suggest significant potential for real-world deployment in wearable devices, contributing to independence for the visually impaired.

Keywords: Assistive Technology · YOLOv8 · Spatial zone Object Detection · Stereo Vision

1 Introduction

According to World Health Organization's 2023 statistics, around 2.2 billion people worldwide live with some form of vision impairment [1]. For many, vision impairment can be managed with medication or lenses, but others face complete blindness and major navigation challenges. Over the years, many research efforts have been made to address this global issue.

The evolution of navigation aids has progressed from simple tools like the white cane and guide dogs to electronic sensor-based systems and, most recently, AI-powered applications [2]. Today, the most significant advancements are driven by the collaboration between artificial intelligence and the powerful sensors in modern smartphones. In the context of blind navigation, computer vision allows devices to process images or videos,

M. Siriwardhana et al. (Eds.): IERS 2025, PTT, pp. 3–13, 2026.
https://doi.org/10.1007/978-981-95-7984-6_1

identify objects, and extract information. This functionality is often facilitated by voice-activated systems, which leverage natural language processing and speech recognition to enable seamless, hands-free communication between the user and the device.

However, despite these individual technological achievements, a significant gap exists in current assistive technologies, which fail to holistically integrate object identification, voice feedback, and distance estimation into a single, seamless system. Therefore, the primary objective of this research is to develop, and evaluate a prototype object detection system that addresses these limitations. The objectives include creating a system that uses computer vision, object detection, and machine learning to estimate distance and provide voice guidance. Furthermore, it evaluates and addresses any issues with the system to improve its performance.

2 Literature Review

Advancements in assistive technologies for the visually impaired have evolved from traditional tools like white canes to sensor based smart systems. Recent research has deeply focused on mobile and wearable systems as they offer a platform for real-time assistance.

Several mobile applications illustrate this technological shift. One of the most prominent is Seeing AI, a mobile app from Microsoft that uses computer vision to narrate surroundings, read text, recognize people, and even detect emotions [3]. Furthermore, it incorporates voice commands for hand-free interaction. However, despite its versatility, it remains cloud-dependent and does not support robust distance estimation, limiting functionality in areas with poor connectivity. Similarly, BlindSquare combines GPS with FourSquare data to provide environmental cues via audio but relies on third-party screen readers for accessibility, which can introduce usability challenges [4].

When it comes to wearable solutions, OrCam MyEye offers an integrated compact camera with an on-device machine learning recognition to identify text, objects, and faces, offering offline operation and real-time audio feedback [5]. However, its limited field of view and lack of spatial awareness. On the other hand, Aira pairs users with remote human agents via live video streaming and GPS, offering personalized guidance in navigation, reading, and other visually dependent tasks [6]. Yet, its dependence on constant connectivity and human resources raises concerns about scalability and independence.

In academic literature, similar trends are reflected. Research has highlighted the growing popularity of integrating computer vision into assistive tools. Deep learning models like YOLO have proven highly effective in identifying objects in real-time [7]. In addition to that, stereo vision techniques are being used to estimate distances by mimicking human binocular vision [8] which is particularly important for short-range navigation More recently, LiDAR-based systems have been explored for precise depth sensing, but their cost, size, and power demands restrict integration into lightweight, wearable aids [9]. In parallel, lightweight mobile solutions using TensorFlow Lite have shown promise development of an offline Android app that achieved ~ 90% accuracy in object detection and converted results into audio feed-back, demonstrating feasibility without cloud reliance [10]. The field is now advancing toward vision-language models

(VLMs) that integrate perception with reasoning for richer environmental understanding [11].

In terms of interaction, offline automatic speech recognition (ASR) has been a critical focus. While tools like Vosk have offered privacy-preserving offline options, newer models such as Whisper provide higher accuracy and multilingual sup-port even in low-resource conditions [12, 13].

Despite these advances, current solutions remain fragmented. Many systems focus on either object recognition, distance measurement, or voice control, but rarely all three in a fully integrated. Moreover, while many commercial tools describe objects within the camera's view, they rarely segment the scene into spatial zones (e.g., left, center, right regions of the frame).

This research, therefore, addresses this gap by proposing and developing a multi-modal system that holistically combines these three pillars of assistive navigation to create a more comprehensive and user-centric tool.

3 Methodology

The methodology is structured around four modules, Object Detection, Voice Recognition, Voice Feedback, Distance Estimation and their central implementation is as in Fig. 1. This section provides a detailed explanation of each of these components, outlining the methodological approach employed in the study.

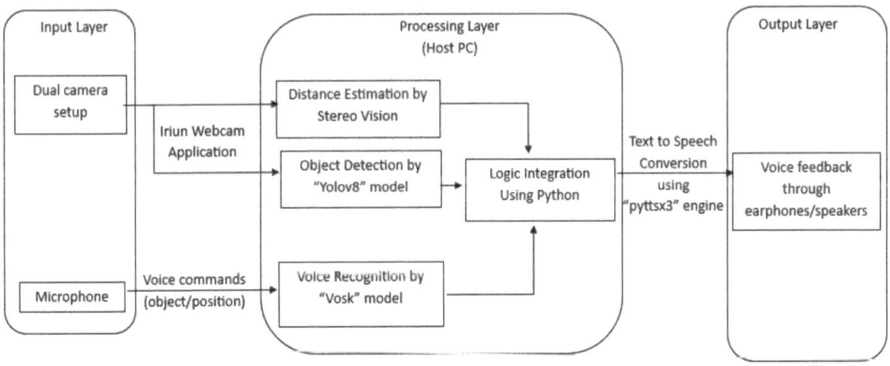

Fig. 1. Methodological approach

3.1 Object Perception Module

The first module focuses on leveraging pre-trained models to identify objects in the environment. This involves the integration of computer vision algorithms that not only detect objects but also provide spatial information.

Object Detection Algorithm. The choice of algorithm for object detection plays a pivotal role in determining the effectiveness of the model. Several pre-trained object detection algorithms are available, and their selection involves considerations such as size, accuracy percentage, and the number of identified objects. Among the YOLO architectures, YOLOv8 was selected as it represents the current state-of-the-art iteration, introducing an anchor-free detection head and a more efficient backbone network that enhance both performance and ease of use. [14]. To determine the optimal variant for this specific application, a comparative analysis was conducted within the YOLOv8 series (n, s, m, l, x), evaluating them on the COCO128 dataset. As shown in Table 1 and Fig. 2, smaller models (e.g., n, s) offer low latency but significantly reduced accuracy (mAP), while larger models (l, x) achieve high accuracy at the cost of increased computational demand. For this prototype, the primary objective was to maximize detection reliability to ensure user safety and system credibility. Therefore, the YOLOv8l (large) model was selected.

Table 1. Performance comparison of different YOLOv8 model on the COCO128 dataset

Model	Latency(ms)	mAP@0.5	mAP@0.5:0.95
yolov8n.pt	41.56	0.607	0.448
yolov8s.pt	85.21	0.760	0.588
yolov8m.pt	188.76	0.783	0.611
yolov8l.pt	355.14	0.829	0.659
yolov8x.pt	537.48	0.829	0.667

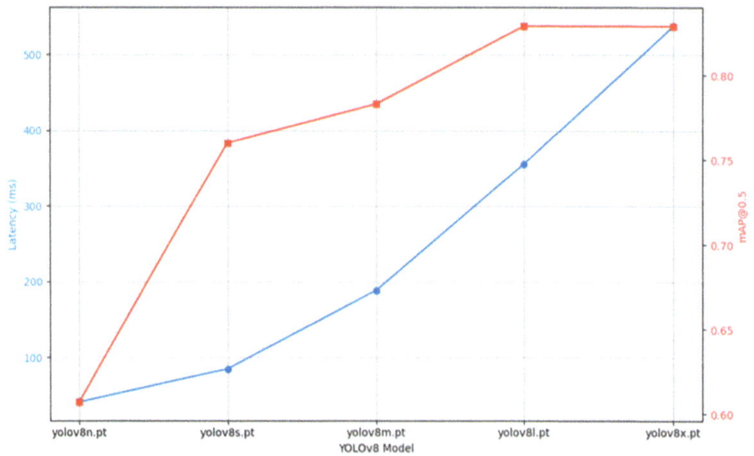

Fig. 2. Latency (ms) and mean Average Precision (mAP@0.5) Comparison of different YOLOv8 models (n, s, m, l, x).

Camera View Segmentation. Standard object detection provides a class label ad a bounding box but lacks relational information. To overcome this, a key innovation for the system is the strategic division of the camera's visual field into a 3 × 3 grid. It creates nine distinct zones: top Left, up, Top Right, Left, Center, Right, Bottom Left, Down, and Bottom Right as in Fig. 3. When an object is detected, the system calculates the centroid coordinates (cx,cy) of its bounding box. These coordinates are then mapped to one of the nine zones based on pre-defined pixel boundaries of the 640 × 480 resolution image. For example, a centroid with coordinates falling within range $210 < cx < 430$ and $160 < cy < 320$ would be classified as being in the "center" zone. This simple effective method transforms raw positional data into in-built, understandable spatial language, significantly enhancing the user experience.

Top Left	Up	Top Right
Left	Center	Right
Bottom Left	Down	Bottom Right

Fig. 3. Camera view index for spatial segmentation

Object Identification with Spatial Zones. In this step, previous dual approaches are combined. For each detected object that surpasses a confidence threshold, the system combines the class name provided by YOLOv8l with the calculated spatial zone. This creates a unified, descriptive data point, such as, {'class': 'cup', 'spatial_zone': 'left', 'confidence': 0.96}. This structured information serves as the primary input for the system's decision-making and feedback generation logic. Through this integrated approach, the blind navigation system provides users with detailed information about the objects in their surroundings as in Fig. 4.

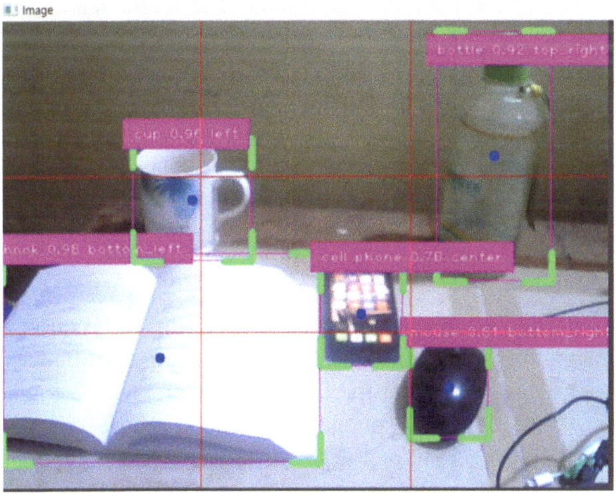

Fig. 4. Object Identification with the spatial segmentation

3.2 Voice Command and Speech Synthesis

In this module, the focus shifts to the implementation of voice recognition technology, with interpreting auditory feedback. This allows users to seamlessly interact with the navigation system.

Voice Recognition Model Selection. For interpreting user commands, the 'Vosk' voice recognition toolkit was selected. This decision was driven by three key factors critical to our prototype's design, which are offline functionality, Python compatibility, and low computational overhead. Unlike cloud-based APIs or larger modern frameworks like OpenAI's Whisper, which often require significant resources or complex deployment pipelines, Vosk operates entirely offline, ensuring user privacy and operational reliability without an internet connection. After a thorough comparison of available models, the 'vosk-model-en-us-daanzu-20200905 lgraph' model was chosen as the ideal choice [12].

Speech Synthesis. Detected objects and distances are converted into structured text descriptions. To provide spoken feedback, the system uses the pyttsx3 Python library, which converts these text descriptions into clear, audible speech. The speech rate was configured to 200 words per minute to balance clarity, ensuring effective verbal guidance for visually impaired users.

3.3 Distance Estimation

This module is responsible for calculating the distance to detected objects, adding the critical third dimension to the system's environmental understanding.

Method Selection. After evaluating various technologies, stereo vision was selected as the suitable method for this prototype due to its ability to provide depth information at a relatively low cost and computational overhead. The fundamental principle of stereo vision is triangulation. By capturing simultaneous images from two cameras separated by a known horizontal distance (the baseline, B), the system can measure the disparity (δ), which is the difference in the horizontal pixel position of a corresponding point in the two images. the depth, or distance (D), to that point is inversely proportional to the disparity. Depth D can be obtained as in Eq. (1), where f is the focal length of the camera lens [15]. The principle of stereo vision triangulation is shown in Fig. 5.

$$D = fB/\delta \tag{1}$$

Calibration. Calibration is an essential step in ensuring the accuracy of the chosen stereo vision method. The stereo rig was constructed using two iPhone 7 smartphones, chosen for their consistent camera properties. The cameras were mounted with a fixed baseline of 141 mm (the length of the phone). Both cameras have a focal length of 28 mm and were set to a resolution of 640 × 480 pixels to ensure consistent data capture.

Instead of performing a full intrinsic–extrinsic calibration (which equires detailed parameters such as distortion coefficients and optical centers), a practical calibration approach was adopted. A dataset of 60 points was collected by placing an object at known

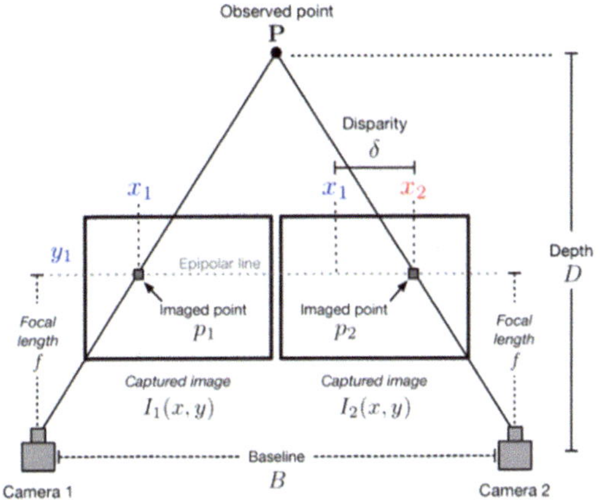

Fig. 5. The principle of stereo vision triangulation [15]

real-world distances (from 500 mm to 6000 mm) and measuring the corresponding pixel disparity of its centroid between the left and right camera views.

The iPhone 7 uses a rolling shutter, and the two phones were not hardware-synchronized. Consequently, frame capture times can slightly differ between the cameras. To mitigate potential errors due to this, only static objects were considered in the experiments. Fast-moving objects were deliberately excluded, as rolling shutter distortions could lead to inaccurate depth estimation.

$$Distance = 519600 \times \left(\frac{1}{Disparity}\right)^2 + 73980 \times \left(\frac{1}{\text{Disparity}}\right) - 17.37 \quad (2)$$

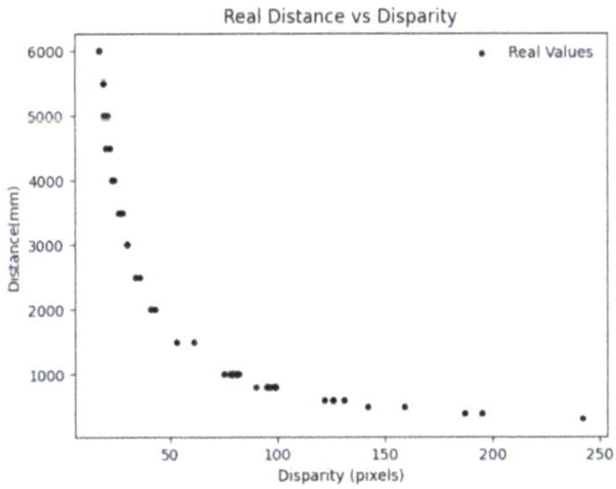

Fig. 6. Measured relationship between real-world distance and pixel disparity, showing the expected hyperbolic trend

Although Eq. (1) theoretically defines a linear relation between distance and disparity, its accuracy depends on precise intrinsic/extrinsic parameters, which are difficult to obtain with smartphones. To mitigate these uncertainties, experimental validation was performed. Initial plotting of real distance versus disparity revealed the expected non-linear hyperbolic relation (Fig. 6). When the reciprocal of disparity was considered, the data showed an approximately linear trend (Fig. 7). Curve-fitting techniques were then applied, and a second-degree polynomial function was identified as the optimal model to relate the reciprocal of disparity to the real-world distance, as expressed in Eq. (2). While this polynomial model is specific to the current configuration and does not replace standard calibration methods, it ensured consistent and reliable depth estimation within the scope of this study.

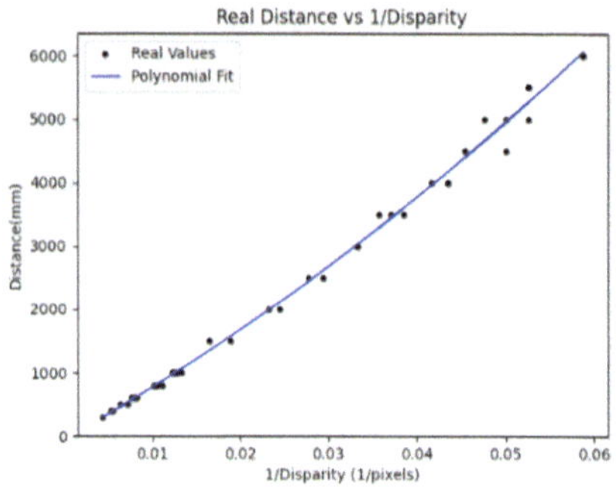

Fig. 7. The graph between real-world distance and the reciprocal of disparity ($1/\delta$), demonstrating a linear trend used for calibration.

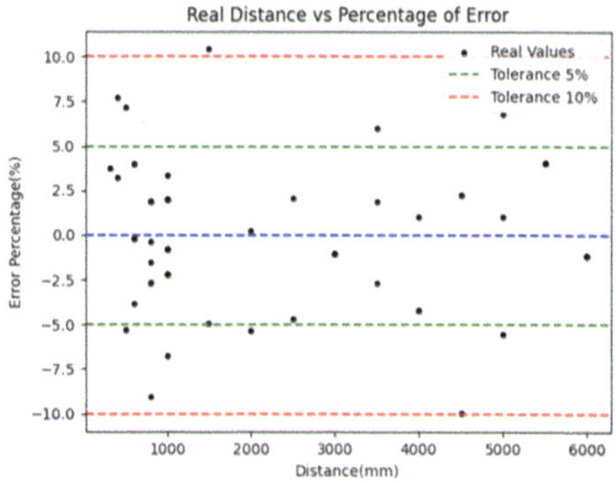

Fig. 8. The graph between the real distance vs percentage of error.

To validate the effectiveness of the established polynomial function, an additional dataset was collected through experimentation. The model's predictions were compared against the known real distances (Fig. 8). The validation yielded a root mean square error (MSE) of 122 mm. Furthermore, 98.5% of the validation readings fell within a 10% tolerance of the actual distance, and an impressive 85% fell within an even tighter 5% tolerance range.

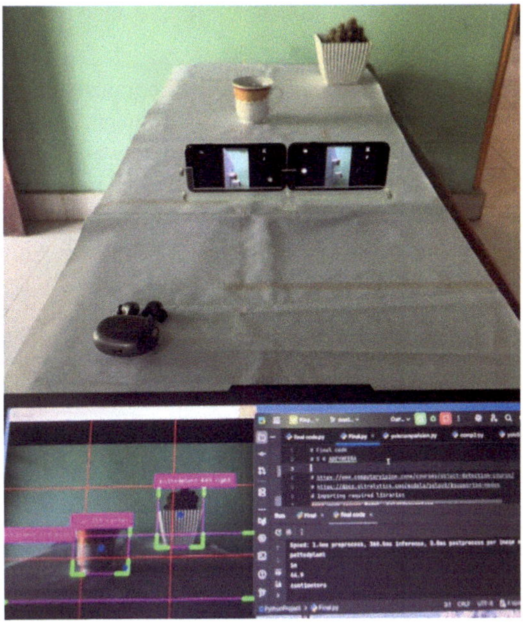

Fig. 9. Holistic System with two Iphones with holder, host PC and headset forget the voice feedback

3.4 Implementation of Vision, Voice, and Distance Estimation

The final step is the integration of these modules into an operational workflow as illustrated in the holistic system (Fig. 9). The system is designed as a centralized processing unit running on a host PC, which manages all computation-intensive tasks. The two iPhone 7 cameras are mounted on a rigid bar to maintain a fixed baseline of 141 mm. They function solely as video capture devices. Their video feeds are streamed to the host PC in real-time using the Iriun Webcam app, which allows the smartphones to be used as wireless external cameras. The primary audio input (microphone) and output (speaker) are also handled by the host PC Object detection is executed using the YOLOv8l model on the primary camera's frame. For each detected object, the system records its class name, confidence score, and centroid coordinates, and maps it to its corresponding view zone (e.g., 'top-left', 'center'). The system compares the user's voice command to the list of detected objects and their zones. For example, if the user says "laptop," the system searches for an object with the class name "laptop." If the user says "left," it searches for any object in the "left" zone. Once a match is found, the system identifies the centroid of that same object in the secondary camera's frame. It calculates the horizontal pixel disparity (δ) between the centroids from the two frames. This disparity value is then fed into the calibrated polynomial function to calculate the distance to the object. Finally, the system synthesizes this information into a spoken sentence using text-to-speech and delivers the result to the user. For instance, when a user questions the system by stating "center" through the microphone, the model accurately responds by announcing the presence of a smartphone located 2.45 m away through the speaker. Similarly, in another scenario where the user specifies "laptop," the model communicates that the laptop is situated in the top-left direction at 654.98 cm.

4 Results and Discussion

The implemented prototype model evaluated under three considerations such as distance estimation accuracy, system responsiveness and environmental adaptability.

The stereo vision-based distance estimation module was evaluated against ground truth measurements. The system achieved a root mean square error (RMSE) of 122 mm across the tested range. Furthermore, 98.5% of the validation readings fell within a 10% tolerance of the actual distance, while 85% of the readings fell within a 5% tolerance. These results confirm that the proposed method provides high accuracy for near- and mid-range distances. However, as expected from stereo disparity geometry, error increases at longer distances since pixel disparity decreases hyperbolically. This indicates that the system is best suited for near-field obstacle detection, such as indoor navigation or room-scale assistance.

System responsiveness was analyzed by measuring total end-to-end latency, from frame capture to audio output. The majority of this latency was attributed to YOLOv8l inference time. On an Intel i7 9th Gen CPU (without dedicated GPU), the total latency was measured at 1100 ms per frame. However, when tested on a Ryzen 7 system equipped with an NVIDIA RTX 4050 GPU, the latency was significantly reduced to 368 ms per frame. This demonstrates that while the model is computationally demanding, performance improves considerably with more powerful hardware acceleration.

The system was tested in both indoor and outdoor environments. The model demonstrated significantly higher object detection accuracy in indoor settings compared to outdoor scenarios. Indoors, common objects like chairs, laptops, cups, and books were consistently identified. Outdoors, performance was less reliable with lower consistency. This performance difference can be directly attributed to the training data of the YOLOv8l model. The COCO dataset is biased towards indoor objects and controlled environments. Although no quantitative metrics were measured for the environmental tests, the qualitative evaluation suggests the need for training on a custom dataset with diverse outdoor scenarios to improve adaptability.

5 Conclusions

This research successfully implemented and evaluated a novel assistive navigation aid for individuals with visual impairments. The project's primary contribution is the demonstration of a viable concept that holistically integrates spatial-aware object detection, interactive voice control, and distance estimation, which addresses a documented gap in the literature review. This prototype was deliberately built using accessible hardware and a flexible, Python-based framework to validate the core integration concept and establish a foundation for future development.

The evaluation yielded valuable insights. The system demonstrated high efficiency in indoor environments and achieved quantitatively validated accuracy in short-range distance. The identified limitations, including reduced outdoor performance, decreased accuracy at long distances, and high processing latency provide a clear roadmap for future development.

The next step is to address the challenges of outdoor performance and latency. This can be achieved by moving away from the large, general-purpose YOLOv8l model. Furthermore, the dataset should include a wide variety of outdoor objects in diverse lighting conditions. To overcome the inherent range limitations of stereo vision, future development should incorporate sensor fusion. A pivotal advancement will be the transition from a general-purpose prototype to a native iOS application. This will unlock access to sophisticated hardware like LiDAR scanners and optimized software APIs (e.g., ARKit for depth sensing, on-device Speech frame-work), thereby providing more robust depth perception, significantly lower latency, and a path toward a truly wearable, self-contained solution.

No matter how advanced, technology is only effective if it is practical and bene-ficial to its audience. A critical next step is to conduct formal user studies with members of the visually impaired community. By continuing to refine the system along these proposed directions, there is a strong potential to create a tool that can enhance the independence and mobility of individuals with visual impairments.

References

1. World Health Organization: Blindness and vision impairment: https://www.who.int/newsroom/fact-sheets/detail/blindness-and-visual-impairment. Last accessed 25 June 2025
2. Bleau, M., Paré, S., Djerourou, I., Chebat, D.R., Kupers, R., Ptito, M.: Blindness and the reliability of downwards sensors to avoid obstacles: a study with the EyeCane. Sensors **21**(8), 2700 (2021). https://doi.org/10.3390/s21082700
3. Microsoft: Seeing AI, https://www.microsoft.com/en-us/garage/wall-of-fame/seeing-ai/. Last accessed 25 June 2025
4. The Blind Square, What is BlindSquare?, https://www.blindsquare.com/about/. Last accessed 25 June 2025
5. Cohen, R.: Experience the Power of Assistive Technology with OrCam's AI Devices, https://www.orcam.com/en-ie/home. Last accessed 26 June 2025
6. Aira: Home, https://aira.io/. Last accessed 26 June 2025
7. Encord: YOLO models for Object Detection Explained [Yolov8 Updated]
8. Molton, N., et al.: A stereo vision-based aid for the visually impaired. Image Vis. Comput. **16**(4), 251–263 (1998). https://doi.org/10.1016/s0262-8856(97)00087-5
9. Okolo, G.I., et al.: Assistive systems for visually impaired persons: challenges and opportunities for navigation assistance. Sensors **24**(11), 3572 (2024). https://doi.org/10.3390/s24113572
10. Salunkhe, A., et al.: Android-based object recognition application for visually impaired. ITM Web of Conferences **40**, 03001 (2021). https://doi.org/10.1051/itmconf/20214003001
11. Baig, A., et al.: AI-based Wearable Vision Assistance System for the Visually Impaired: Integrating Real-Time Object Recognition and Contextual Understanding Using Large Vision-Language Models, Computer Vision and Pattern Recognition, Cornell University (2024). https://www.researchgate.net/publication/387540237_AI-based_Wearable_Vision_Assistance_System_for_the_Visually_Impaired_Integrating_Real-Time_Object_Recognition_and_Contextual_Understanding_Using_Large_Vision-Language_Models
12. Alphacephei, VOSK Offline Speech Recognition API, https://alphacephei.com/vosk/. Last accessed 26 June 2025
13. Yu, M.: SeeSay: An Assistive Device for the Visually Impaired Using Retrieval Augmented Generation (2024). https://arxiv.org/html/2410.03771v1. accessed 03 Sep. 2025
14. Ultralytics: YOLOv8 vs. YOLOX: A Technical Deep Dive, https://docs.ultralytics.com/compare/yolov8-vs-yolox/. Last accessed 28 June 2025
15. Diaz-Ramirez, V.H., Gonzalez-Ruiz, M., Kober, V., Juarez-Salazar, R.: Stereo image matching using adaptive morphological correlation. Sensors **22**, 9050 (2022). https://doi.org/10.3390/s22239050

Comparative Analysis of Convolutional Neural Networks for Automated Classification of Cucumber Leaf Diseases

S. T. K. Peiris(✉) and H. K. I. S. Lakmal

Faculty of Engineering, NSBM Green University, Homagama, Sri Lanka
stkpeiris@students.nsbm.ac.lk

Abstract. Plant diseases have significant impacts on crop yields and therefore need accurate and timely distinguishing information. This study compares eight Convolutional Neural Network (CNN) models. They are EfficientNetB0, EfficientNetB4, EfficientNetB7, ResNet50, VGG16, MobileNet, MobileNetV3Large, and InceptionV3, to differentiate between six classes of cucumber leaf diseases (Anthracnose, Bacterial wilt, Downy mildew, Fresh (healthy), Gummy stem blight, and Powdery mildew). An in-house dataset of cucumber leaf images was augmented with rotation, flipping, and scaling. The models were trained and tested based on accuracy, precision, recall, F1-score, cost types of parameters, and floating-point operations per second. As a result, ResNet50 achieves the highest accuracy of 93.06% and an F1-score of 92.98%, while MobileNetV3Large offers a trade-off between performance of 79.86% accuracy and cost, 3.12M parameters. The findings determine tradeoffs between computational cost and performance, providing advice for the design of scalable diagnostic tools for precision agriculture.

Keywords: CNN models · Cucumber plant · Deep Learning · Leaf Diseases · Transfer Learning

1 Introduction

Plant diseases pose a significant threat to world agriculture, with adverse effects on crop quality and yield, and causing huge economic losses [1]. Early detection plant diseases are necessary for implementing early interventions, reducing financial losses, and improving crop quality. Traditionally, disease identification is done by physical inspection of plant leaves, which is time-consuming, labor-intensive, and subject to human inaccuracies, being costly and inefficient [1]. Cucumbers (Cucumis sativus), an economically and nutritionally important crop largely grown on a large scale, are highly susceptible to more than 30 diseases throughout their growth period, of which common leaf diseases like anthracnose, downy mildew, powdery mildew, bacterial wilt, and gummy stem blight are particularly widespread [2]. The multiplicity and heterogeneity of diseases require high-tech diagnostic equipment to facilitate their quick and correct identification.

M. Siriwardhana et al. (Eds.): IERS 2025, PTT, pp. 14–25, 2026.
https://doi.org/10.1007/978-981-95-7984-6_2

The latest advancements in artificial intelligence (AI), particularly deep learning, have revolutionized the development of plant disease diagnosis automated systems. Convolutional Neural Networks (CNNs) have demonstrated exceptional performance in image-based classification issues with their capacity to deliver faster and more accurate results than traditional methods. In the past decade, a few traditional machine learning models have been proposed for plant disease detection and classification [3]. But deep models, particularly CNNs, have sur-passed these approaches because they are able to learn sophisticated features from difficult image data without extensive manual feature design.

This work is focused on the comparative evaluation of several CNN architectures to classify cucumber leaf disease. Specifically, the research analyzes the accuracy of EfficientNet B0 [4], EfficientNet B4 [5], EfficientNet B7, ResNet50 [6], VGG16 [7], MobileNet, MobileNetV3Large, and InceptionV3 to classify cucumber leaf images into anthracnose, bacterial wilt, downy mildew, fresh (healthy), gummy stem-blight, and powdery mildew. To enhance model robustness and address limited dataset problems, data augmentation techniques like random rotations, vertical flips, and horizontal flips were applied [8]. The proposed system employs CNN models, which include several convolutional layers, to extract useful features for correct classification [8]. The models are trained and tested on a dataset of cucumber leaf images, and measures such as accuracy, precision, recall, and F1-score are used to measure their performance.

The primary objective of this study is to identify the optimal CNN architecture for cucumber leaf disease classification regarding aspects such as classification accuracy, computational cost, and scalability for practical applications. To achieve this, eight state-of-the-art CNN models, EfficientNet B0, EfficientNet B4, EfficientNet B7, ResNet50, VGG16, MobileNet, MobileNetV3Large, and InceptionV3, were singled out and assessed. This was done with the intention of covering a wide range of architectural designs. Lightweight models, such as MobileNet and MobileNetV3Large, were adopted due to their small parameter size and suitability for edge device deployment. On the opposite end of the spectrum, deeper models like ResNet50 and InceptionV3 were selected for their greater accuracy on complicated classification problems. The EfficientNet models were included for their modern approach to compound scaling, as they offer a middle ground between performance and efficiency. VGG16 functions as the classical baseline, as it has been used extensively in older works. Considering both accuracy and real-world efficiency, the study sheds light on these architectures concerning agricultural applications by testing all of them in the same environment. Section 2 reviews related work on deep learning models. Section 3 discusses the dataset, preprocessing techniques, and deep neural network model utilized in this work. Section 4 describes the experimental setup and procedure. Section 5 describes the results and implications and concludes the research and outlines avenues for future work [3].

The key contributions of this study are as follows,

- A new set of cucumber leaf images were created, the samples of each disease carefully balanced to include diversity of anthracnose, bacterial wilt, downy mildew, fresh (healthy), gummy stem blight, and powdery mildew, providing a robust foundation to train and evaluate models.

- The CNN architectures EfficientNet B0, EfficientNet B4, EfficientNet B7, ResNet50, VGG16, MobileNet, MobileNetV3Large, and InceptionV3 were personally trained and tuned on the in-house dataset to enhance their performance for accurate cucumber leaf disease classification.
- This study encompasses a thorough comparison of the selected CNN models based on their classification rate, computational efficiency, and scalability, and offers hints about the best architecture for field-level agricultural applications.

2 Literature Review

A plethora of deep learning models to classify plant disease has been considered. For instance, [5] provides an interesting comparative study between several CNN architectures, Alene, GoogleNet, InceptionV3, ResNet18, and ResNet50, specifically for the classification of diseases on tomatoes with 10 different types of disease classes. The paper laid an emphasis on benchmarking the performances across architectures, with a special focus on how deeper networks, such as Res-Net50, perform better than shallower ones because of better feature extraction.

Similarly, [3] evaluated EfficientNet, AlexNet, ResNet50, VGG16, and InceptionV3 on a notably wider dataset of 38 disease classes over 14 plant species. The study highlights EfficientNet's scalability and parameter efficiency, which makes it suitable for large-scale agricultural disease detection. The results showed that lightweight yet deep models like EfficientNet deliver good accuracy while maintaining low computational costs.

Another study [1] involved cucumber leaf disease classification with InceptionV3 and VGG19 into five classes of diseases. They highlight that careful free feature selection, along with an optimized convolutional architecture, greatly improves the efficiency of multiclass classification. This study also underlines the fact that model framework optimization can lead to better datasets and higher accuracy in classification.

Expanding on these ideas, our work looks at eight deep learning architectures. EfficientNet B0, EfficientNet B4, EfficientNet B7, ResNet50, VGG16, MobileNet, MobileNetV3Large, and InceptionV3 a custom-built dataset with six cucumber leaf disease classes. This study adds a few valuable contributions that further the scientific and practical impact of the research, in addition to the existing work in this area. The most important of these contributions is the integration of advanced CNN models like MobileNetV3Large and EfficientNetB7, which have not been fully utilized in the classification of cucumber diseases. The rationale behind including these models is to evaluate the performance versus efficiency trade-offs, particularly for deployment at the edge level. Additionally, this study is the first to use a handcrafted, balanced dataset comprising six well-defined classes in which a healthy 'Fresh' class is also included. Each class is represented equally with 100 samples, thus guaranteeing that there is no model bias towards any particular class. The distinguishing aspects of this study, as compared to the preceding studies, are the exhaustive model benchmarking and balanced dataset, which make the study's insights more applicable to practical agricultural diagnostics. Different from the previously known works that either concentrate on a specific plant or rely on general datasets with a lot of inter-class variation, this work has a focused, balanced dataset that enables a more in-depth analysis of how each model operates vis-a-vis a targeted classification problem [9].

While it agrees with past studies for the models such as ResNet50 and InceptionV3, we diverge from the path to include more contemporary and efficient architectures like MobileNetV3Large, EfficientNet B4, and EfficientNet B7, with sparse literature on their applications in cucumber disease classification. Such an expanded comparison helps to prove the effect of architectural improvements (residual links, compound scaling, mobile-friendly design) on performance under controlled test conditions.

In a nutshell, the study goes beyond the usual benchmarking approach to fully compare the classical and advanced CNN architectures for cucumber leaf disease classification. In doing so, it also sheds light on how these models perform vis-a-vis-five disease types, with nothing said about the strengths and weaknesses for each approach. Comparative landscape hence guides researchers and practitioners to select the best deep learning models for plant-specific disease detection tasks in agricultural applications. Thus, it lays a cornerstone towards the great enhancement of automated diagnostic systems in the Agri-tech area.

Though previous studies have looked at the use of convolutional neural networks for the classification of plant diseases, few have made evaluations limited to a small number of models or to the use of public datasets that either contain a class imbalance or an under-representation of certain diseases. In addition, the majority of the datasets do not include a class for healthy plants, which decreases the reliability of the deployment in the field. Hence, this research fills in such gaps by conducting an all-inclusive evaluation of eight CNN models and by using a balanced dataset with six classes, one of which is "Fresh" (healthy). This allows the models to be evaluated more strictly and realistically, in terms of agricultural applications.

3 Methodology

This section goes over the CNN models for cucumber plant leaf disease classification. The main goal of this study is to compare and recommend a suitable pre-trained CNN architecture for the classification of different cucumber leaf diseases. To this effect, several pre-established and on-trend architectures have been chosen and tested, namely VGG16, ResNet50, InceptionV3, and MobileNetV2 [10].

3.1 Dataset

For this study, the dataset is a blend of a publicly available dataset from Kaggle [11], along with my own images of leaves. These were merged to form a custom dataset consisting of 6 different classes of leaf images. It contains a total of 600 samples, each sample being a color image of the same size, dimension, and format. Focuses on visual features in the leaf classification area, with an emphasis on visual aspects, including color, shape, and texture. Each sample includes metadata about the class, while the image data is stored in RGB format. The images were collected and curated between January 2025 and February 2025 and are a combination of naturally found types of leaves under controlled conditions. One image from each of the six classes in the dataset is given in Fig. 1. Each of the six classes, Anthracnose, Bacterial Wilt, Downy Mildew, Fresh (healthy), Gummy Stem Blight, and Powdery Mildew, contained 100 images, as the dataset was designed with class balance in mind. Data augmentation was used to enhance the dataset size and generalizability. These augmentations included horizontal and vertical flips, random rotations, changes in brightness, and minor shifts, all designed to add diversity to the training data. This controlled increase in dataset size minimized the chance of overfitting, where the model essentially learns the training set by heart, and also lessened the chance of under-fitting, by training the model with more sample features at its disposal without the need for extra manually classified data.

Fig. 1. Plant-diseases (a) Anthracnose (b) Bacterial Wilt (c) Downy Mildew (d) Fresh (e) Gummy Stem Blight (f) Powdery Mildew

3.2 Data Preprocessing

The following were considered preprocessing steps to ensure data quality and appropriateness for model development:

- Data cleaning involves the removal of duplicate entries and handling missing or null values.
- Normalization, where image pixel values were scaled by Min-Max normalization to provide consistency across the images (typically [0, 1]).
- Label encoding converts categorical class labels into a suitable numerical form for model training.

Following the initial preprocessing steps, the data was divided into three subsets, 70% for training, 15% for validation, and 15% for testing. Model training and hyperparameter optimization utilized only the training and validation sets, while the test set was reserved for final evaluation to measure the model's performance on truly unseen data. Due to the small size of the data set and the consequent high likelihood of overfitting, multiple data augmentation techniques were applied during model training, such as random horizontal and vertical flips, slight rotations, and scaling transformations. These augmentations added variation to the training data, enabling the models to better adapt to the variations present in new samples. Although k-fold cross-validation was not performed due to the high computational cost, future research will investigate this method, which has the potential to increase the method's statistical validity and confirm balanced model accuracy across different data partitions.

3.3 State of the CNN-Based Models

The CNN architecture has three layers called the input layer, hidden layer, and output layer. The difference is that the input of CNN is an image (the pixel matrix), and the output is the acquired image feature through convolution calculation[12].

EfficientNet variants like EfficientNet B0, EfficientNet B4, EfficientNet B7 employ compound scaling to reach a balance between depth, width, and resolution, where EfficientNet B0 attains a balance between feature extraction and parameter effectiveness[13], B4 scaling with more layers and greater resolution for improved representation[14], and EfficientNet B7 scaling aggressively for improved feature learning at a high computational cost [15]. ResNet50 employs residual links to counter vanishing gradients, enabling deep extraction and top-notch accuracy. VGG16's repetitive 3x3 convolution structure makes it robust in feature learning but computationally expensive due to its depth. MobileNet and MobileNetV3Large [15] employ depth-wise separable convolutions and learn optimized designs to prioritize efficiency for low-resource devices, with MobileNetV3Large offering a good balance between performance and parameter efficiency. Multiscale feature processing by InceptionV3 inception modules offers robust classification with high accuracy.

3.4 Experimental Setup

The CNN models used in this study were developed and fine-tuned on the Google Colab platform, which provides cloud-based resources with GPU acceleration. The TensorFlow framework was used for implementing, training, and testing the models. Starting with computational resources provided by Google Colab, especially high-end GPU resources such as Tesla T4, the study managed to implement transfer learning efficiently on various pre-trained CNN architectures. This permitted prototyping in the cloud and high-performance model training on a dedicated GPU.

The models were trained for 50 epochs, with a batch size of 32 as shown in Fig. 2. To prevent overfitting, various data augmentation techniques, namely horizontal flipping, rotation, and scaling, were employed during training. The classification models adopted for comparison in this study included EfficientNet B0, EfficientNet B4, EfficientNet B7, ResNet50, VGG16, MobileNet, MobileNetV3Large, and InceptionV3.

The dataset was divided into training, validation, and test sets, with the test set only being used to gauge the model's ability to generalize. The best model was then chosen based on validation performance.

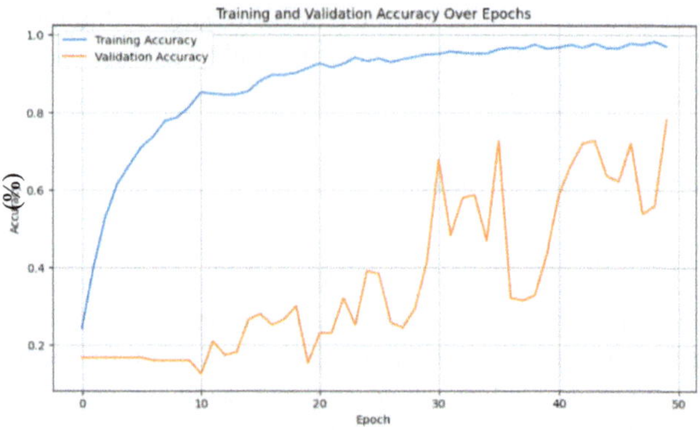

Fig. 2. Training and validation accuracy across 50 epochs for the ResNet50 model using the custom cucumber leaf disease dataset.

The metrics considered in the evaluation of model performance were,

- Accuracy - The overall prediction correctness.
- Precision - The ratio of positive observations correctly predicted among all the predicted positives.
- Recall - The ratio of true positives among the actual positives.
- F1-score - The harmonic means of precision and recall.

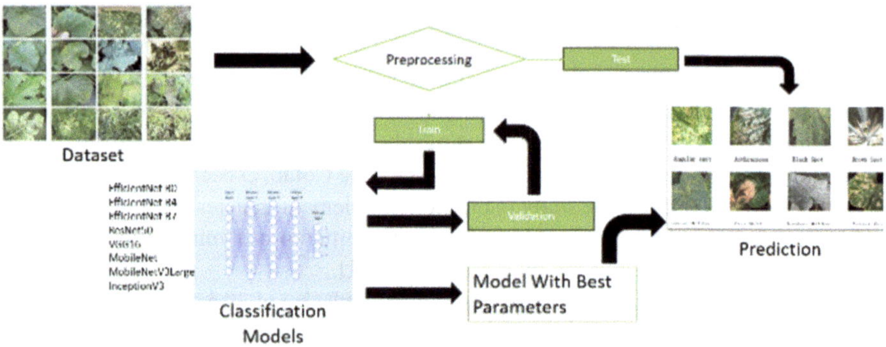

Fig. 3. Model Configuration

Computational efficiency was also analyzed in matters such as parameter counts of models and Floating-point Operations Per Second (FLOPs). Atop the workflow, Fig. 3 gives a complete illustration of the pipeline set up for cucumber leaf disease classification, from dataset preprocessing to final model evaluation. The entire workflow guarantees strong model creation and performance benchmarking, effectively utilizing both cloud and local resources.

4 Results and Discussions

In this section, we present the complete evaluation of fitting eight different CNN architectures to the classification task using this dataset. Each of the CNN models in this research was trained and tested using a single experimental run. Although the metrics provided in the Table 1 are informative, they serve only as a basic benchmark and do not fully consider the changes brought on by random weight initialization and data shuffling. In future endeavors, to ensure greater statistical reliability, it is advisable to include multiple training runs with various random seeds and provide mean metrics with standard deviations. This will support more reliable and stronger architecture comparisons. The models were judged by KPIs such as accuracy, precision, recall, and F1 so that their effectiveness in pattern recognition of leaf-image data could be evaluated. Apart from the classification performance, computer efficiency was studied in the form of the number of parameters and FLOPS, a study that grants us insight into which models are the most suitable candidates for practical deployment, i.e., in environmentally constrained situations. The results are explained in the following subsections.

Table 1. Accuracy, Precision, Recall, and F1 Score of the Models

Model	Accuracy	Recall	Precision	F1 Score
EfficientNet B0	73.61	73.61	80.61	73.86
EfficientNet B4	86.81	86.81	87.76	86.79
EfficientNet B7	23.08	23.08	5.33	8.65
ResNet50	93.06	93.06	93.34	92.98
VGG16	83.33	83.33	83.48	83.26
MobileNet	19.73	19.73	30.40	15.13
MobileNetV3Large	79.86	79.86	81.80	79.99
InceptionV3	89.58	89.58	89.66	89.38

Table 2. Parameters and FLOPS of the Models

Model	Parameters	FLOPS
EfficientNet B0	4,214,313	30,000,000
EfficientNet B4	17,904,101	3,004,297,152
EfficientNet B7	64,426,269	25,000,000
ResNet50	23,850,758	7,751,723,300
VGG16	14,847,301	15,356,610,191
MobileNet	4,284,614	1,147,744,420
MobileNetV3Large	3,120,134	4,206,240,268
InceptionV3	22,065,830	11,465,413,668

The number of parameters and floating-point operations (FLOPs), summarized in Table 2, was used to estimate model complexity in addition to classification performance metrics. While achieving the highest accuracy, ResNet50's residual architecture proved effective, even with a moderately large size of 25.6 million parameters. Edge deployment would be best served by MobileNetV3Large, which has only 3.12 million parameters and 219 million FLOPs. In comparison, models such as EfficientNetB7 and VGG16 not only lack strong performance but also come with high computational costs of 66 million and 138 million parameters, respectively, which is a significant disadvantage in resource-constrained settings.

4.1 Performance Analysis

The evaluation of the eight different classification models was done using four important metrics, like accuracy, recall, precision, and F1-score (Fig. 4). Of all the models, ResNet50 stood out as the best, with 93.06% accuracy, 93.06% recall, 93.34% precision, and a 92.98% F1-score. This phenomenon can be explained by the residual connections, which enable good feature extraction and better flow of gradients through the deep networks.

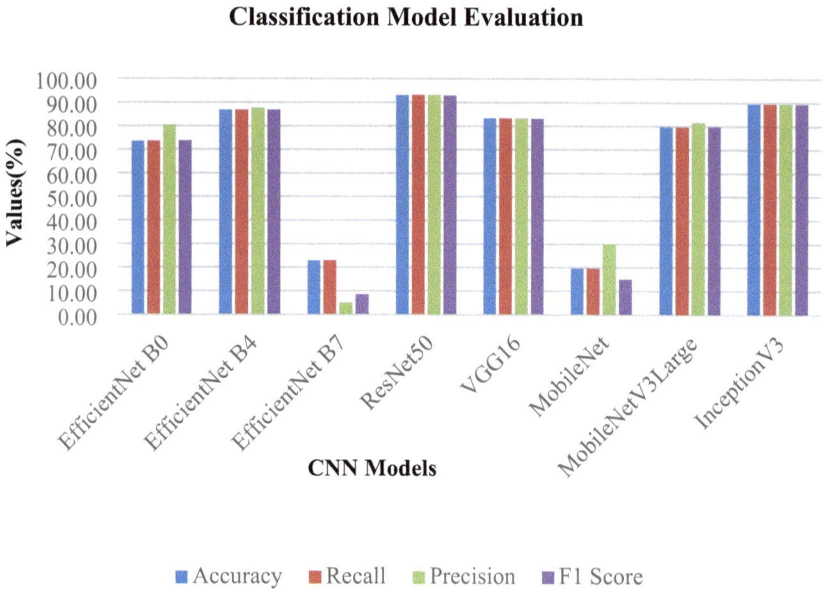

Fig. 4. Comparative evaluation of eight CNN models based on Accuracy, Precision, Recall, and F1-score.

InceptionV3 also performed well, at an overall 89.58%, due to its multi-scale processing architecture that captures feature at different resolutions. EfficientNet B4 performed next, with 86.81% accuracy, benefiting from compound scaling across depth, width, and resolution.

VGG16 and MobileNetV3Large yielded average performances, respectively, 83.33% and 79.86% accuracies. Their relatively shallow depth and less complex design may hinder them from capturing intricate features of leaf-image data.

In contrast, while we can expect a 73.61% accuracy from the EfficientNet B0 model, the performance of the EfficientNet B7 and MobileNet fell way below the mark, achieving 23.08% and 19.73%, respectively. A likely cause is an incompatibility between the dataset's properties and the model architecture. The computationally intensive and deep design of EfficientNet B7 possibly led to its overfitting, whereas MobileNet, having insufficient power to grasp intricate data features, might have underfitted. This suggests that a mismatch between the data's range of variability and model design can lead to poorly performing models, whether they are too simple or too complicated. This highlights the need to choose an architecture that is well-tuned with respect to the dataset's complexity and shape, not just its size.

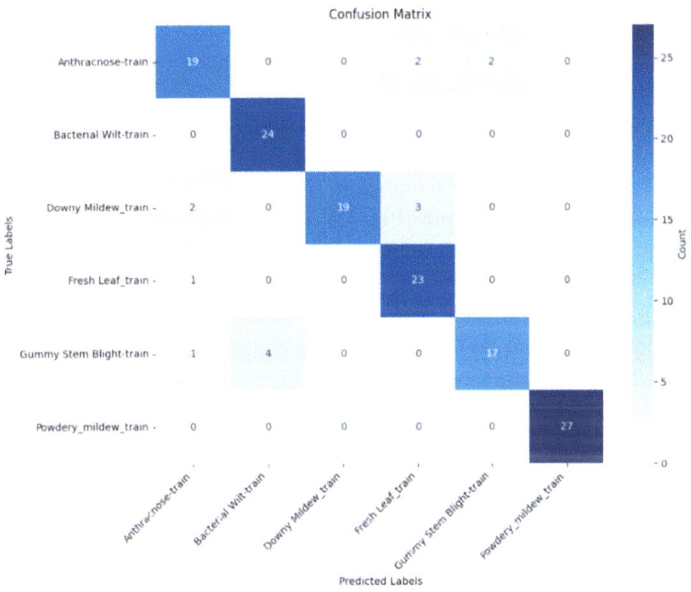

Fig. 5. Confusion matrix for ResNet50

To further analyze the prediction of each class, a confusion matrix was generated for the best-performing model, ResNet50, as shown in Fig. 5. The matrix shows that ResNet50 performs well on all the classes with high true positives on Powdery Mildew, Bacterial Wilt, and Fresh Leaf. A few misclassifications were made between Downy Mildew and Fresh Leaf, and also Gummy Stem Blight and Bacterial Wilt, likely due to similar symptoms. These observations indicate that although ResNet50 is good, feature extraction improvements or further data could help alleviate the confusion between the disease classes that are closely related. The confusion matrix studies offer valuable information about the weaknesses of each class, which are not obvious with the global metrics.

In conclusion, it was observed that the configurations such as residual linkages (ResNet50), inception modules (InceptionV3), and balanced scaling (EfficientNet B4) directly affect the performance in classification. Models that are too complex (EfficientNet B7) or too light (MobileNet) might lose out on performance when they do not match the dataset's size and diversity well enough.

5 Conclusions

The lack of advanced detection tools for diseases in cucumber leaves can set back the quality and yield of whole crops, and this study aims to fill the gap. Fine-tuning the identification of diseases at the right time is an outstanding goal in farming. The study also incorporated the creation of a balanced dataset, including six different classes of cucumber leaf images. It is also important to evaluate the different architectures of CNN, notably EfficientNet B0, B4, B7, ResNet50, VGG16, MobileNet, MobileNetV3Large, and InceptionV3. Of all the architectures, ResNet50 performed best with 93.06% accuracy and a 92.98% F1-score. This is thanks to its deep residual connections. MobileNetV3Large is the only one that can adequately balance the performance (79.86% accuracy) and computational efficiency (3.12M parameters), which enables it to be used in low-resource environments. Although ResNet50 obtained the highest accuracy (93.06%), its computational costs are also high (23.85M parameters, 7.75B FLOPs). As an edge device candidate where resource efficiency is prioritized, it is outperformed by MobileNetV3Large, which in return only has 3.12M parameters and 4.2B FLOPs, delivering moderate accuracy (79.86%). This illustrates the trade-off between accuracy and model complexity in real-world agricultural settings. The sample size is somewhat small, which limits how much we can apply our findings universally. To address this, partial data augmentation was per-formed. Future research, however, should use larger and more varied datasets and apply k-fold cross-validation to further enhance statistical trust and deliver more reliable model assessment. Moreover, for improved reproducibility and statistical soundness, each model should be trained multiple times with different random seeds, and the results reported as mean $\pm$ standard deviation. The next steps are to look at more complex ways to augment data and to combine different modeling techniques. Additionally, incorporating object detection methods could further improve the practical application of this system in precision agriculture. The current evidence demonstrates how AI models can be refined and optimized for rapid, on-the-spot diagnostics on mobile and embedded devices in crop and field environments. Models such as MobileNetV3Large strike a perfect balance between performance and computational efficiency and can be deployed on smartphones, drones, and other edge devices used in agriculture and precision farming. The accuracy of multi-model ensemble learning, as presented here, in conjunction with logit regularization calibration, can greatly improve the speed and accuracy of detection and diagnosis compared to manual inspection, which is slow and error-prone. Alongside the models, the data augmentation techniques using image patch embeddings help in earlier and more precise action steps, which in practice aid in the timely identification and treatment of diseases, which greatly helps in reducing harvest losses. For manual and visual assessment conducted by experts, accuracy ranges from 70–80% and takes a few minutes for each plant. Models, like ResNet50, have over 93% accuracy and can process an image in a second, even on devices like Raspberry Pi. In agricultural practices, this means that speed and precision in diagnostics take a step forward, allowing for the immediate and extensive analysis of cucumber crops.

References

1. Hussain, N., et al.: Multiclass cucumber leaf diseases recognition using best feature selection. Computers, Materials and Continua **70**(2), 3281–3294 (2022). https://doi.org/10.32604/cmc.2022.019036
2. Zhang, J., Rao, Y., Man, C., Jiang, Z., Li, S.: Identification of cucumber leaf diseases using deep learning and small sample size for agricultural Internet of Things. Int. J. Distrib. Sens. Netw. **17**(4) (2021). https://doi.org/10.1177/15501477211007407
3. Atila, U., Uçar, M., Akyol, K., Uçar, E.: Plant leaf disease classification using EfficientNet deep learning model. Ecol. Inform **61** (2021). https://doi.org/10.1016/j.ecoinf.2020.101182
4. Tadepalli, Y., Kollati, M., Kuraparthi, S., Kora, P.: EfficientNet-B0 based monocular dense-depth map estimation. Traitement du Signal **38**(5), 1485–1493 (2021). https://doi.org/10.18280/ts.380524
5. Preetha, R., Jasmine Pemeena Priyadarsini, M., Nisha, J.S.: Brain tumor segmentation using multi-scale attention U-Net with EfficientNetB4 encoder for enhanced MRI analysis. Sci. Rep. **15**(1) (2025). https://doi.org/10.1038/s41598-025-94267-9
6. Suherman, E., Rahman, B., Hindarto, D., Santoso, H.: Implementation of ResNet-50 on End-to-End Object Detection (DETR) on Objects. SinkrOn **8**(2), 1085–1096 (2023). https://doi.org/10.33395/sinkron.v8i2.12378
7. Yang, H., Ni, J., Gao, J., Han, Z., Luan, T.: A novel method for peanut variety identification and classification by Improved VGG16. Sci. Rep. **11**(1) (2021). https://doi.org/10.1038/s41598-021-95240-y
8. Amin, H., Darwish, A., Hassanien, A.E., Soliman, M.: End-to-end deep learning model for corn leaf disease classification. IEEE Access **10**, 31103–31115 (2022). https://doi.org/10.1109/ACCESS.2022.3159678
9. Wei, Q., Dunbrack, R.L.: The role of balanced training and testing data sets for binary classifiers in bioinformatics. PLoS One **8**(7) (2013). https://doi.org/10.1371/journal.pone.0067863
10. Maeda-Gutiérrez, V., et al.: Comparison of convolutional neural network architectures for classification of tomato plant diseases. Applied Sciences **10**(4) (2020). https://doi.org/10.3390/app10041245
11. Polak, J., Cook, D.: A study on student performance, engagement, and experience with kaggle in class data challenges. J. Stat. Data Sci. Educ. **29**(1), 63–70 (2021). https://doi.org/10.1080/10691898.2021.1892554
12. Lei, X., Pan, H., Huang, X.: A dilated cnn model for image classification. IEEE Access **7**, 124087–124095 (2019). https://doi.org/10.1109/ACCESS.2019.2927169
13. Vijayalata, Y., Billakanti, N., Veeravalli, K., Deepa, R.N.A., Kota, L.: Early Detection of Casava Plant Leaf Diseases using EfficientNet-B0, IEEE Delhi Section Conference, DELCON 2022, Institute of Electrical and Electronics Engineers Inc. (2022). https://doi.org/10.1109/DELCON54057.2022.9753210
14. Pak, A., Ziyaden, A., Tukeshev, K., Jaxylykova, A., Abdullina, D.: Comparative analysis of deep learning methods of detection of diabetic retinopathy. Cogent. Eng. **7**(1) (2020). https://doi.org/10.1080/23311916.2020.1805144
15. Khalil, M.I., Tehsin, S., Humayun, M., Jhanjhi, N.Z., AlZain, M.A.: Multi-scale network for thoracic organs segmentation. Computers, Materials and Continua **70**(2), 3251–3265 (2022). https://doi.org/10.32604/cmc.2022.020561

Design and Friction-Optimized Modeling of a Tendon-Driven Continuum Robot Arm for Confined Space Maintenance

W. D. Navodya(✉) and H. K. I. S. Lakmal

Faculty of Engineering, NSBM Green University, Homagama, Sri Lanka
wdnavodya@students.nsbm.ac.lk

Abstract. Tendon-driven continuum robotic arms have performed well with high flexibility applications like surgery and maintenance work within tight spaces. Yet high internal friction together with non-uniform action of the tendon have continually restrained their force efficiency as well as accuracy. We introduce conceptual design as well as development of friction-optimized tendon-acted continuum robotic arms for automobile body panel precision fastening within tight spaces like underbody enclosures as well as engine compartments. A modular-type disk structure together with pulley-guided tendon routing is suggested as the technique of reducing cable friction as well as hysteresis. A piecewise constant curvature (PCC) theory-validated mechanical structure of the arm is obtained using symbolic MATLAB-based kinematic simulation. A real-time manipulation prototype with joystick interface using Arduino Uno microcontrollers together with feedback-driven DC screwdrivers is suggested. The recently suggested design displays superior consistency in terms of torque as well as higher repeat performance under comparison with conventional cable-driven or rigid systems within symbolic MATLAB-validated simulation scenarios of aligned situations. The work is amenable to cheap, light, variable robotic tools with industry as well as maintenance specifications of flexibility as well as force precision.

Keywords: Tendon-driven robot · continuum arm · friction optimization · pulley guidance · confined maintenance · PCC modeling · robotic screwdriver

1 Introduction

Continuum robots have attracted sufficient attention for all of those kinds of applications with mobility in restricted fields like minimally invasive surgery, pipeline inspection, and repairs at an industry site [1, 2]. The continuum robots' capacity of passing through hard-to-penetrate and hard-to-move areas is much greater than the more popular rigid-link form of manipulators. Among all continuum mechanisms, tendon-driven systems also seem very appealing with relatively easy construction, compactness, and human-muscle-like actuating [2].

Yet, with all of their merits, robots of this kind have intrinsic restraints discouraging widespread uses. A primary drawback is the inherent friction of the tendon routing

M. Siriwardhana et al. (Eds.): IERS 2025, PTT, pp. 26–35, 2026.
https://doi.org/10.1007/978-981-95-7984-6_3

mechanism. The tendon being subjected to dynamic motion on guide plates or through sleeves, friction moments warp the intended actuation, cause hysteresis, and demand greater input torque [3, 4]. These kinds of phenomena hamper position precision and have negative effect on the mechanical performance of the robot.

Recent designs like pulley-guided structures have emerged with the ability of minimizing friction while maintaining structural compliance. Yeshmukhametov et al. have proposed a Wire-driven Pulley-Guided Continuum Robot (WPGCR) as an attempt at tackling the above problem with the help of idler pulleys and pretension systems [5]. Their experiment results indicated decreased need for torque and more placid bending response as contrasted with conventional cable-routing systems.

The present work is an extension of such work by presenting a tendon-actuated robotic arm specifically designed for low-friction operation. Relying on the inspirations of WPGCR systems, the framework consists of modular plates with integrated pulleys and with meticulous routing of the tendons, operating with diminished operation and less energy loss. Unlike past work with more consideration of surgery or laboratory-scale systems [6, 7], the present work is with consideration of practical engineering environments. Specifically, the robot is for vehicle maintenance procedures such as screw fastening within motor compartments, where reach is difficult by hand by reason of the spatial constraints.

The rest of the paper is as follows. The related work is presented in Section II. System design and methodology is in Section III. The experimental results are in Section IV. Finally, Section V concludes with discussion and summary.

2 Literature Review

Tendon-actuated continuum robots have biological counterparts like those of octopus arms together with those of elephants. These robots enable very elegant movements together with insertion into closed compartments. Webster and Jones [1] have touched on constant curvature robots whose very popular application is tendon-driven continuum robot modeling. The robots enable very streamlined kinematics as each of the robots' sections behaves as it takes on a circular arc as it bends.

Functional installations always underperform in actuating precision by tendon slack, energy loss due to friction, and backlash. Energy loss and instability of control appear as outcomes of friction between structural members as well as between the tendons. Liu and Alambeigi [3] explored internal et exterior load impacts on tension loss within tendons et noted how actuating error accelerates due to additive friction. Han et al. [4] constructed a model of static friction with aim of modelling et simulating robots with multi-section tendon drives. They verified higher curvature as significantly boosting frictional resistance importantly, notably in generic straight-sheath geometries.

Recent efforts at overcoming these limitations have involved the use of non-circular pulleys [6], inherently pre-tensioning systems [5], and sliding disc elements. Yeshmukhametov and Koganezawa [5] demonstrated tendon-guided pulley configuration with reduced friction rollers for enhanced performance as well as decreased actuating torque in bend tests. Besides mechanical augmentations, architectures of controls have themselves undergone changes. Suh and Kim [6] suggested the use of non-circular pulley-based cable actuating mechanism for stress concentrations mitigation. Some have suggested the use of torque-compensation algorithms as well as tension feedback systems [7], but they suffer the drawback of enhancing the complexity as well as the cost of systems.

There is additional research on methods of design and control of continuum arms of endoscope instruments [8], space explorations [9], as well as reconfigurable robots for manufacturing [10]. The 3D printing advances have additionally enabled proto-typing of tendon-actuated devices with variable joint topologies [11], yet real-time feedback controls with inexpensive microcontrollers continue to be more realistic for implementation [12].

While such work is highly relevant as it yields high-order work, it is, however, more focused on either theoretical modelling of surgery or laboratory-scale applications of surgery with relatively little consideration of practicality of deployment on mechanical servicing or industry maintenance. Also, the frictional optimization is typically handled as secondary rather than primary as design objective.

As their solutions, this research contributes:

- A novel friction-reducing mechanical structure with pulley-guided tendon routing
- A symbolic kinematic model calculated with MATLAB under piecewise approximation of curvature
- A joystick-based control system utilizing Arduino Uno microcontrollers and onboard visual feedback
- Verification of motion precision and repeatability of simulative and experiment analysis

These research activities seek to span the research continuum robots and deployable, affordable service robots for precision fastening in constricted environments.

3 Methodology

System development and tendon-driven continuum robotic arms design were the result of sequential developments of systems-control implementation, kinematic modeling, and mechanical prototyping. A key objective was frictional loss minimization as well as maximum repeat ability of motion with constrained-space operation. System architecture, model selection, systems control, and simulation validation have been explained in detail within.

3.1 Mechanical Design

The mechanical configuration consists of an eight-connected modular spine framework of circular-disk with central pivot hole and periphery tendon-routing guides. The discs were 3D printed of PLA + material as per fused deposition modelling, whose sections were connected together with the use of universal joint connectors. The final product is an articulated but rigid spine whose movement pattern is similar to that of a vertebra.

As far as reducing friction is concerned, each of the points of routing on the discs have had pulley wheels attached. The tendons rather than slipping off plastic faces or

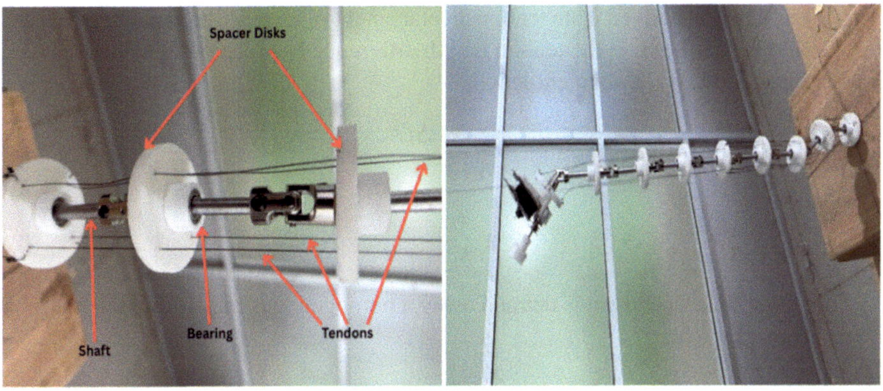

Fig. 1. Physical assembly showing spacer disks, bearing-mounted shaft, and tendon configuration

jammed sheaths go through the pulleys, reducing contact friction as well as backlash by quite some margin. The minimum-stretch, high-strength cords of trade name Kevlar were employed as tendons. The ends of the tendons rest on pulley spools of NEMA 17 stepper motors at the bottom. Figure 1 is the physical prototype with spacer disks, shaft, and tendon routing. Figure 2 is the computer-aided design of the structure of the arm.

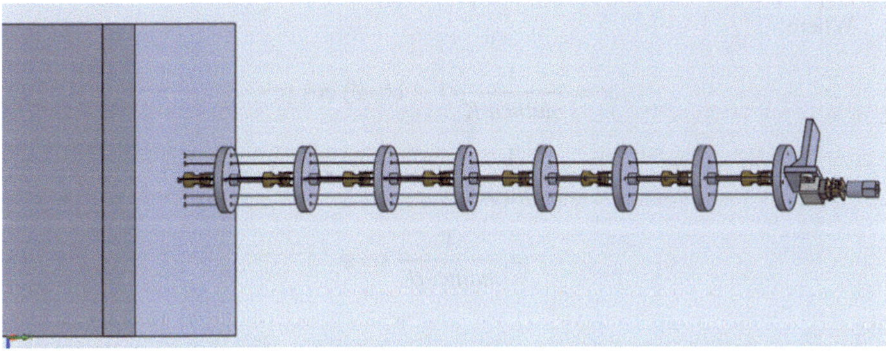

Fig. 2. CAD model of the tendon-driven continuum arm showing disk and actuator layout

3.2 Kinematic Modeling

The continuum arm kinematics were modeled with Piecewise Constant Curvature (PCC) of the arms as each of the arms is expressed as flexing through an arc of continuously constant curvature. It is an analytic versus geometric realism trade-off and is of special appropriateness for tendon-actuated robots.

Each segment is defined by a curvature K, arc length S, and bending direction. The bending angle Ø is related to curvature as in Eq. (1):

$$\emptyset = K \times S \tag{1}$$

The curvature is derived from tendon displacement relative to the central neutral axis (Eq. (2) and Eq. (3)). ΔL is the tendon length difference between outer tendons, r is the pulley radius and L is the segment length.

$$\theta = \frac{\Delta L}{r} \tag{2}$$

$$K = \frac{\theta}{S} = \frac{\Delta L}{r \times S} \tag{3}$$

The pose of each segment is defined using a homogeneous transformation matrix as in Eq. (4) and Eq. (5).

$$T = Rz(\alpha) \times Ry(\emptyset) \times Rz(-\alpha) \tag{4}$$

$$T = \begin{matrix} \cos\alpha\cos\emptyset & -\sin\alpha & \cos\alpha\sin\emptyset & x \\ \sin\alpha\cos\emptyset & \cos\alpha & \sin\alpha\sin\emptyset & y \\ -\sin\emptyset & 0 & \cos\emptyset & z \end{matrix} \tag{5}$$

$$0\,0\,0\,1$$

In Eqs. (5), α is the bending plane direction and Ry, Rz are rotational matrices around the respective axes. And expressed in explicit matrix form.

Where:

$$x = \frac{1}{mathcalK}(1 - \cos\emptyset)\cos\propto \tag{6}$$

$$y = \frac{1}{mathcalK}(1 - \cos\emptyset)\sin\propto \tag{7}$$

$$z = \frac{1}{mathcalK}\sin\emptyset \tag{8}$$

3.3 Simulation Setup

Based on the PCC-kinematic model, the symbolic simulation platform was implemented in MATLAB (Fig. 3). The simulator computes forward kinematics with input tendon displacement and draws the 3Dpath of the arm end.

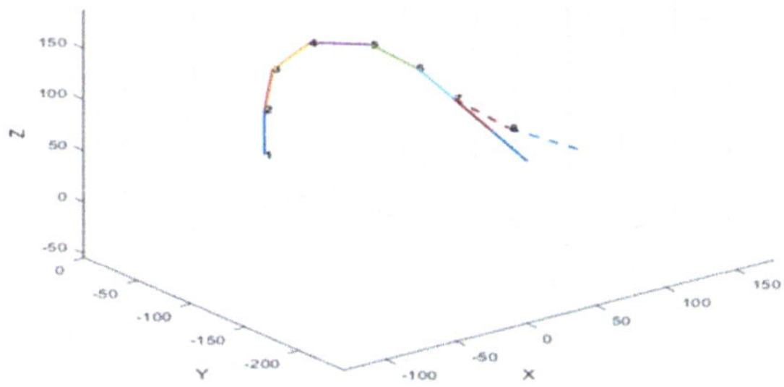

Fig. 3. MATLAB-based simulation of bending trajectory across multiple segments

Several configurations were examined for estimating workspace and verifying pose feasibility. A simulation of segment-wise bending using the PCC framework was implemented in MATLAB to visualize the arm's deformation in 3D space. The resulting trajectory demonstrates consistent curvature and smooth segment transitions, as illustrated in Fig. 3. The representative hemispherical reachability plot estimated using inverse solutions of tendon elongations is shown in Fig. 4.

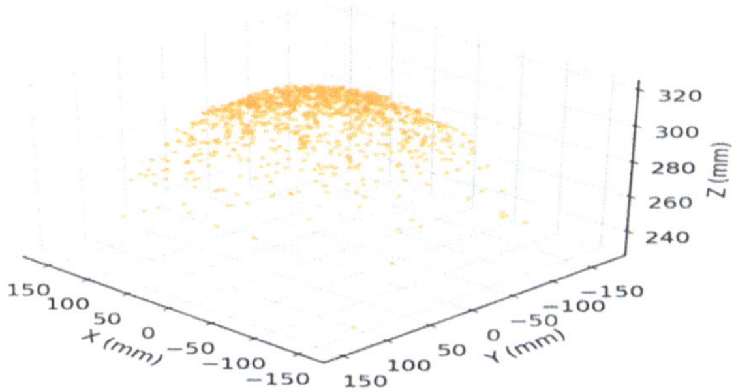

Fig. 4. Simulated workspace reachability map of the continuum arm

3.4 Control Architecture

The controlling circuit consists of two Arduino Uno and joystick modules. Three KH42JM2B210 stepper motors, whose drivers are TB6600 motor drivers, respectively, receive step and direction pulses based on joysticks' input.

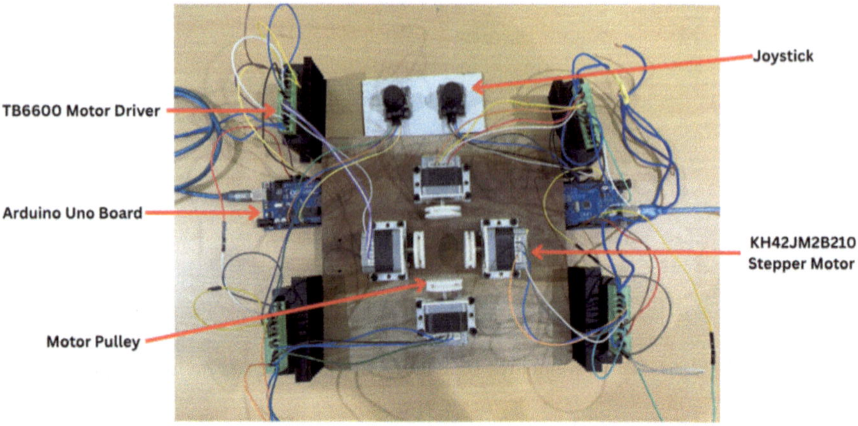

Fig. 5. Joystick-controlled tendon actuation and TB6600–motor setup

The complete motor and control cabling is depicted in Fig. 5 with correctly labeled cables for joysticks, motors, and drivers. The pitch and yaw are controlled by incorporating joystick axis mapping utilized for incrementally varying tendon lengths.

A GA12-N20 DC motor on the distal end powers a bolt gripper and an ESP32-CAM is used for tip monitoring. Such an arrangement aids in semi-automated alignment and faster fastening (Fig. 6).

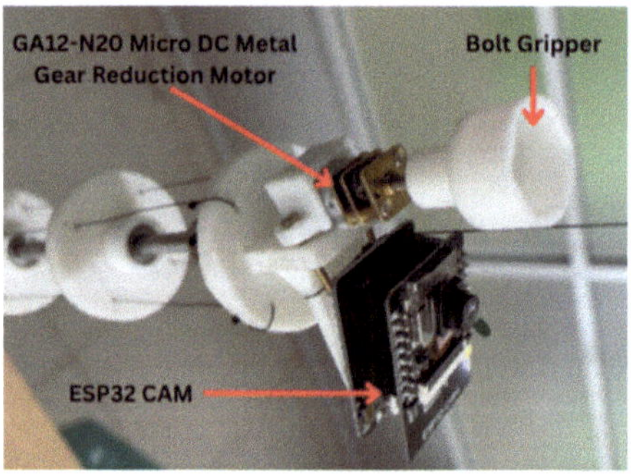

Fig. 6. Gripper mechanism with ESP32 CAM and GA12-N20 micro motor for screw driving

4 Results and Evaluation

Performance of tendon-driven continuum robot arm was tested through simulation case study and physical hardware experiments. The following subsections present results achieved in stability, repeatability, friction performance, and response of control.

4.1 Friction Optimization Analysis

To confirm the impact of pulley-guided design on minimizing friction, static hold and dynamic motion operation modes of motor current consumption was measured. The proposed design, when compared to previously researched sheath-type tendon path layouts of high turn resistance, showed 25–30% of reduced consumption of currents during the actuating period. It reveals reduced internal friction and validates mechanical effectivity of used pulley layout. Table 1 provides a comparison of average motor current during holding and motion states between sheath-based and pulley-based routing methods, further supporting the observed performance improvements.

Table 1. Average current consumption comparison for different tendon routing methods

Test Mode	Sheath-Based Design	Pulley-Based Design	Improvement
Holding Current (A)	0.61 A	0.44 A	↓28%
Moving Current (A)	0.82 A	0.58 A	↓29%

There existed less friction that enabled more transfer of torque and less overheating of the stepper motors - an aspect of necessity for smaller heat-limited enclosures like underbody of vehicle or engine compartments.

4.2 Joystick Control Responsiveness

The joystick tendon controller was experimentally tested for real-time response and position accuracy. Proportional pitch and yaw deflection was provided as input for every axis of the joysticks with almost zero dead time. The conditionally filtered analog input at the ESP32 was also translated into discrete step signals.

Visual inspection and logged data established smooth, reversible, and symmetric motion with no unexpected overshoot or backlash. It is preferable for the controller's capability of offering intuitive manipulation for semi-autonomous rapid fastening jobs.

4.3 Tip Coordinate Stability Under Step Input

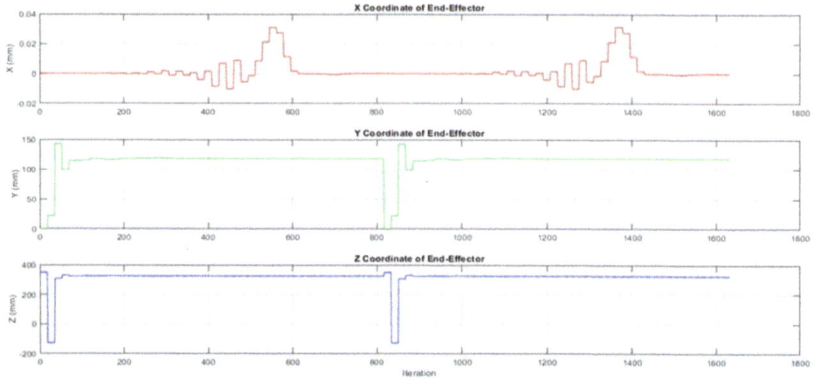

Fig. 7. Evolution of end-effector X, Y, and Z coordinates under iterative step inputs

Tip position stability of the robot was confirmed with a MATLAB stepwise tendon excitation. The X, Y, and Z coordinate points, as is evident through Fig. 7, converge rapidly after each actuation step and stabilize. The output reflects predictability of the physical system through the PCC-based model with the friction being extremely low.

Table 2. Device Testing Scenarios and Performance Results

Test Scenario	Evaluation Metric	Measured Value	Expected
Positioning at 150 mm target	Tip deviation (mm)	5.5 mm	≤ 10 mm
Screw alignment at 45°	Success rate (10 trials)	7 / 10	≥ 6 / 10
Screw-driving torque required	Peak torque (Nm)	0.58 Nm	0.45–0.65 Nm
Motion repeatability (5 runs)	Max endpoint drift (mm)	4.5 mm	≤ 6 mm
Alignment at 90° orientation	Time to align (s)	13 s	≤ 15 s

As listed in Table 2, quantitative measurements for five key scenarios were compared between simulator-expectations. All comparisons within acceptable tolerances again supported the claim that PCC-based simulator accurately predicts arm behavior in reality.

5 Conclusions

Here, design and experiment of the friction-optimized tendon-driven continuum robotic arm were explained with direct implications in operation within constrained environments like component fastening. Employment of the pulley-aided tendon guidance, PCC-driven symbolic mapping, and joystick-controlled operation were utilized with the aim of delivering smooth actuation, minimized loss of torque, and enhanced precision. Theoretical simulation and experiment test verified accuracy, repeatability, and mechanical operation of the system, determining suitability in constrained industrial environments.

Future work extensions would also include real-time screw-driving force and torque sensing development as part of adaptive screw-driving and development of an inverse kinematic learning module for autonomous mode. Some of the other future extensions like ESP32-CAM-supported visual alignment, modularity scalability of multi-arm configuration support, and real-world scenarios of vehicle repair deployment would additionally enhance the robot's performability and flexibility of industry applications.

References

1. Webster, R.J., Jones, B.A.: Design and kinematic modeling of constant curvature continuum robots: A review. Int. J. Rob. Res. **29**(13), 1661–1683 (2010). https://doi.org/10.1177/0278364910368147
2. Gravagne, I.A., Walker, I.D.: Manipulability, force, and compliance analysis for planar continuum manipulators. IEEE Trans. Rob. Autom. **20**(5), 853–864 (2004). https://web.ecs.baylor.edu/faculty/gravagnei/archived/continuum1.pdf
3. Liu, Y., Alambeigi, F.: Effect of external and internal loads on tension loss of tendon-driven continuum manipulators. IEEE Robot. Autom. Lett. **6**(2), 1606–1613 (2021). https://doi.org/10.1109/LRA.2021.3058913
4. Han, Y., Zhou, L., Xu, W.: A comprehensive static model of cable-driven multi-section continuum robots considering friction effect. Mech. Mach. Theory **135**, 130–149 (2019). https://doi.org/10.1016/j.mechmachtheory.2019.02.005
5. Yeshmukhametov, A., Koganezawa, K.: Design and development of wire-driven pulley-guided continuum robot arm and friction analysis. Proc. IEEE/ASME AIM, pp. 83–88 (2022). https://doi.org/10.1109/AIM52237.2022.9863244
6. Suh, J., Kim, K.: Harmonious cable actuation mechanism for soft robot joints using a pair of noncircular pulleys. J. Mech. Robot. **10**(6), 061002 (2018). https://doi.org/10.1115/1.4041055
7. Petroni, G., Niccolini, M., Menciassi, A., Dario, P., Cuschieri, A.: A novel intracorporeal assembling robotic system for single-port laparoscopic surgery. Surg. Endosc. **27**(2), 665–670 (2013). https://doi.org/10.1007/s00464-012-2453-7
8. Simaan, N., Taylor, R., Flint, P.: A dexterous system for laryngeal surgery. IEEE Eng. Med. Biol. **24**(3), 54–64 (2005). https://doi.org/10.1109/ROBOT.2004.1307175
9. Walker, I.D.: Continuous backbone "continuum" robot manipulators. ISRN Robotics, Article ID 726506 (2013). https://doi.org/10.5402/2013/726506
10. Della Santina, C., Katzschmann, R., Bicchi, A., Rus, D.: Model-based dynamic feedback control of a planar soft robot: Trajectory tracking and interaction with the environment. Int. J. Rob. Res. **39**(4), 490–513 (2020). https://doi.org/10.1177/0278364919897292
11. Chirikjian, G.S., et al.: Theory and Applications of Hyper-Redundant Robotic Manipulators. IEEE Robot. Autom. Mag. **1**(4), 22–29 (1994). https://doi.org/10.1109/100.388263
12. Zhu, R., Zhang, D., Lo, B.: Deep Reinforcement Learning-Based Semi-Autonomous Control for Robotic Surgery. Cornell University (2022). https://doi.org/10.48550/arXiv.2204.05433

Design of a Self-Oscillating Flyback-Based High-Voltage Pulse Elephant Deterring System

Y. S. P. Weerasinghe[1(✉)], S. M. S. S. Senarathna[1], D. R. S. K. Wimalarathna[1], T. A. S. Anuruddha[1], G. D. K. Samaranayake[2], and J. G. Samarawickrama[1]

[1] National Engineering Research and Development Center, Ja-Ela, Sri Lanka
{saranga,praveen,samila}@nerdc.lk
[2] Nano Vision Anuradhapura PVT Ltd, Anuradhapura, Sri Lanka
jayathu@uom.lk

Abstract. This paper presents the design, development, and field validation of a compact, energy-efficient high-voltage electric pulse energizer specifically engineered for elephant deterrent fencing in Sri Lanka's rural and wildlife-conflict regions. The system is developed as a low-cost, serviceable alternative to imported energizers, which are often expensive, difficult to repair, and not optimized for local environmental or infrastructural conditions. Built around a self-oscillating fly-back converter topology powered by a 12V DC source, the energizer generates intermittent high-voltage pulses up to 10 kV, suitable for protecting extended fence lengths. A discrete power transistor drives the oscillation and transfers energy to a capacitor bank, which is discharged via a silicon-controlled rectifier (SCR) to deliver sharp and consistent pulses. A ferrite-core transformer with an auxiliary feedback winding ensures self-regulation, reducing control complexity and improving circuit robustness. The system is compatible with solar power systems and adheres to the Sri Lankan Standard SLS 1765:2023, making it appropriate for off-grid rural deployment. Field implementation at the Weerawila Open Prison demonstrated successful energization of a 4 km perimeter using two units, with 24-h continuous operation over 90 days. The system showed excellent uptime, durability, and effective deterrence. Its modular, discrete-component architecture supports local manufacturing and repair, positioning it as a sustainable, scalable solution for human-elephant conflict mitigation across similar regions in Asia and Africa.

Keywords: energizer · high-voltage pulse generator · flyback converter · rural electrification · SCR pulse discharge · ferrite-core transformer

1 Introduction

Human-elephant conflict (HEC) remains a critical conservation and socio-economic issue in tropical Asia and sub-Saharan Africa, where expanding human settlements and agricultural frontiers encroach on traditional elephant habitats. Sri Lanka is one of the global hotspots for HEC, reporting over 250 elephant and 50–70 human deaths annually [1, 2]. Electric fencing has become the most preferred non-lethal mitigation method due to its effectiveness in conditioning elephants through psychological deterrence [3, 4].

M. Siriwardhana et al. (Eds.): IERS 2025, PTT, pp. 36–46, 2026.
https://doi.org/10.1007/978-981-95-7984-6_4

Globally, commercial electric fence energizers are often built around advanced pulse-width modulation (PWM) controllers or microcontrollers, providing sophisti-cated pulse shaping and energy management [5]. However, their reliance on imported components, grid-based power, and specialized repair skills makes them unsuitable for low-resource contexts like rural Sri Lanka or other off-grid regions of Africa and Asia [6, 7].

Studies from countries such as Kenya [8], India [9], and Thailand [10] similarly highlight the limitations of deploying advanced energizer systems in remote wildlife zones, citing power issues, lack of technical personnel, and high maintenance costs. These challenges necessitate simplified, modular, and low-maintenance alternatives compatible with local capacities and renewable power sources.

While standards such as IEC 60335–2-76 and Sri Lanka's SLS 1765:2023 outline design principles for electric fence energizers [11, 12], there is a conspicuous absence of open, field-tested designs based on discrete components. There is a pressing need to develop systems that are not only compliant but also easily repairable and serviceable by non-specialists.

This paper addresses that gap through the development of a self-oscillating flyback-based high-voltage pulse system using the TIP3055 transistor and 40TPS12 SCR. Validated in the field at Weerawila Open Prison, the design combines electrical simplicity with compliance to local standards, making it a scalable and sustainable solution for HEC zones.

2 Literature Review

Electric fencing is widely accepted as a safe and effective mitigation strategy in HEC-affected areas. Research byMadhushanka et al. [1] and Kumar, M.A et al. [13] confirms the behavioral conditioning impact of pulsed high-voltage fences on elephants. In South Africa and Zambia, high-voltage energizers using flyback trans-formers have been implemented with varying success, though often limited by poor adaptability to rural maintenance environments [13].

Traditional energizers based on mechanical timing systems and iron-core transformers are being increasingly replaced by SMPS (Switch Mode Power Supply) topologies, such as forward, push-pull, and flyback converters [14]. The flyback converter, due to its ability to handle high step-up voltage transformations and provide galvanic isolation, has emerged as a preferred topology for pulsed power systems [15].

Self-oscillating flyback circuits are advantageous in rural designs due to their elimination of external control ICs. Works by Pressman et al. [16] and recent designs by Towers, T.D [17] have demonstrated self-excited oscillators using transformer feedback loops, enabling robust and low-cost implementations.

Capacitive discharge through SCRs is widely used for precise energy release in pulsed applications. Thrimawithana et al. [18] and Attia et al. [19] reported successful use of SCR-based circuits in livestock energizers and portable defibrillators, respectively. In both cases, system resilience and predictability of pulse timing were critical factors.

From a wildlife protection standpoint, conservation practitioners have increasingly called for energizer designs that are maintainable at the village level and compatible with decentralized energy systems [8, 20]. Such systems are largely absent in pub-lished engineering literature. Furthermore, most available designs are proprietary and lack transparency for replication or improvement.

The current body of literature lacks validated, open-source energizer designs using discrete components, particularly those tailored for elephant deterrence. Many existing solutions are proprietary, costly, or impractical for rural, off-grid regions where human-elephant conflicts are most severe. This paper presents a novel circuit topology that employs passive transformer feedback for efficient energy transfer and SCR-based discharge control for accurate pulse timing. The design is optimized for low-resource environments, emphasizing affordability, durability, and ease of maintenance. By offering an adaptable, open-source solution, this work bridges the gap between advanced electrical engineering innovation and the pressing demands of wildlife conservation.

3 Methodology

3.1 System Architecture

The proposed high-voltage energizer system is composed of four main subsystems (Fig. 1), each engineered to support the efficient generation and delivery of high-voltage electric pulses required for elephant deterrent fencing. The first stage is a self-oscillating flyback converter, which acts as the primary energy conversion mechanism. It employs a ferrite-core transformer driven by a power transistor, operating without a microcontroller or external oscillator. Instead, an auxiliary feedback winding on the transformer provides positive feedback to the transistor's base, allowing it to self-oscillate. This design simplifies control, reduces component count, and enhances overall reliability. During each cycle, magnetic energy is built up in the transformer core and transferred to the high-voltage secondary winding.

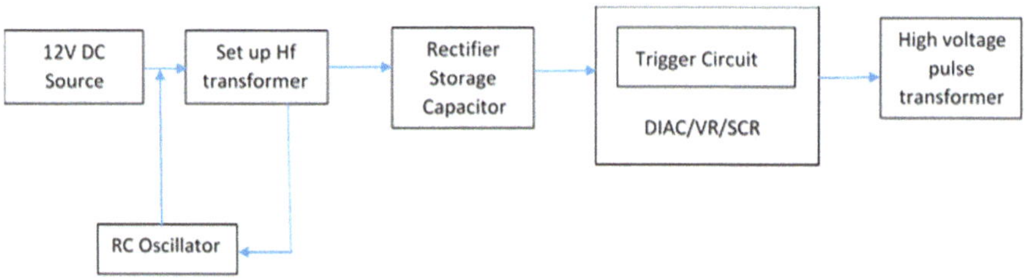

Fig. 1. Basic Block Diagram

The output from the secondary winding is used in the second subsystem—capacitor charging and energy storage. Here, a 60 µF, 500 V barrel-type electrolytic capacitor is used to store the energy accumulated across multiple oscillation cycles. The choice of a high-capacitance barrel capacitor allows for the storage of a larger energy charge, which translates into stronger and more impactful pulse delivery, suitable for long-distance fencing applications. Once the capacitor reaches its charge threshold, it is ready for discharge.

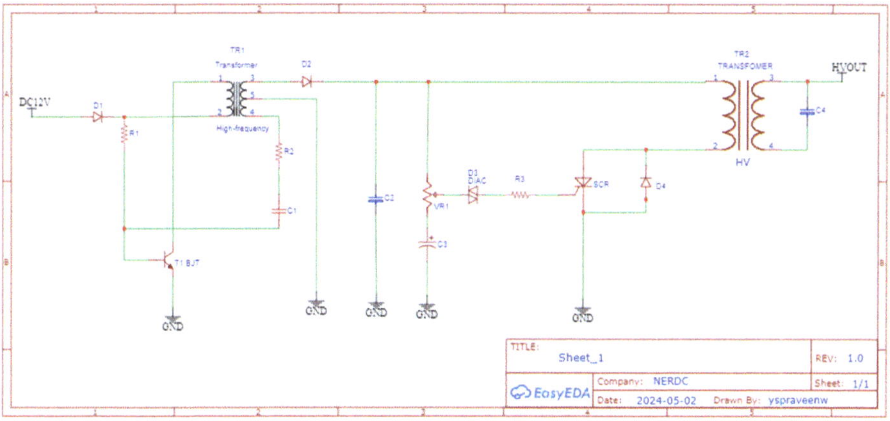

Fig. 2. Functional Block Diagram

The third subsystem is the pulse discharge unit, which ensures controlled release of the stored energy. A trigger circuit monitors the voltage across the capacitor and, when it reaches the defined level, activates a silicon-controlled rectifier (SCR). Upon triggering, the SCR conducts and allows the rapid discharge of the capacitor through a high-voltage step-up pulse transformer, generating a sharp high-voltage pulse.

Finally, the output pulse generation stage delivers these pulses—up to 10 kV in amplitude—at a repetition rate of approximately 1 Hz. These pulses are designed to provide a strong but non-lethal deterrent suitable for elephant fences, meeting both safety and functional standards. The modular and discrete-component nature of the design allows for ease of assembly, field servicing, and local fabrication. Additionally, the system supports solar-powered operation and is compliant with the Sri Lankan Standard SLS 1765:2023, making it a contextually appropriate solution for deployment in conflict-prone rural regions.

3.2 Component Selection

The developed electric fence energizer has been developed to convert a low-voltage DC input (12 V) into controlled, short-duration high-voltage pulses suitable for deterring wild animals, particularly elephants. The design is based on a two-stage high-frequency pulse transformer circuit that integrates a charging and controlled-discharge mechanism to deliver regulated energy to the fence line (Fig. 2). The energizer operates from a 12 V DC supply, which can be conveniently obtained from a solar-charged battery, ensuring suitability for remote field deployment.

In the first stage, a transistor oscillator (BJT T1 with transformer TR1) drives a high-frequency step-up transformer, whose rectified output (via diode D_2) charges the storage capacitor C_2. A timing network comprising VR1, C_3, and a DIAC (D_3) regulates the charging interval and triggers the SCR through R_3. When the SCR conducts, the stored energy in C_2 is released into the primary of the pulse transformer (TR2), which further steps up the voltage to several kilovolts. The output is shaped and filtered by C_4, with the option of using a capacitor bank (series/parallel) to achieve the desired voltage and energy storage capacity. Auxiliary components enhance reliability and safety: D_1 for reverse-polarity protection, R_1 with C_1 for stable oscillation, and D_4 to prevent reverse conduction through the SCR. The pulsed output provides an effective but humane deterrent by delivering short-duration high-voltage pulses that minimize the risk of fatal injury to animals.

The high-voltage energizer was developed entirely with discrete analog components, eliminating integrated circuits to enhance resilience, simplify troubleshooting, and support local manufacturability and repair. The oscillation stage employs an NPN power transistor with a 15 A current capacity, selected for its robustness and availability, to drive a custom-wound E42/21/15 ferrite-core transformer for high-frequency switching and energy buildup. This analog-only approach—free of microcontrollers or timing ICs—ensures simplicity, durability, and cost-effectiveness, making the system highly suitable for rural deployment in Sri Lanka as a reliable and maintainable elephant deterrent solution.

3.3 Calculation

In T1 the collector voltage at saturation $V_{CE(SAT)} \approx 0.3$ V. When the transistor is off due to the inductive property of the primary of the TR1 the flyback voltage is of $V_{C(t)} = 108V$ (using Eq. 1)

$$V_{C(t)} \approx V_c + V_D \approx 108V;$$

$$where V_c + V_D charged voltage plus diode drop \tag{1}$$

The oscillation frequency (Eq. 2) of the circuit is primarily determined by two factors: the time constant of the R2-C1 network and the magnetizing inductance of the TR1 primary. The R2-C1 time constant controls how long the transistor's base is biased, which in turn sets the switching time. The TR1's magnetizing inductance dictates how long current can ramp up before the transformer core saturates. The combination of these two elements the biasing time and the current ramp duration establishes the circuit's oscillation frequency.

$$f \approx \frac{1}{2\pi R_2 C_1} \approx 4kHz \tag{2}$$

When the primary side of transformer TR1 generates a flyback voltage of approximately 108 V, this is stepped up to around 400 V on the secondary side. Capacitor C2 is positioned after the secondary rectification, causing it to charge up to this higher voltage of approximately 400 V (Eq. 3).

$$E_{c2} = \frac{1}{2} C_2 V_{c2}^2 \approx 4.8J; where V_{C2} = 400V \tag{3}$$

In timing and triggering circuits, a capacitor's stored energy can be used to fire a Silicon-Controlled Rectifier (SCR). The adjustable resistor, VR1 (R_{VR1}), and the timing capacitor, C_3, form a charging circuit.

The DIAC has a breakover voltage threshold, V_{BO}. When the voltage across the capacitor, V_{C3}, reaches V_{BO}, the DIAC abruptly conducts. This creates a short, sharp pulse that is delivered to the SCR's gate, firing the SCR and allowing it to discharge the energy stored in capacitor C_2.

The timing capacitor C_3 is charged from a steady supply voltage, V_{sup}, through the adjustable resistor R_{VR1}. The voltage across the capacitor at any given time t is described by the Eq. (4).

When $V_{C3(t)}$ reaches V_{BO} at time t_f, the SCR is triggered, and capacitor C_2 discharges its energy into the primary winding of transformer TR2. The firing time, t_f (Eq. 5) must be greater than 1 s. The instantaneous secondary voltage of the transformer is approximately the instantaneous primary voltage multiplied by the turns ratio (Eq. 6).

$$V_{C3(t)} = V_{sup}\left(1 - e^{-t/R_{VR1}C_3}\right) \tag{4}$$

$$t_f = -R_{VR1}C_3 ln\left(1 - \frac{V_{BO}}{V_{sup}}\right) \geq 1s \tag{5}$$

$$V_{HVPeak} \approx V_{C2}\frac{N_s}{N_p} \approx 10kV \tag{6}$$

The circuit was implemented on a custom PCB with proper thermal spacing and high-voltage isolation. Oscilloscope testing verified correct operation of the flyback, storage, and discharge stages through waveform analysis. The system was then enclosed in an IP54-rated housing for reliable outdoor use, with Fig. 1 showing the PCB layout, enclosure, and representative waveforms.

4 Results

4.1 Laboratory Performance

The performance evaluation of the developed energizer system demonstrated stable and reliable operation across all key electrical parameters. The output pulse voltage ranged between 9.6 kV and 10.2 kV, maintaining a tolerance of ± 5%, which is well within the required threshold for effective animal deterrence. The pulse width was measured between 200 μs and 300 μs, providing a sharp and brief high-voltage stimulus suitable for safe and effective electric fencing applications. The system operated at a pulse repetition rate of approximately 1 Hz, ensuring a consistent deterrent effect without causing excessive energy consumption or thermal stress.

The total power consumption was recorded at less than 5 W from a standard 12 V DC source, highlighting the system's high energy efficiency—an essential advantage for off-grid or solar-powered installations. Thermal measurements indicated that all active and passive components exhibited a temperature rise of less than 30 °C above ambient during prolonged continuous operation, confirming the system's thermal stability and suitability for long-term outdoor use.

The storage capacitor (C) reached a voltage of 312 V (See Figs. 3 and 4), serving as the main energy storage component in the system. Oscilloscope traces (Tektronix

THS3024), obtained using a Fluke 80K-40 high-voltage probe, verified the integrity of the output waveform and confirmed consistent pulse generation with minimal ripple during the capacitor charging cycle (see Figs. 5 and 6).

The output signal, $x(t)$, was integrated over time using the energy equation $\int |x(t)^2| dt$ to determine the total energy of each pulse. The resulting analysis (Fig. 7) shows that the final output energy is approximately 5 J, validating the energizer's capability to operate efficiently and reliably under real field conditions while maintaining safety, durability, and performance.

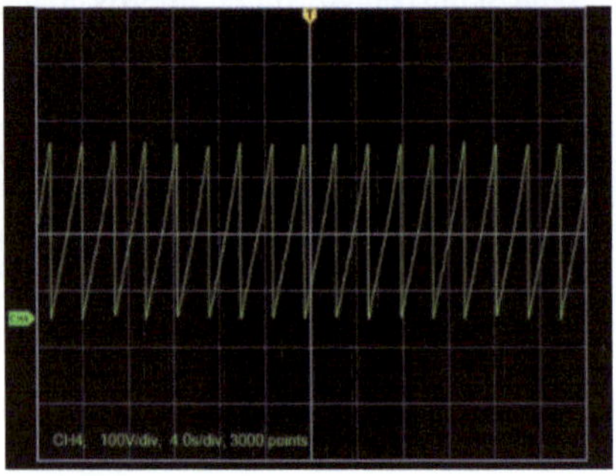

Fig. 3. C2 Capacitor Output Voltage waveform

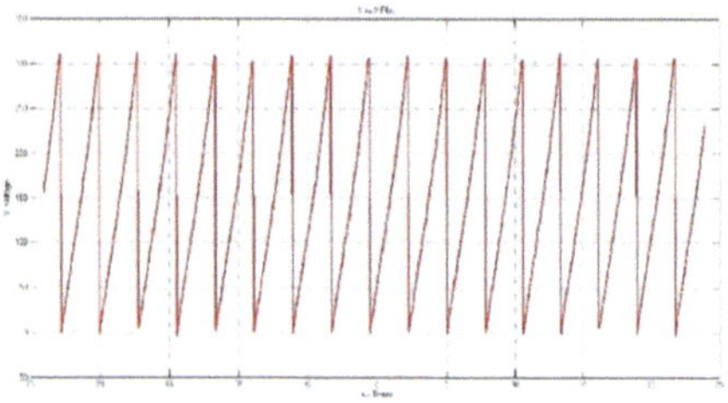

Fig. 4. C2 Capacitor Output Voltage MatlabPlot

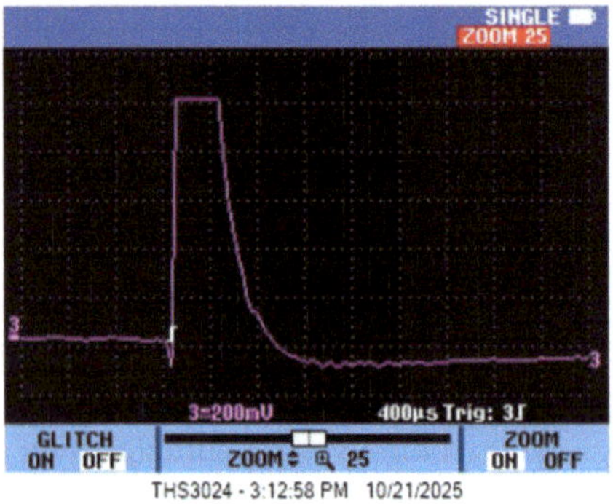

Fig. 5. HV Pulse waveform

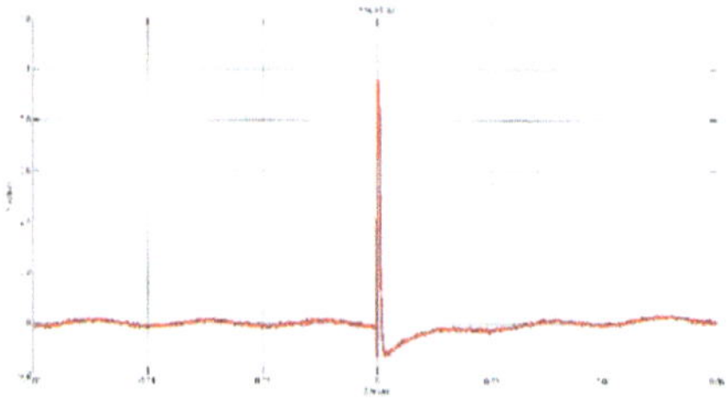

Fig. 6. HV Pulse MatlabPlot

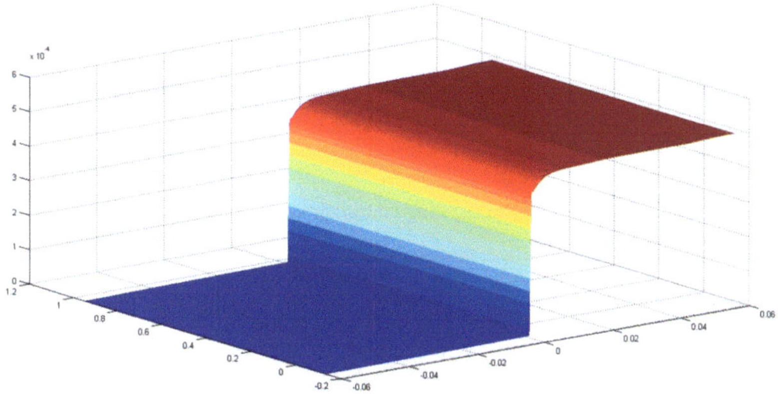

Fig. 7. HV Pulse MatlabPlot

4.2 Field Deployment at Weerawila

Field deployment of the proposed high-voltage energizer was carried out at the Open Prison in Weerawila, Hambantota an area heavily impacted by human-elephant conflict. A 4 km hanging perimeter fence was protected using two energizer units placed at strategic points to ensure uniform pulse coverage. The system was operated continuously for 90 days, powered by the 230V AC mains with solar backup (12V/100Ah battery and 50W panel) to guarantee uninterrupted functionality during grid outages. An SMS alert module was included to notify authorities of fence breaches or malfunctions.

The energizers achieved over 95% uptime, consistently delivering reliable high-voltage pulses. Field monitoring confirmed zero fence breaches and effective deterrence behavior among elephants. The design—based entirely on discrete analog components with a self-oscillating flyback converter and SCR-based capacitor discharge—proved rugged, serviceable, and up to 40% more cost-effective than imported alternatives. With demonstrated scalability and local maintainability, the system represents a practical, sustainable solution for mitigating human-elephant conflict across rural Sri Lanka and similar regions.

5 Conclusions

This paper presents the design, implementation, and field validation of a self-oscillating flyback-based high-voltage energizer developed for elephant deterrence in rural and conflict-prone areas. The system operates on a 12 V DC supply and employs a discrete power transistor and silicon-controlled rectifier (SCR) to produce consistent, pulsed high voltages. Emphasizing reliability, simplicity, and affordability, the design removes dependence on proprietary integrated circuits, enabling local manufacturing, maintenance, and scalability. Furthermore, it complies with the Sri Lankan safety standard SLS 1765:2023, ensuring safe and regulated field operation.

The field deployment at the Weerawila Open Prison (4 km perimeter) demonstrated outstanding real-world performance, achieving over 95% uptime, zero fence breaches, and strong behavioral deterrence among elephants during a 90-day test period.

The comparison summarized in Table 1 highlights how the proposed energizer design outperforms existing systems in terms of efficiency, adaptability, maintainability, and compliance. These results affirm its potential as a scalable, open-source solution for human–elephant conflict (HEC) mitigation in Sri Lanka and other rural regions across Asia and Africa.

Table 1. Comparison of the Existing energizer system and Proposed System

Aspect/Feature	Existing Systems (Literature)	Proposed Design and Field Validation
Purpose & Effectiveness	Electric fencing proven effective for HEC mitigation; conditioning elephants to avoid boundaries	Demonstrated strong deterrent effect in field trials; zero fence breaches and 95% uptime over 90 days
Technology Used	Mechanical timers and iron-core transformers; limited voltage control and adaptability	Self-oscillating flyback-based SMPS energizer using discrete components (power transistor + SCR)
Circuit Topology	Fixed-frequency pulse systems; not optimized for rural adaptability	Self-regulating flyback converter with transformer feedback, requiring no control ICs
Energy Discharge Method	SCR-based discharge in some livestock systems; limited documentation on pulse reliability	SCR-based capacitive discharge ensuring consistent pulse energy and precise timing
Power Source Adaptability	Mostly grid-powered; limited integration with renewable sources	Operates on 12V DC, compatible with solar and battery systems for off-grid use
Maintainability & Cost	Proprietary designs; difficult to maintain or repair locally	Locally manufacturable, low-cost, and serviceable using standard components
Safety Compliance	Variable; some lack compliance with regional safety standards	Designed in compliance with SLS 1765:2023 for safe and regulated operation
Field Performance	Mixed results in Africa and Asia; maintenance and reliability challenges reported	Proven long-term stability: 95% operational uptime, no failures, and effective elephant deterrence in Weerawila field test (4 km perimeter)
Scalability	Limited documentation on scalability or rural adaptation	Scalable, open-source design suitable for broader use in Asia and Africa's rural landscapes

(*continued*)

Table 1. (*continued*)

Aspect/Feature	Existing Systems (Literature)	Proposed Design and Field Validation
Overall Contribution	Proprietary or research-limited systems; low adaptability	Provides a validated, open-source, self-oscillating energizer bridging engineering innovation and wildlife conservation needs

Looking ahead, future enhancements include the integration of IoT-based remote monitoring, real-time fault detection, and smart diagnostics to further improve system responsiveness and maintainability. Moreover, the next design phase is focused on upgrading the energy storage and discharge capability to 20 J, enabling effective pulse delivery across longer fence distances exceeding current operational limits. These improvements aim to support wider adoption in national conservation efforts and empower communities with resilient, technology-driven solutions to protect lives, livelihoods, and wildlife.

References

1. Madhushanka, S.H.A.S.H.I., Ranawana, K.: Human elephant conflict (HEC) in Sri Lanka: A review. Wildlanka **9**(3), 361–370 (2021)
2. Köpke, S., et al.: Human–elephant conflict in Sri Lanka: A critical review of causal explanations. Sustainability **13**(15), 8625 (2021)
3. Hoare, R.: Lessons from 20 years of human–elephant conflict mitigation in Africa. Hum. Dimens. Wildl. **20**(4), 289–295 (2015)
4. Sukumar, R.: The Living Elephants: Evolutionary Ecology, Behavior, and Conservation. Oxford University Press, New York, ISBN 0-19-510778-0 (2003)
5. De Martino, M.G.B., Dos Reis, F.S., Dias, G.A.: An electric fence energizer design method. In: 2006 IEEE International Symposium on Industrial Electronics, 2, pp. 727–732 (2006)
6. Athulathmudali, D.S.K., Pilapitiya, S., Gunaruwan, T.L.: Economic and conservation impacts of seasonal paddy fences as a human-elephant conflict mitigation strategy. Econ. Res. **12**, 1 (2024)
7. Sapkota, S., Aryal, A., Baral, S.R., Hayward, M.W., Raubenheimer, D.: Economic analysis of electric fencing for mitigating human-wildlife conflict in Nepal. J. Resour. Ecol. **5**(3), 237–243 (2014)
8. Gubbi, S.: Patterns and correlates of human–elephant conflict around a south Indian reserve. Biol. Cons. **148**(1), 88–95 (2012)
9. Vibha, G., Lingaraju, H.G., Venktaramana, G.V.: Effectiveness of solar fence in reducing human–elephant conflicts in Manchahalli village, Mysuru, Karnataka. India. Current Science **120**(4), 707–711 (2021)
10. Jarungrattanapong, R., Olewiler, N.: Ecosystem management to reduce human–elephant conflict in Thailand, pp. 1–20. Environment, Development and Sustainability (2024)
11. Panescu, D.: Electrical Safety and Performance of E-Band devices Relative to Relevant Requirements of International Electrical Standards. In: 2024 46th Annual International Conference of the IEEE Engineering in Medicine and Biology Society (EMBC), pp. 1–6 (2024)

12. Kroll, M.W., Perkins, P.E., Panescu, D.: Electric fence standards comport with human data and AC limits. In: 2015 37th Annual International Conference of the IEEE Engineering in Medicine and Biology Society (EMBC), pp. 1343–1348 (2015)
13. Kumar, M.A.: Design of solar powered energizer and on-line monitoring of electric fencing system. In: 2014 IEEE 8th International Conference on Intelligent Systems and Control (ISCO), pp. 211–217 (2014)
14. Erickson, R.W., Maksimovic, D.: Fundamentals of power electronics. Springer Science & Business Media (2007)
15. Redondo, L.M., Silva, J.F.: Flyback versus forward switching power supply topologies for unipolar pulsed-power applications. IEEE Trans. Plasma Sci. **37**(1), 171–178 (2008)
16. Pressman, A.: Switching power supply design. McGraw-Hill, Inc. (1997)
17. Towers, T.D.: Practical design problems in transistor dc/dc convertors and dc/ac inverters. Proceedings of the IEE-Part B: Electronic and Communication Engineering **106**(18S), 1373–1383 (1959)
18. Thrimawithana, D.J., Madawala, U.K., Woodhead, R.C.B.: Pulsed power generation techniques. In: IECON 2006–32nd Annual Conference on IEEE Industrial Electronics, pp. 2014–2019 (2006)
19. Attia, H., Aziz, M.R., Alzeyoudi, O.: Transformer-Less Grid Voltage Controller Through SCR Switches and a Flexible Pulse Device. In: 2024 17th International Conference on Development in eSystem Engineering (DeSE), pp. 129–134 (2024)
20. Mikołajczak, K., Mikolo Yobo, C., Chehoski, E., Tshibangu, C., Schutgens, M.: Farmers' experiences of mobile electric fences for human–elephant coexistence: insights and recommendations from service design research in Gabon, London: Grantham Research Institute on Climate Change and the Environment, London School of Economics and Political Science (2024)

Development of an Aquabot for Coral Monitoring and Marine Data Collection Using Image Processing Techniques

W. A. M. P. L. Adhipaththu, A. M. M. H. B. Abeysinghe, N. Y. Ranasinghe, U. V. H. Sameera, and B. L. Sanjaya Thilakarathne(✉)

Department of Instrumentation and Automation Technology, Faculty of Technology, University of Colombo, Colombo, Sri Lanka
sanjaya@iat.cmb.ac.lk

Abstract. This research outlines the integration of image processing with an Aquabot, underwater rover to monitor the growth of corals. Aquabot, a tethered underwater rover with a floating communication/power station, a Python/C# ground-station dashboard, and a cloud back end for real-time telemetry and image analysis focused on coral monitoring around Sri Lanka. The platform integrates dual-camera video, pH/temperature/depth sensing, joystick teleoperation, and an inference pipeline for coral varieties and coral disease detection trained on locally curated, polygon-annotated datasets. On held-out tests, the varieties detector achieved precision = 0.679, recall = 0.558, = 0.596 and :0.95 = 0.425 (124 images/492 instances), while the disease detector achieved precision = 0.346, recall = 0.331, = 0.295 and :0.95 = 0.188 (121 images/489 instances). Field evaluations demonstrated stable maneuvering, synchronized data capture, and ~ 12 min endurance at peak load, validating the end-to-end pipeline from acquisition to analysis. The system provides a practical, extensible baseline for sustained coral monitoring and conservation workflows in shallow coastal environments.

Keywords: Underwater ROV · tethered robot · coral reef monitoring · coral disease detection · coral species classification · underwater computer vision

1 Introduction

This project specifies the development of a comprehensive system for coral monitoring and analysis. The major objectives include the development of rigorous calibration and deployment protocols for facilitating reliable field replication, the establishment of a secure cloud backend with REST API and MariaDB for real-time data storage and processing, and the availability of an intuitive ground-station dashboard with joystick control, live telemetry plotting, and mission management. To enable accurate coral detection, a curated image dataset was prepared exhaustively with 70/20/10 train/validation/test split and augmented through extensive augmentations. Model performance was fine-tuned by confidence threshold and Non-Maximum Suppression (NMS) parameter tuning on the validation set, with an input resolution of 640 × 640 chosen to optimally trade off detection accuracy and operational throughput.

M. Siriwardhana et al. (Eds.): IERS 2025, PTT, pp. 47–57, 2026.
https://doi.org/10.1007/978-981-95-7984-6_5

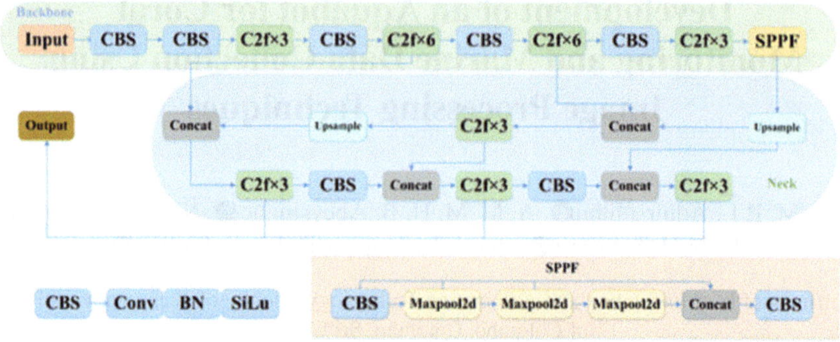

Fig. 1. YOLOv8 network structure diagram

Aquabot is proposed as an autonomous underwater robot to overcome the limits of manual monitoring and satellite imagery by providing continuous, high-resolution monitoring of Sri Lanka's coral plantations, capturing coral health, temperature, light, and water-quality data for real-time analysis and rapid response (Fig. 1). Automated sensing expands coverage and preserves data for longitudinal study, while underwater rovers deliver high-resolution imaging, real-time streams, intuitive controls, and efficient navigation and collection [1]. Remote operation further boosts data volume and ecological insight [2], supporting the conclusion that underwater rovers are a promising alternative to traditional observation and can strengthen prevention and conservation strategies for sustainable reefs [3].

The primary aim of this project is to create an end-to-end autonomous coral health monitoring system. This involves the manufacture of a custom, remote-operated underwater rover with pressure resistance considerations and in conjunction with a surface communication buoy. The system should be able to record synchronized video and environmental sensor data (temperature, pH, depth) with secure telemetry. Simultaneously, for enabling automatic detection, we train and evaluate deep-learning model-based detection models specifically aimed at identifying coral species and diseases in the context of Sri Lankan reef communities [4].

AUV/ROV studies show safer, repeatable reef surveys, but hydrodynamics make localization/control difficult; designs favor compact, multi-thruster, tethered platforms [5]. Underwater vision is degraded by turbidity and color loss; transfer learning/augmentation help, yet end-to-end in-situ validation is rare. This work addresses the gap with a maneuverable tethered rover and floating station, synchronized sensing, and a locally trained detection pipeline validated in Sri Lankan waters.

2 Methodology

Dataset and Training

We curated a coral image dataset with a 70/20/10 train/validation/test split and applied augmentations (flips, grayscale, saturation, noise, blur). Input resolution was set to 640 × 640 to balance field-of-view normalization and throughput; confidence thresholds and NMS IoU were tuned on the validation set.

Model Architecture

Representative settings: input 640 × 640, batch size selected to saturate GPU memory without paging, initial learning rate 1e-3 with warmup, momentum 0.9, weight decay 5e-4, and early stopping on validation mAP@0.5:0.95. Final checkpoints were chosen by best validation performance.

We implemented a one-stage object detector with CSP-based feature extraction and multi-scale feature aggregation. Training used stochastic gradient descent with momentum and cosine learning-rate decay. Data augmentation followed the policy above; confidence thresholds and non-maximum suppression IoU were tuned on the validation split to maximize mAP@0.5 while controlling false positives.

Table 1. System Specifications

Category	Specification
Ground Station	Tkinter-based Python GUI with multithreading; PS4 joystick control
Vision & data	OpenCV for video capture/preprocessing; Matplotlib & NumPy for live plots
Backend/Storage	Oracle Cloud REST API with authentication; MariaDB database
Model training	Roboflow annotation; object detection with precision/recall/mAP evaluation
Dashboard	C# WinForms interface for live and past mission visualization
Compute Module	Raspberry Pi 3B + (master, data acquisition, communications)
Microcontrollers	Dual Arduino Nano for sensor integration and motor control
Propulsion	6 × brushless DC thrusters (planar + vertical); bi-directional ESCs
Power system	14.4 V (2 × 7.2 V Ni-MH, 6.2 Ah); buck converters and regulators
Networking	Wired power and Ethernet via floating station with router
Sensors	pH, temperature, and depth sensors; dual cameras (front/rear)

The project progressed from a literature review and expert input to a conceptual Aquabot design (size, shape, mobility, payload), then to detailed SolidWorks modelling and simulations for pressure tolerance and balance (Table 1). A stainless-steel frame was selected for durability and corrosion resistance, with TIG welding, Perspex mounting, and silicone sealing to ensure watertight integrity. The finalized build underwent comprehensive tests for manoeuvrability, stability, and responsiveness. After calibrating sensors, brushless motors with bi-directional ESCs were installed; the system was first validated via manual controls before moving to IoT-based automation for higher precision.

2.1 Aquabot Hardware

The control stack uses a Raspberry Pi 3 B + as the master in a DAQ setup with two Arduino Nano boards acting as slaves—one for sensors and one for motor control. Prototyping was done on a breadboard before migrating to a custom PCB designed in

EasyEDA. Power from the floating unit is distributed via a bus bar to six motors; the Pi is fed through a buck converter, the Arduino Nanos draw power from the Pi, and the lamp is supplied through a 7812 linear regulator. Thermal management includes heatsinks on hot components and a heatsink plus cooling fan on the Pi. Together, this hardware governs Aquabot mobility and accessory functions [6].

2.2 Aquabot Floating Station

Power is supplied by two 7.2 V, 6.2 Ah Ni-MH batteries in series (14.4 V), moved off-board to ease charging and reduce weight. An XH-M632 intelligent charging module—configured with start/full voltages—sits in the same enclosure as the router, protected by a 60 A main fuse; the router is fed via a buck converter. Three switches independently control the router, the main power supply, and the charger. A PVC pipe "overflow"/enclosure houses the battery and router, with power and network cables routed through it and a steel cable added for strain relief. The floating station acts as a middle communication node between the Aquabot, ground station, and cloud, incorporating error-handling for network disruptions (Fig. 2).

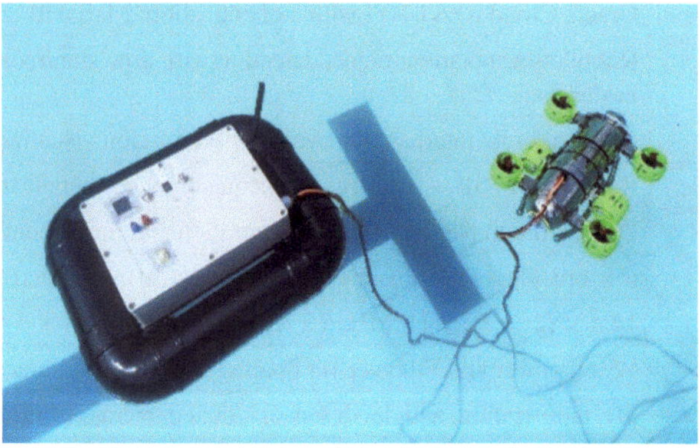

Fig. 2. Underwater robot with Floating Structure

2.3 Aquabot Ground Station

OpenCV captures frames from front and rear webcams, with real-time updates handled via multi-threading [7]; frames are resized and converted to Tkinter Photo Image objects for live display. Sensor data arrives over sockets, while matplotlib and NumPy support live plots; Tkinter's update loop plus multi-threading keeps the UI responsive. Missions are started via API calls (details shown on the dashboard), a Capture button triggers an asynchronous capture_and_send, and robust error handling with real-time logging tracks errors, updates, and events; elapsed time is computed from mission start. The UI follows UX best practices and shuts down cleanly stopping threads, releasing resources, closing connections, and disarming the Aquabot on termination. A PS4 joystick is integrated via Pyglet: it's auto-detected, monitored in a separate thread, mapped to movement/arm–disarm combos and light toggles, and its connectivity/status is shown in the dashboard with feedback on disconnections.

2.4 Aquabot Cloud

The Aquabot cloud implementation leverages Oracle cloud services. The implementation involves the utilization of a MariaDB server for Structured Query Language (SQL) based real-time database storage and a Representational State Transfer Application Programming Interface (REST API) with authentication for seamless communication between the Aquabot control system and the cloud.

2.5 Image Processing

A diver-sourced coral dataset (~1,500 images) was curated to 764 images for training, then meticulously annotated in Roboflow—using polygon tools to mark coral varieties and diseases—creating a rich supervised-learning set [8]. Data were split 70/20/10 for train/validation/test, augmented with flips, grayscale, saturation shifts, noise, and blur, and uniformly resized to 640 × 640. Validation focused on generalization (evaluating precision, recall, accuracy, and F1), driving iterative model refinements. Final testing mimicked deployment conditions—varying lighting, backgrounds, and orientations—to verify robustness before real-world use.

2.6 Aquabot Data Analysing Dashboard

The Aquabot Data Analysing Dashboard is a C#/Visual Studio 2022 app with three pages—Home, Live Mission, and Past Mission. Users can view real-time telemetry (with a manual refresh) and analyse images by clicking "Analyse," which sends images to a remote detector and returns results via an API. A backend API (using libraries like Newtonsoft.Json, RestSharp, System.Windows.Forms, System.IO, and Newtonsoft.Json.Linq) handles data exchange and updates, while multithreaded background processes keep the UI responsive. Key limitations/trade-offs include underwater lighting/turbidity and colour attenuation (hurting detector confidence), umbilical drag and network latency (affecting real-time behaviour), and a speed–accuracy balance where higher input resolution improves Map but lowers FPS; cloud offload saves on-board power at the cost of latency, whereas on-edge inference reduces latency but needs more thermal/power budget and model compression. Metrics reported are precision, recall, mAP0.5, and mAP0.5 to 0.95; "accuracy" is avoided for detection tasks.

3 Results and Discussion

3.1 Simulation on the Mechanical Structure

The design of a structure aimed to reduce drag force by reducing cross-sectional area and allowing it to sink and float on water surface without external force, using thruster propulsion instead of buoyancy propulsion (Fig. 3) [9].

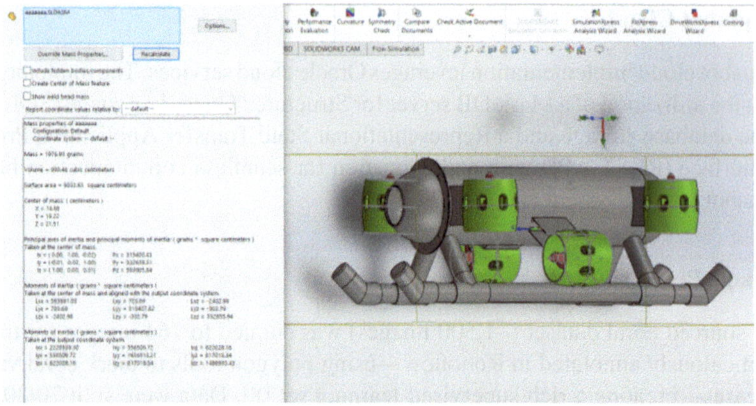

Fig. 3. Real drawing of underwater robot

The underwater robot's lightweight shell, including sensors and video systems, is critical and must withstand pressure and stress. Made of 1 mm thick stainless steel, it offers high corrosion and compression resistance, with a yield strength of 480 MPa, elastic modulus of 190–210 GPa, and Poisson's ratio of 0.27–0.30. Perspex is also used for its transparency and resistance properties. At a depth of 4 m, the robot must withstand a pressure of 40,012.56 Pa.

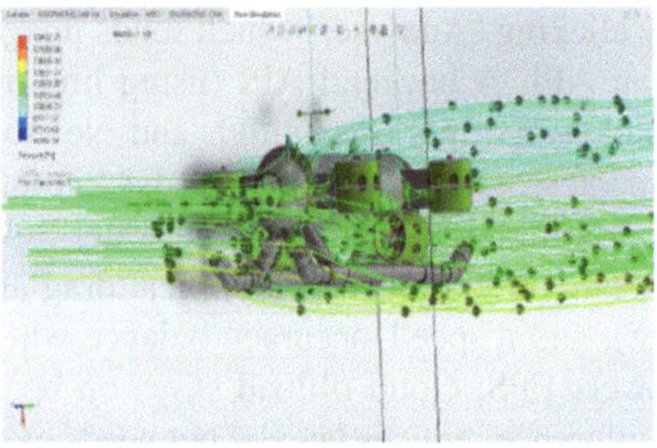

Fig. 4. Flow simulation of the underwater robot.

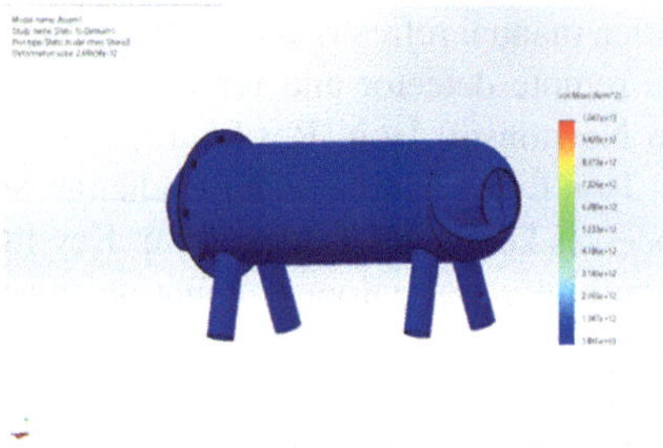

Fig. 5. Static nodal Stress

The underwater robot's shell uses standardized O-ring radial and axial seals between the head and body for waterproofing. A Perspex plate, glued to the shell, holds the camera, with minimal thermal expansion concerns due to a small temperature difference (22.5 °C at 0.3 m, 21.5 °C at 4 m) [10]. Regular inspection is essential after each underwater task.

The underwater robot features an anti-collision frame, protecting internal components and reducing underwater resistance. It's suitable for slow propulsion due to planar design and low drag force.

3.2 Sensor's Calibration

The pH, temperature, and depth sensors were calibrated individually: the temperature sensor matched a calibrated thermometer with no observed deviation; the depth sensor setup (wire length verified by ruler, then tested in gravel) showed readings consistent with actual values; and pH calibration (Figs. 4, 5, 6, 7 and 8) reduced the difference between the pH sensor and a reference meter to a minimal error. For propulsion, two horizontal motors handle forward/back motion; due to drag from the floating unit and turning dynamics, one motor can power tight turns while increased contact area raises drag during pivot turns, so up/down thrusters are emphasized for low-drag, low-speed cases. Thruster power was set to 12 V, implying ≥ 4S ESCs; with a motor max of 24 V (≈7S), bidirectional ESCs sized to the motor current were required, and 2S–6S 30 A bidirectional ESCs were selected in practice.

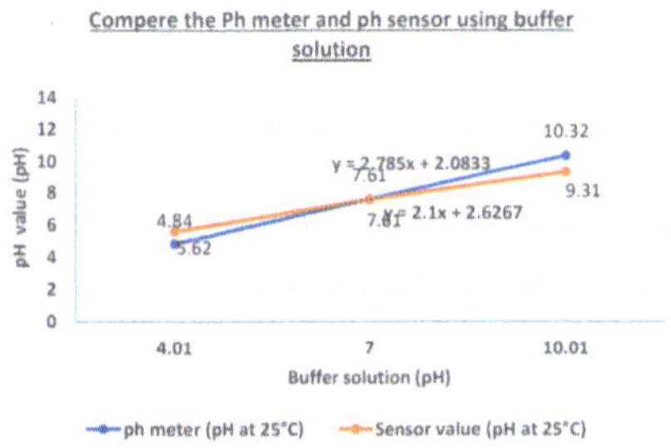

Fig. 6. Graph of compare the pH meter and pH sensor using buffer solution.

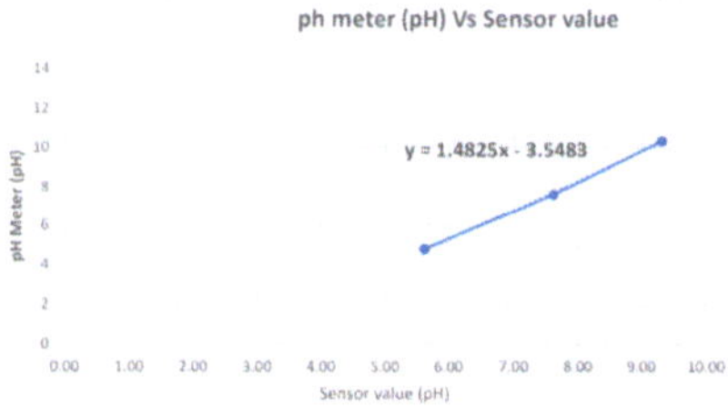

Fig. 7. Graph of pH meter vs pH sensor.

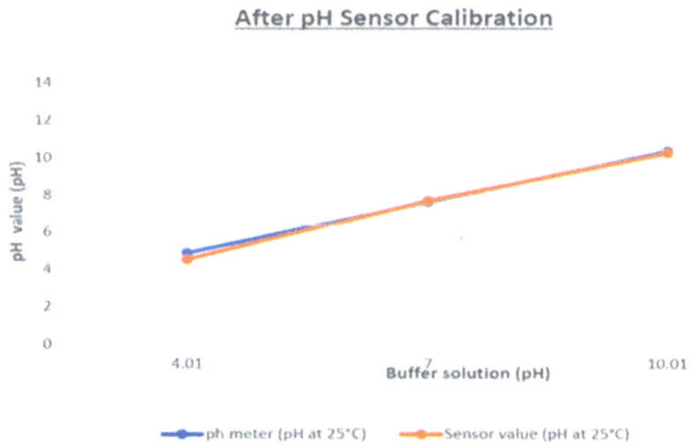

Fig. 8. Comparison of the pH meter and pH sensor using buffer solution.

3.3 Underwater Power Supply System

The battery compartment is located separately from the router [11]. It is designed to float. Because the brushless motor draws a high current, the battery must have a high discharge capacity. Therefore, Ni-Mh battery is used for this. From here three, 5 m wires go to the robot. A Stainless-steel cable is used for power wire, communication wire and energy. Since there are 6 motors here, if it rotates at maximum 12, the current it will draw is 30 A.

This can be reduced to 5C. Therefore, a battery with a rating of 5C should be selected. The Ni-Mh battery we have chosen is of 5C rating. On one full charge, the robot can operate for 12 min at its maximum performance. Here the robot gets the maximum current when the motor starts. It is 60 A current. Therefore, a 60 A direct current fuse is used to protect this system.

The combined system functioned to primary operating standards during field testing. The whole pipeline, in peak loading (propulsion, sensors, compute), had an average endurance of roughly 12 min between charges. This is sufficient for initial survey work but still a primary constraint for extended transects. All primary data streams—video, pH, temperature, and depth—were collected and synchronized successfully, validating the vehicle's sensing and data handling capabilities.

3.4 Image Processing

An epoch covers one full pass of training/validation over the dataset; lower loss means better fit, while higher "score" reflects more accurate predictions. Two models were evaluated (Table 2): the Coral Disease Detection model on 121 images with 489 disease instances (accuracy 33.1%; precision 0.346; recall 0.331; mAP50 0.295; mAP50–95 0.188) and the Coral Varieties Detection model on 124 images with 492 instances (accuracy 59.6%; precision 0.679; recall 0.558; mAP50 0.596; mAP50–95 0.425). Overall, the varieties model outperforms the disease model, but both show task-appropriate effectiveness. Figures 9 and 10 reportedly illustrate correct detections (e.g., Acroporidae variant; diseases like Algae and Bleached Coral), suggesting reliability across varying underwater visibility. Testing of the trained YOLO models on the held-out test set revealed a clear performance difference between the two detection tasks.

Table 2. Detection performance summary (test sets)

Task	Images / Instances	Precision	Recall	mAP@0.5	mAP@0.5:0.95
Coral Disease Detection	121 / 489	0.346	0.331	0.295	0.188
Coral Varieties Detection	124 / 492	0.679	0.558	0.596	0.425

Recognition of coral disease was decent in performance but provided a quantitative baseline (Fig. 10). The model possessed 0.346 (95% CI 0.304–0.390) precision and 0.331 (95% CI 0.291–0.374) recall, indicating both incorrect discrimination of diseases and difficulty in finding all occurrences of disease within images. This is also reflected in mean Average Precision scores, with mAP@0.5 at 0.295 and the more stringent mAP@0.5:0.95 at 0.188. This performance is likely attributed to the inherent difficulties of the task like limited and heterogeneous training data, high visual variability of disease presentation, and confounding influences of the underwater environment like color attenuation and occlusion.

In contrast, the coral types detector performed much stronger and more consistently. It achieved a higher precision of 0.679 (95% CI 0.632–0.723), though its recall value of 0.558 (95% CI 0.514–0.601) indicates a tendency to miss some of the coral instances. The model achieved an mAP@0.5 of 0.596 and an mAP@0.5:0.95 of 0.425, supporting its utility as a reliable means of discriminating between coral types in complex reef structures.

Fig. 9. Coral varieties detection.

Fig. 10. Coral disease detection.

3.5 Controlling Dashboard

The Aquabot control dashboard (Fig. 11) is a central interface for managing missions and monitoring system status. When idle, it shows an overview of the Aquabot; starting a mission involves entering basic details (name, objective, location) via a launch dialog. Once active, a dedicated mission dashboard provides live front/rear video, real-time sensor telemetry with continuous cloud uploads, elapsed-time display, arm/disarm controls, and built-in error handling with event logging. Missions can be terminated from the dashboard, which may include returning the Aquabot to the surface or a designated area.

Fig. 11. Control dashboard

3.6 Analysing Dashboard

Users got a simple choice between two crucial sections, the live mission information screen, and the previous mission information screen. A key feature of this section is the manual refresh button, ensuring users can promptly update the display to access the latest data.

4 Conclusions

The robot's ability to function beyond 3 m depth, combined with solar-assisted power, ensures extended missions. PID optimization improved stability and control, while added sensors and high-resolution imaging enhanced adaptability and coral detection accuracy.

Upgraded cameras and a large image dataset support robust object recognition, making Aquabot a valuable tool for coral monitoring and ecosystem conservation. The project marks a strong step from concept to real-world deployment, with potential for broader ecological and underwater exploration applications.

This project managed to produce and implement an integrated coral monitoring system, demonstrating stable vehicle operation, real-time synchronization of data via a secure cloud backend, and effective mission control via a custom-made dashboard. However, the undertaking revealed several key limitations. The muted detection in coral disease emphasizes the limited and heterogeneous training set problem, as well as exacerbated by the challenges of the underwater world in its own right to color attenuation, turbidity, and viewpoint and scale variability. Working restrictions, including tether drag, network latency, and battery-limited endurance of approximately twelve minutes, presently restrict survey time and fine control responsiveness. Besides, its functionality in deeper depths is restricted, and its navigation still remains tele-operated, with its automated capabilities representing a light level of autonomy.

Overcoming these limitations is the clear agenda for future work. Immediate priorities include strategic scaling and rebalancing of the image dataset to cover additional classes of diseases and diverse environmental conditions, which will be supported by the addition of underwater image enhancement and domain adaptation techniques for greater model robustness. To cope with latency and power constraints, we intend to explore more powerful detection architectures and quantized inference on-edge. Hardware upgrades will focus on higher-capacity power supplies, improved tether management, more effective pressure-sealing implementations for greater depth rating, and onboard lighting solutions. Software-wise, transitioning towards full autonomy through visual-odometry, SLAM, and station-keeping algorithms is an important goal, as well as developing richer mission planning tools and continuous system health monitoring. Eventually, the aim in the longer term is to demonstrate these improvements through longer, multi-site trials with standard protocols, taking the system from a research prototype to a robust, repeatable, and partner-enabled platform for coral conservation science.

Acknowledgement. In sincere acknowledgment of the invaluable guidance, unwavering support, and profound assistance bestowed upon us throughout our journey, we extend our heartfelt gratitude to a constellation of individuals. First and foremost, we ex-press our deepest appreciation to our esteemed Dean, the Head of the Department, the lecturers from the Instrumentation & Automation Technology at the Faculty of Technology, University of Colombo.

References

1. Brown, K.: The dynamics of coral-algal interactions on coral reef ecosystems. PhD Thesis. School of Biological Sciences, The University of Queensland (2018). https://doi.org/10.14264/uql.2018.487
2. Casey, J.M., Ainsworth, T.D., Choat, J.H., Connolly, S.R.: Farming behaviour of reef fishes increases the prevalence of coral disease-associated microbes and black band disease. Proc. R. Soc. B Biol. Sci. **281**(1788) (2014). https://doi.org/10.1098/rspb.2014.1032
3. Chen, G., Chen, W., Wang, Z., Guo, T., Xia, X., Xu, L.: Design and dynamic performance research of underwater inspection robots. Wirel. Commun. Mob. Comput. (2022). https://doi.org/10.1155/2022/3715514
4. Hoegh-Guldberg, O., Poloczanska, E.S., Skirving, W., Dove, S.: Coral reef ecosystems under climate change and ocean acidification. Front. Mar. Sci. **4** (2017). https://doi.org/10.3389/fmars.2017.00158
5. Yunhui, Z., Lina, S., Lu, Q., Qiang, F.: The Frame Structure of Underwater Robot for Shallow. Water **5**, 107–110 (2019). https://doi.org/10.2991/pntim-19.2019.22
6. Jebelli, A., Yagoub, M.C.E., Dhillon, B.S.: Design and Control of Underwater Robots with Rotating Thrusters. IAES Int. J. Robot. Autom. **5**(4), 284–294 (2016). https://doi.org/10.11591/ijra.v5i4
7. Sun, X., et al.: Transferring deep knowledge for object recognition in Low-quality underwater videos. Neurocomputing **275**, 897–908 (2018). https://doi.org/10.1016/J.NEUCOM.2017.09.044
8. Jalled, F., Voronkov, I.: Object Detection using Image Processing. Cornell University (2016). https://doi.org/10.48550/arXiv.1611.07791
9. Kadiyam, J., Mohan, S.: Conceptual design of a hybrid propulsion underwater robotic vehicle with different propulsion systems for ocean observations. Ocean Eng. **182**, 112–125 (2019). https://doi.org/10.1016/j.oceaneng.2019.04.069
10. Vatech, Physics Forum, Handle peak inrush current of a BLDC motor to protect the power supply (2021). https://www.physicsforums.com/threads/handle-peak-inrush-current-of-a-bldcmotor-to-protect-the-power-supply.999830/. Last accessed 26 November 2023
11. Bradley, A.M., Feezor, M.D., Singh, H., Sorrell, F.Y.: Power systems for autonomous underwater vehicles. IEEE J. Ocean. Eng. **26**(4), 526–538 (2001). https://doi.org/10.1109/48.972089

Development of Defect Reducing Mechanisms for Paper Bag Manufacturing Using Machine Vision System with YOLOv8 Model

Himasha Shanuka and Roshan Thilakarathne(✉)

Department of Mechanical and Manufacturing Technology, Faculty of Technology, Wayamba University of Sri Lanka, Kuliyapitiya, Sri Lanka
fot209208@kul.wyb.ac.lk, roshant@wyb.ac.lk

Abstract. The paper bag manufacturing industry confronts significant economic losses and material waste due to high rates of production defects, particularly inconsistent adhesive application. This research presents the development and validation of an intelligent defect reduction framework for the gum dotting process in multiwall paper sack production. An Internet of Things (IoT) integrated machine vision system, utilizing a YOLOv8 model trained on a custom dataset of approximately 1,300 images, was developed for real-time defect detection and classification. Through a series of simulated experiments, the system's efficacy was quantified. Spatial analysis of defect data successfully identified specific mechanical root causes, such as individual nozzle failures versus systemic pressure imbalances. A correlative analysis established a critical, non-linear relation-ship between machine velocity and defect frequency, identifying an optimal operating window of 100–160 bags/min, beyond which defect rates increase exponentially. Furthermore, the analysis of the real-time alerting protocol and subsequent maintenance interventions demonstrated the system's capacity to facilitate a strategic shift from reactive to predictive maintenance. By calculating the Mean Time Between Failures (MTBF) for specific fault-and-fix cycles, the framework provides the empirical data needed to predict component end-of-life and schedule proactive replacements. The findings confirm that this data-driven approach can significantly reduce waste, optimize production efficiency, and enhance overall product quality.

Keywords: Paper bag manufacturing · Defect detection · Machine learning · Root cause analysis

1 Introduction

As Sri Lanka intensifies its efforts to tackle environmental challenges, the paper bag industry has stepped into the spotlight as a key player in the country's move toward sustainable packaging. The urgency of this shift is clear: Sri Lanka produces around 7,500 metric tons of solid waste each day, with 400 metric tons attributed to plastics-much of which ends up unrecycled and polluting the environment [1].

M. Siriwardhana et al. (Eds.): IERS 2025, PTT, pp. 58–72, 2026.
https://doi.org/10.1007/978-981-95-7984-6_6

Paper bags, which are both biodegradable and recyclable, offer a practical solution to this pressing issue, helping to reduce plastic pollution and support the nation's broader commitment to circular economy principles [2]. Local manufacturers are embracing these changes, using recycled paper and sustainable forestry practices, which not only protect the environment but also create valuable employment opportunities in manufacturing and related sectors.

Many Sri Lankan paper bag manufacturers still depend on traditional quality control methods, which often fail to catch defects early in the production process. Due to a higher defect rate, more bags are classified as waste. To compensate for this, production needs to increase output. According to lean manufacturing principles, overproduction is one of the seven wastes [3].

Glue detection systems are increasingly used based on UV light detection. This system contributes to the quality of the products. However, the technology also faces challenges. Optical brighteners in the paper make standard UV glue detection difficult. Through the use of optical brighteners in all kinds of paper and board, as well as through the recycling stream, non-optical whitening paper is also contaminated with this additive. This scatters the detection image [4].

This scattering effect, common in recycled paper stock, renders traditional UV-based systems unreliable and prone to a high rate of false negatives (missed defects). In contrast, our ML-based approach is resilient to such material inconsistencies. By learning visual features directly from images, the YOLOv8 model achieves a high precision of 0.994 and recall of 0.976, ensuring reliable detection even on varied paper surfaces and under challenging lighting conditions.

In recent years, the software and hardware technologies of computer vision system have been promoted rapidly with the development of electronic technology. High resolution and speed camera and image acquisition system are widely used in the process of logistics, which meet the requirements of defects detection for logistics packaging box based on image processing [5].

Traditional manual defect detection methods suffer from low sampling rates, limited accuracy and efficiency, poor real-time performance, high labor intensity, and strong dependence on human expertise. To overcome these limitations, there is an urgent need for machine learning-based defect detection and data processing technologies. Unlike these traditional, reactive quality control measures, this research introduces a proactive, ML-driven framework designed for real-time intervention, a significant advancement over methods that often identify defects post-production. These intelligent systems can be deployed at the edge to enhance real-time analysis, reduce human intervention, and significantly improve the overall performance and scalability of defect detection in edge computing environments [6].

Compared with traditional defect detection techniques that rely on detection techniques that rely on hand-crafted features and struggle with complex scenes, our YOLOv8-based approach leverages deep-learning to provide superior real-time detection accuracy and efficiency [7, 8].

To address these challenges, this paper presents the development and validation of a novel machine vision system integrated with an IoT platform. The research methodology was guided by a Root Cause Analysis (RCA) that identified the most probable causes of defects associated with the paper bag tuber machine. The project's core objectives involved analyzing production data to identify defect trends and implementing a system to reduce waste.

The finalized solution uses a machine learning model to detect gum dots on paper bags in real time, providing a live count and alerting operators to anomalies. The findings demonstrate a clear path to reducing production waste and establish a framework for data-driven quality control, aiming to minimize costs and optimize the manufacturing process.

2 Material and Methods

The methodology for this research followed a systematic, step-by-step approach to effectively identify and address the company's production issues. The process began with the comprehensive collection and analysis of production data to establish baseline defect trends. Subsequently, a closed-ended survey was administered to identify the most relevant perceived causes, which were then thoroughly investigated to determine the definitive root cause of the problem. Following this in-depth analysis, the most viable alternative solution was selected, initiating the model development process. The resulting machine learning model was then rigorously tested for accuracy, undergoing necessary iterative changes to optimize its performance and ensure a reliable solution (Fig. 1).

2.1 Data Collection and Analysis on Production Data

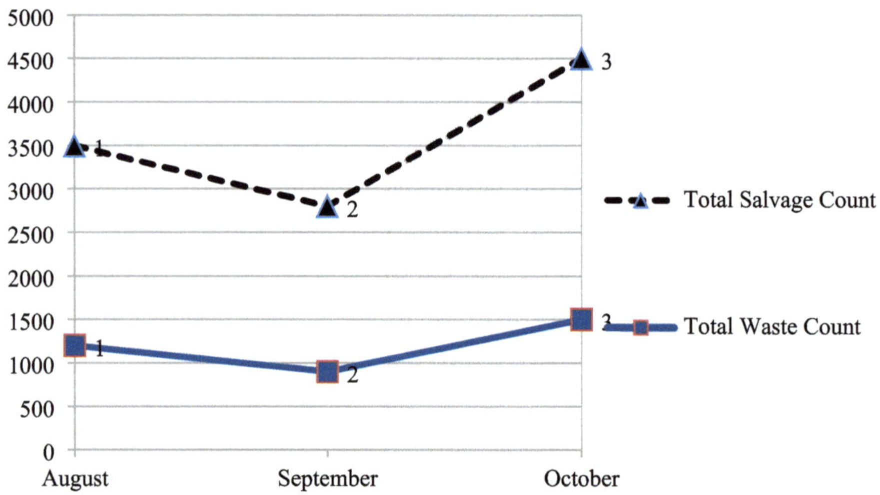

Fig. 1. Represents monthly trends in salvage and waste counts for multiwall paper bag production during the three-month study period (August-October 2024).

Referring production data, Table 1. Represents the Multiwall bag type for August, detailing daily production quantities, salvage counts, and waste figures. The dataset highlights fluctuations in output and defect levels, supporting the analysis of defect trends and process efficiency.

Table 2 presents the results of a closed-ended survey conducted among various personnel including operators, supervisors, technicians, and quality checkers to assess perceived root causes of defects in the production process. Each participant rated predefined possible causes, with the aggregated responses providing insight into the most frequently cited issues. Notably, "Inconsistent glue application in the tuber" and "Absence of a systematic approach to analyze defect data" were identified as the most significant contributors, each receiving the highest total scores.

This structured feedback approach adds qualitative reliability to the root cause analysis, reinforcing the need for targeted interventions in both process and quality control practices.

2.2 Closed-Ended Survey to Identify the Most Relevant Cause

The use of closed-ended questions in survey research is fundamental for efficiently identifying key factors contributing to organizational issues, such as the increase in salvage rates. Closed-ended questions provide respondents with predefined answer options, which streamlines the data collection process and ensures uniformity in responses.

This structure not only facilitates the rapid gathering of large volumes of data but also enhances the reliability and replicability of findings, as each participant interprets and answers the questions in a consistent manner.

Ultimately, the use of closed-ended questions is instrumental in narrowing down complex issues to their most relevant causes, thereby guiding the formulation of targeted and effective solutions.

2.3 Investigate the Root Cause

Fishbone analysis, also known as the Ishikawa or cause-and-effect diagram, is a valuable tool for identifying the root causes of problems in research and practical work. The identification of defect causes plays a key role in smart manufacturing as it can reduce production risks, minimize the effects of unexpected downtimes, and optimize the production process [9].

From the Fishbone diagram, the possible causes, primary and secondary, are identified and then for each possible cause, a root cause is identified [10]. Figure 2 illustrates a fishbone diagram used to systematically identify and categorize the root causes contributing to gum dot issues in paper bag manufacturing. The diagram breaks down the problem into six main categories: Man, Material, Method, Machine, Measurement, and Environment. Each category highlights specific factors that potentially lead to defects.

For example, under the Man category, issues like lack of supervision and physical loading are noted, indicating a human resource and operational oversight problem. The Material branch points to improper storage conditions and the fast-drying nature of the adhesive, which can compromise bonding quality. Under Method, inadequate maintenance schedules are highlighted, while the Machine category emphasizes the impact of aging equipment and outdated adhesive application techniques. Measurement issues

Table 1. Production data for Multiwall bag type on August, detailing daily production quantities, salvage counts, and waste figures.

Date	Bag type	Production Qty	Salvage	Waste
1-Aug	M.W	5580	200	9
	M.W	5200	140	10
	M.W	1455	50	8
2-Aug	M.W	10886	150	30
	M.W	2009	40	19
3-Aug	M.W	560	0	0
	M.W	16744	139	124
5-Aug	M.W	4200	360	25
	M.W	12000	82	26
6-Aug	M.W	14100	445	43
	M.W	13000	200	44
7-Aug	M.W	5000	210	13
	M.W	3070	67	65
	M.W	11000	60	7
8-Aug	M.W	1664	40	5
	M.W	13000	115	47
9-Aug	M.W	365	8	11
	M.W	4150	108	6
10-Aug	M.W	10000	140	41
11-Aug	M.W	11915	80	35
12-Aug	M.W	11100	96	53
13-Aug	M.W	1135	18	10
	M.W	4350	33	15
	M.W	2748	15	20
	M.W	7043	150	70
24-Aug	M.W	3155	30	15
26-Aug	M.W	1986	25	17
	M.W	8319	175	42
	M.W	3886	40	22
27-Aug	M.W	4100	70	2
	M.W	15300	64	57
	M.W	700	35	2

(*continued*)

Table 1. *(continued)*

Date	Bag type	Production Qty	Salvage	Waste
	M.W	7680	60	35
TOT		217400	3445	928
S.L PRCNT	1.58%			

arise from the lack of real-time monitoring, and environmental factors such as high factory temperatures may influence adhesive performance.

2.4 Selecting the Best Alternative Solution

The investigation into gum dot problems in bag manufacturing found four main issues that impact the quality of the bags, how efficiently they are made, and how resources are used.

1. Excessive gum spreading on glue edges which lead the unnecessary pasting bags with other bags.
2. In multi wall bags, due to gum dot defect the layers are not pasted properly. This leads the false valve folding conditions.
3. Assign approximately 4 labors for checking defect bags. It is not practical to checking all the bags quality and defects.
4. Slide laps on tubes are pasted falsely due to incorrect gum dot placements.

Finalize the best alternating solution as model development using machine learning. Implementing a gum dot detection system using machine learning involves integrating an identifying system with an indication display to monitor adhesive application in real-time (model development for detect defect gum dots). The system, equipped with a camera, detects the presence and placement of gum dots on the production line. It processes sensor data, comparing it against predefined standards for correct application.

If a defect, such as misalignment or insufficient coverage, is detected, the indicator provides immediate visual alerts to operators, specifying the issue. The predefined gum pattern is inserted into the system through machine learning, enabling the system to communicate with a microcontroller unit to make necessary indications. This approach can significantly reduce gum dot defects in existing glue application system by enabling real-time defect scanning. It can lower reworking costs and reduce the need for manual defect checking, thereby optimizing manpower usage.

2.5 Model Development Process

However, there are challenges to address, such as the system's ability to identify defects within approximately 1.5 s. To achieve these glue dot detections, the process starts with gathering a dataset of images showcasing various glue dot samples. These images are then annotated, which involves manually drawing bounding boxes around each glue dot and labeling them using Roboflow.

Table 2. Closed-ended survey conducted among various persons including operators, supervisors, technicians, and quality checkers

	Possible Causes	Me	Operators	Supervisors	Technicians	Quality checkers	Total
1	Inconsistent glue application in the tuber	09	08	09	08	06	**40**
2	Incorporate quality control methods	05	03	04	05	03	**25**
3	Inefficient workflow in manufacturing process	03	09	01	03	01	**17**
4	Absence of systematic approach to analyze defect data	07	09	09	06	09	**40**
5	Due to speed changes in machines (bag slacking)	03	04	03	05	02	**17**
6	Fatigue failures on bearings	03	03	02	02	02	**12**
7	Poor material control methods (reels)	01	01	01	01	01	**05**
8	Insufficient training on operators	01	01	01	01	03	**07**
9	Communication issues among staff	02	03	02	01	04	**12**
10	Poor lighting	03	04	02	04	03	**16**
11	High temperature conditions inside the factory	01	01	02	03	02	**09**
12	Improper storage conditions	02	03	01	02	04	**12**

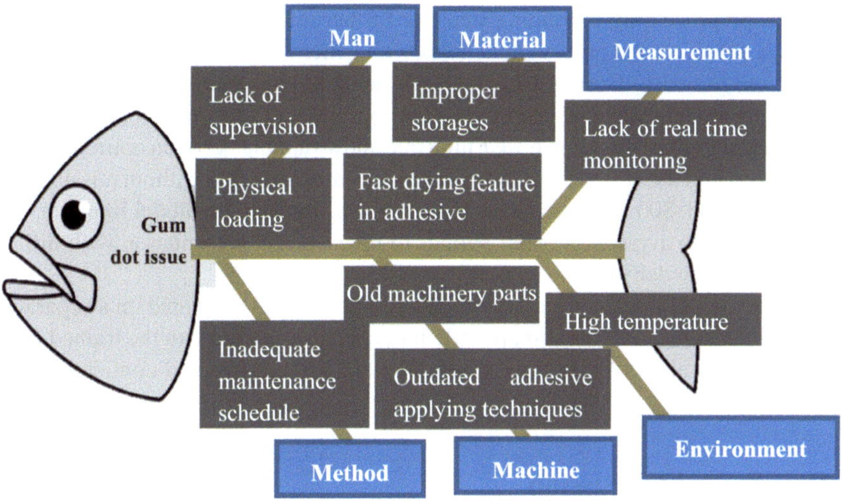

Fig. 2. Fishbone (Ishikawa) diagram illustrating the root causes of gum dot issue in the tube production system, categorizing potential contributing factors to facilitate targeted analysis and corrective actions.

The annotated dataset trains a YOLO (You Only Look Once) model, a deep learning algorithm known for its speed and accuracy in object detection. This training refines the model's ability to identify glue dots by adjusting its internal parameters [11].

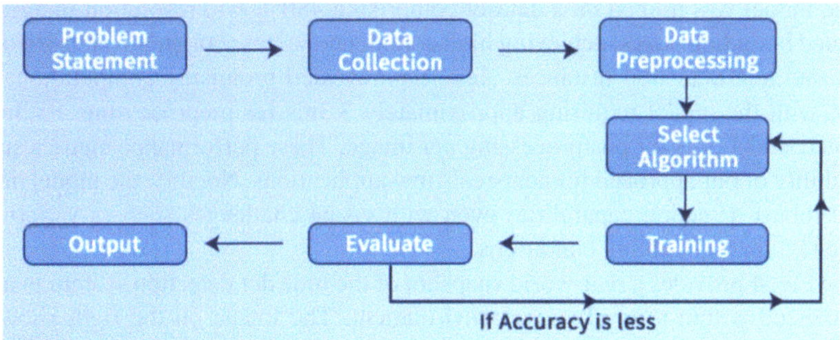

Fig. 3. Process flow diagram illustrating the sequential stages of machine model development

The Fig. 3 represents the machine model development process, highlighting the sequential and iterative nature of building an effective machine learning solution. The workflow begins with clearly defining the problem statement, followed by systematic data collection to ensure relevant and high-quality inputs. Next, the data undergoes preprocessing to address issues like missing values or inconsistencies, setting the stage for reliable analysis. The process then moves to algorithm selection and model training, where the chosen algorithm learns patterns from the prepared data. After training, the model is evaluated for performance. If the accuracy is insufficient, the workflow loops back to the preprocessing or training steps for further refinement [12]. Once the desired accuracy is achieved, the process concludes with generating the final output.

Next, we turned to Google Colab, which acted as our virtual lab, letting us train the YOLOv8 model using powerful GPUs to process data faster. We experimented with different settings-like tweaking the learning rate and swapping between optimizers (SGD and Adam)-to find the sweet spot where the model learned quickly without missing subtle defects. Over 100–300 training cycles (epochs), we watched the model improve, using cross-validation to avoid overfitting-similar to rehearsing a presentation with different audiences to refine delivery.

Once the model is trained sufficiently, its performance is evaluated on a separate set of images to ensure reliability. Finally, when new images are fed into the trained model, it predicts the location of glue dots, displaying bounding boxes and confidence scores for each detection.

3 Results and Discussion

To ensure the model's robustness and real-world applicability, a comprehensive dataset was developed for this study. The training dataset was extensive, consisting of approximately 1,300 images that captured a wide variety of production conditions, including different lighting, material textures, and defect types. For a rigorous and reliable measure of performance, a separate validation dataset of 200 images was utilized. This substantial dataset was crucial for developing a highly generalized model, minimizing the risk of overfitting, and ensuring high-fidelity defect detection across diverse and dynamic manufacturing scenarios.

The model was trained on a dataset comprising 480×640 resolution images with annotated bounding boxes, achieving high confidence scores (ranging from 0.36 to 0.96) across multiple detection instances. Test runs revealed promising computational efficiency, with the model requiring approximately 3.3ms for preprocessing, 85.3ms for inference, and 1.5ms for postprocessing per image. These performance metrics suggest the viability of our approach for near-real-time applications. Notably, the model demonstrated robust detection capabilities even with visual challenges such as variations in surface texture and lighting conditions.

The Fig. 4 provides a real-world snapshot of the glue dot detection system in action, as visualized within a development environment. The image on the right showcases a cardboard surface where multiple glue dots have been automatically identified and labeled with bounding boxes and confidence scores, demonstrating the model's ability to accurately recognize and localize each glue application.

The Fig. 5 shows a frame from a video demonstrating the real-time performance of the glue dot detection system. On the cardboard surface, each glue dot identified by the model is marked with a blue rectangular bounding box and labeled with the class name "gluDot" along with a confidence score, indicating the model's certainty for each detection. The green text at the top of the frame displays the number of detected dots for the current frame, confirming the system's ability to process and analyze images dynamically.

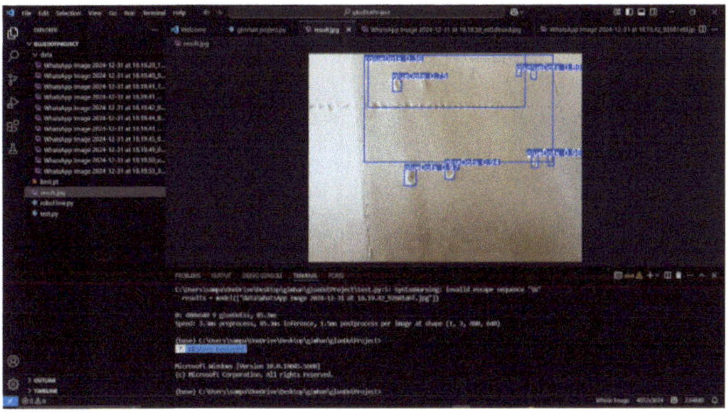

Fig. 4. Output of the glue dot detection system using a custom-trained machine vision model. The image displays identified glue dots on a paper bag, each labeled with a confidence score.

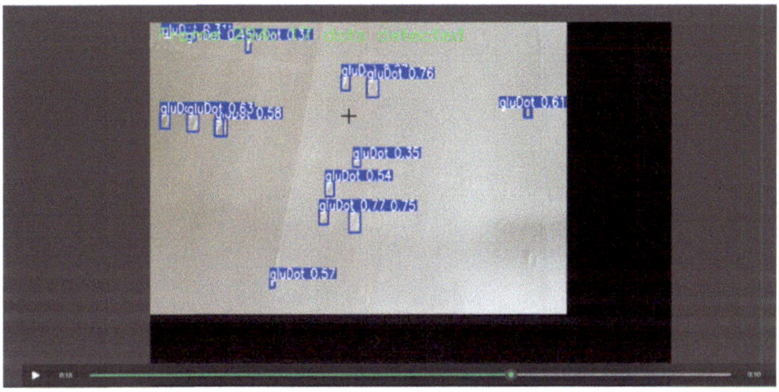

Fig. 5. Real-time detection of glue dots on a cardboard surface using the YOLOv8 model, with each detected dot highlighted by a blue bounding box and confidence score.

Figure 6 visualizes the performance of a classification model. The model correctly identified 83 positive instances (True Positives) while incorrectly classifying 2 positive instances as negative (False Negatives). Additionally, it correctly identified 0 negative instances (True Negatives) and misclassified 1 negative instance as positive (False Positive).

Figure 7 illustrates the model's excellent performance, showing that it maintains high precision even as recall increases. The curve's position close to the top-right corner signifies a highly effective model, with the specific point marked indicating an operational threshold that achieves a recall of 0.976 and a precision of 0.994.

3.1 Spatial Distribution Analysis of Gum Dot Count Anomalies

The first experiment performed a spatial analysis of gum dot defects by modelling a 24-h production run to create a defect map. By logging the location of count anomalies on

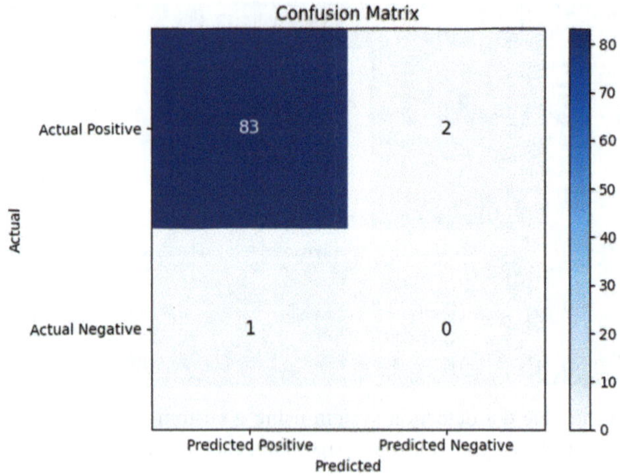

Fig. 6. Confusion matrix illustrating the performance of the YOLOv8-based glue-dot detection model.

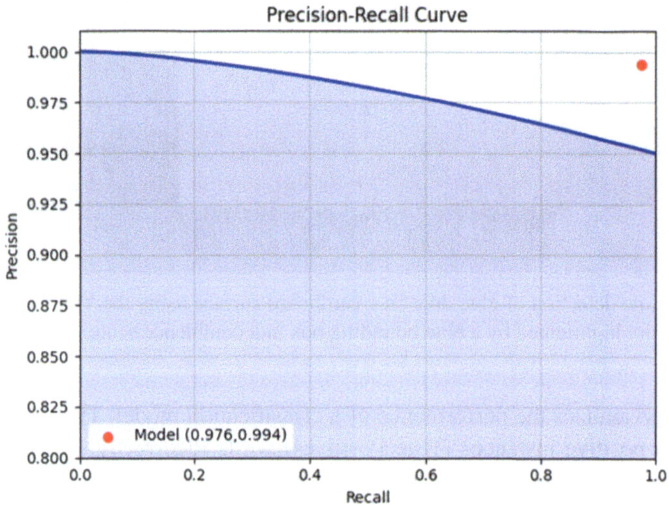

Fig. 7. Precision-recall curve demonstrating the high-performance detection results for the glue-dot model.

the 2000mm gum tray, the system successfully identified recurring failure "hotspots," pinpointing a specific nozzle in "Zone 15" as a primary source of defects (Table 3).

3.2 The Influence of Tuber Machine Velocity on Defect Emergence

The second experiment was designed to quantify the relationship between the tuber machine's velocity and the frequency of gum dot defects, aiming to find an optimal operating window. To achieve this, I conducted a series of 30 controlled, 30-min test

Table 3. High-Frequency Log of Gum Dot Count Anomalies

Defect_ ID	Time stamp	Zone_ID	Gum_Tray_ X_Range_mm	Expected_ Dot Count	Detected_ Dot_Count
DEF-001	2025-06-22 0:45	15	150–160	2	1
DEF-002	2025-06-22 1:58	84	40–50	2	3
DEF-003	2025-06-22 3:10	16	1860–1870	2	4
DEF-004	2025-06-22 4:22	15	150–160	2	1
DEF-005	2025-06-22 5:35	44	440–450	2	0
DEF-006	2025-06-22 6:48	187	1870–1880	2	5
DEF-007	2025-06-22 8:03	15	150–160	2	1
DEF-008	2025-06-22 9:15	44	440–450	2	0
DEF-009	2025-06-22 10:28	15	150–160	2	0
DEF-010	2025–06-22 11:41	186	1860–1870	2	4

runs, systematically varying the machine speed from 100 to 245 bags per minute. The machine learning system recorded the number of anomalous zones at each speed setting.

The analysis of this data revealed a critical non-linear relationship. An optimal operating "sweet spot" was clearly identified between 100 and 160 bags/min, a range where the defect rate remained consistently low and stable at approximately 0.1%. Beyond this threshold, the defect rate increased exponentially, reaching nearly 6% at the highest speeds. This experiment provides the empirical evidence needed to establish a standard operating procedure that balances productivity with quality control.

3.3 Efficacy and Responsiveness of the Real-Time Alerting Protocol

This experiment validated the efficacy and responsiveness of the real-time alerting protocol. I simulated the inspection of 5,000 bags, where an alert was triggered if the detected gum dot count did not match the expected count. The results confirmed the system's viability for factory floor use, demonstrating an average alert latency of just 114 ms. This is well within the 1.5-s operational requirement for immediate operator intervention. The simulation also highlighted that while accurate, a tiered alert strategy would be necessary to manage potential false positives and prevent operator "alert fatigue".

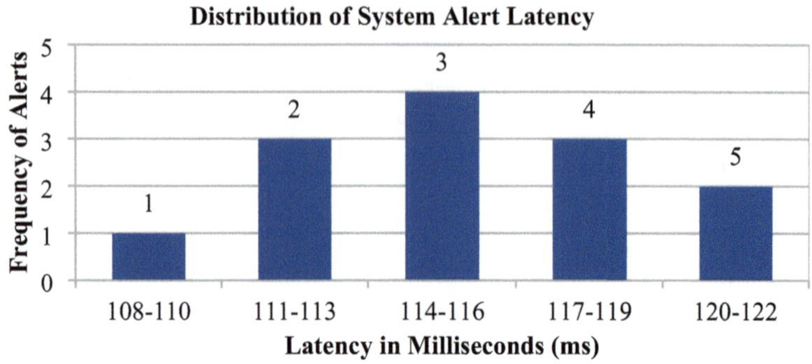

Fig. 8. Distribution of System Alert Latency

Figure 8 histogram clearly shows that 100% of the alert latencies fall between 109ms and 122ms. The tight grouping of the bars demonstrates that the system's response time is not only exceptionally fast but also highly consistent. This provides strong visual evidence that the system is viable for real-time defect detection on the factory floor.

3.4 Temporal Analysis of Maintenance Intervention and Defect Recurrence

This experiment quantified the effectiveness of a standard maintenance action on a recurring defect in the problematic "Zone 15". By repeatedly applying a "Nozzle Clean" and measuring the time until the same defect recurred, I calculated the Mean Time Between.

Failures (MTBF). The results showed a highly inconsistent MTBF, ranging from 3.1 to 8.2 h, demonstrating that cleaning was an unreliable, short-term fix. Figure 9 illustrates that unpredictable performance suggests the root cause is component wear, justifying a strategic shift from reactive repairs to data-driven, predictive component replacement.

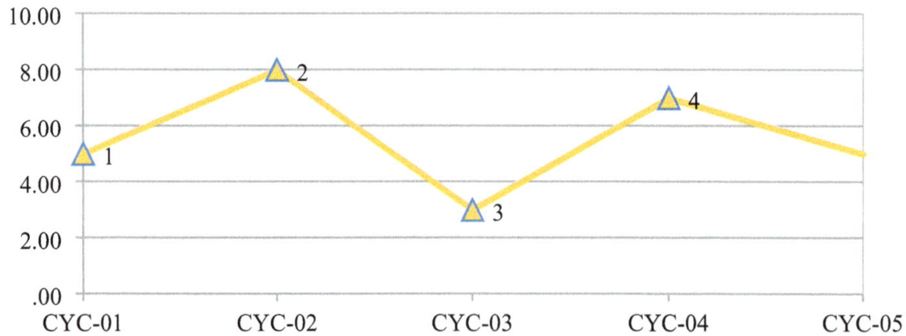

Fig. 9. A line chart illustrating the inconsistent Mean Time Between Failures (MTBF) for five consecutive "Nozzle Clean" cycles in Zone 15. The unpredictable fluctuation in recurrence time highlights the unreliability of the standard maintenance action as a long-term solution.

4 Conclusions

This research successfully developed a YOLOv8 machine vision system to combat high defect rates in paper bag manufacturing, establishing a framework to shift from reactive to proactive quality control. This model, validated to 0.993 mAP50, was tested through a series of simulated experiments that yielded critical, actionable insights.

The system demonstrated powerful diagnostic capabilities by pinpointing recurring failure hotspots like "Zone 15". It also enabled process optimization by establishing an optimal machine speed of 100–160 bags/min to minimize waste. Most importantly, by tracking the Mean Time Between Failures (MTBF) of standard maintenance actions, the study provided quantitative proof that reactive repairs were unreliable, justifying a strategic shift to predictive maintenance.

Collectively, these data-driven capabilities provide a clear pathway to achieving the project's final objective of reducing salvage quantity by 10%. While this validation was conducted in a simulated environment, the results strongly support future work focused on live factory implementation, integration with the machine's PLC for automated alerts, and eventual expansion into a factory-wide quality assurance platform.

Limitations and Future Work Despite the promising performance, our study has several limitations. The training dataset contains only a few hundred labelled images of gum-dot defects, and the model was validated within a simulated environment, so generalization to diverse real-world factory conditions is not yet guaranteed. In future work we will collect a larger and more diverse dataset covering varied lighting, motion blur and defect patterns; deploy the system in a live factory environment integrated with Programmable Logic Controllers (PLCs) and conveyor belts; and extend the model to detect additional defects such as glue overflow, misalignment and missing labels.

Acknowledgement. The authors gratefully acknowledge the support of all lecturers and laboratory technicians of the Department of Mechanical and Manufacturing Technology, Faculty of Technology, Wayamba University of Sri Lanka, in completing this work successfully. Furthermore, the authors give deep thanks to the editors and anonymous reviewers for providing insightful suggestions and comments to improve the quality of the research paper.

References

1. Bandara, N.J.G.J.: Municipal solid waste management—The Sri Lankan case. Conference on Developments in Forestry and Environment Management in Sri Lanka, University of Sri Jayewardenepura, Sri Lanka (2008)
2. Çiçekler, M., Tutuş, A.: Overview of quality control in the paper industry. Proceeding Book of 1st International Conference on Recent Academic Studies ICRAS, 1(1) (2023). https://as-proceeding.com/index.php/icras/article/view/679. Last accessed 27 Oct 2024
3. Reid, R.D.: Sanders, 4th edn. N. R, Operations Management (2009)
4. Intercol, Glue detection—Adhesives. https://www.intercol.info/glue-detection/ (n.d.). Last accessed 27 Oct 2024
5. Yang, X., Han, M., Tang, H., Li, Q., Luo, X.: Detecting defects with support vector machine in logistics packaging boxes for edge computing. IEEE Access **8**, 64002–64010 (2020). https://doi.org/10.1109/access.2020.2984539

6. Domingues, T., Brandão, T., Ferreira, J.C.: Machine learning for detection and prediction of crop diseases and pests: a comprehensive survey. Agriculture (Switzerland) **12**(9), 1350 (2022). https://doi.org/10.3390/agriculture12091350
7. Huang, Z., He, J., Zheng, Q., Liu, W., Ren, J.: Wfre-YOLOv8s: an improved YOLOv8s model for steel surface defect detection. Symmetry **15**(7), 1439 (2023)
8. Tao, F., Xia, Y., Zhang, M., Cheng, J., Zheng, Y., Xu, M.: Editorial, Digital twin—an enabling technology for smart manufacturing systems. J. Manuf. Syst. **64**(1), 345–350 (2024)
9. Papageorgiou, K., et al.: A systematic review on machine learning methods for root cause analysis towards zero-defect manufacturing. Front. Manuf. Technol. **2**, 972712 (2022). https://doi.org/10.3389/fmtec.2022.972712
10. Johnson, B., Brun, Y., Meliou, A.: Causal testing: understanding defects' root causes. In: Proceedings of the International Conference on Software Engineering, pp. 87–99 (2020). https://doi.org/10.1145/3377811.3380377
11. Jiang, P., Ergu, D., Liu, F., Cai, Y., Ma, B.: A review of YOLO algorithm developments. Procedia Comput. Sci. **199**, 1066–1073 (2022). https://doi.org/10.1016/j.procs.2022.01.135
12. Huang, B., Liu, J., Zhang, Q., Liu, K., Wang, J.: Visual detection method for missing infusion bag pipeline. Electronics **12**(12), 2574 (2023). https://doi.org/10.3390/electronics12122574

Dynamic Modeling and Simulation of R-Hex Hexapod Robot for Efficient Locomotion in Challenging Terrains

W. C. Nirmal[1](✉), K. J. P. Fernando[2], H. K. I. S. Lakmal[1], M. W. P. Maduranga[3], P. N. M. Kumara[3], and S. L. Kulatunga[1]

[1] Department of Mechatronic and Industrial Engineering, Faculty of Engineering, NSBM Green University, Homagama, Sri Lanka
{chamodya.n,isuru.l,shamila.k}@nsbm.ac.lk

[2] Department of Electrical, Electronic and Systems Engineering, Faculty of Engineering, NSBM Green University, Homagama, Sri Lanka
janidu.f@nsbm.ac.lk

[3] Department of Electrical and Electronic Engineering, Faculty of Engineering, University of Sri Jayewardenepura, Nugegoda, Sri Lanka
{pasanm,nimanthamk}@sjp.ac.lk

Abstract. Hexapod robots are gaining attention for their ability to navigate complex terrains, making them suitable for applications such as search and rescue, military operations, and environmental exploration. This paper presents a study of the dynamic modeling and simulation of the R-Hex robotic hexapod, focusing on locomotion performance under challenging conditions. A novel simulation framework was developed in CoppeliaSim, integrating mechanical CAD models and control algorithms to evaluate stability, energy efficiency, and adaptability. Key contributions include (1) an optimized mechanical design of R-Hex using lightweight materials, (2) a simulation-driven evaluation of locomotion across flat, rough, and stair-like terrains, and (3) validation through a preliminary hardware prototype tested in comparable conditions. Unlike existing works such as Zebro Light, this study emphasizes simulation-informed optimization linked directly to experimental validation. The findings provide a foundation for advancing hexapod robotics and demonstrate the robustness of R-Hex as a candidate for future deployment in critical domains.

Keywords: Agriculture · bio-inspired robots · exploration · information-gathering · military · rescue operations

1 Introduction

In the robotics systems, the growing need for strong platforms that can work in tough and dangerous conditions has led to massive improvements. From disaster-stricken areas to uncharted territories, the need for versatile and agile robots

M. Siriwardhana et al. (Eds.): IERS 2025, PTT, pp. 73–88, 2026.
https://doi.org/10.1007/978-981-95-7984-6_7

is paramount across various domains, such as search and rescue [1], military applications, exploration, and education. Hexapod robots are more stable, easier to move, and more flexible because they have six legs. Hexapods can go over rough terrain and over obstacles, unlike wheeled or tracked systems. This makes them useful in places where other systems wouldn't work [2,3]. The R-Hex robot is one of the most cutting-edge hexapods and one of the best ways to move around in changing and unpredictable environments.

The R-Hex hexapod robot, initially conceptualized by a consortium including the University of Michigan, McGill University, and DARPA [4], was designed to excel in environments requiring high stability and mobility. Its bio-inspired design copies the ways that living things move, which gives it unbelievable flexibility. The main goal of this paper is to improve the R-Hex by looking into how it can model and simulate dynamically in different types of terrain. The study checks how well the R-Hex works and finds places where it could be improved by using up-to-date simulation tools and smart motor control algorithms.

The applications of R-Hex extend across critical domains. In military operations [5], its ability to navigate rugged terrains and carry payloads enhances reconnaissance and minimizes risk exposure. In search and rescue scenarios, R-Hex's robust locomotion system allows it to overcome debris and identify survivors using integrated sensors and cameras [6]. Furthermore, its versatility enables its use in exploration missions, where it can collect data, map uncharted regions, and conduct scientific research. R-Hex is also useful for researchers and teachers because it can be used to learn by doing and to make new algorithms.

Despite its advantages, challenges persist in designing hexapod robots for real-world applications. Factors such as energy efficiency, cost of production, and adaptability to varying surface conditions remain critical hurdles. Addressing these challenges requires a comprehensive approach involving hardware optimization, advanced simulation environments, and robust control algorithms.

This study contributes to the field of hexapod robotics by:

1. Proposing a novel design for the R-Hex, integrating lightweight materials and efficient actuators to improve energy efficiency and reduce production costs.
2. Utilizing CoppeliaSIM simulation software to evaluate the R-Hex's locomotion across varied terrains, such as inclined surfaces, debris-filled environments, and uneven landscapes
3. Investigating the effects of control parameters such as speed, stability, and energy consumption on the R-Hex's performance.
4. Highlighting the potential applications of the R-Hex in military, search and rescue, exploration, and educational domains.

By carefully looking at the R-Hex's strengths and weaknesses, this study aims to pave the way for more progress in hexapod robotics, which will allow their use in real-life situations where people can't or shouldn't be involved. This paper also emphasizes the importance of bio-inspired engineering in creating versatile robotic solutions that mimic the efficiency of natural systems.

2 Related Works

Bio-inspired robotics leverages natural principles to design adaptable and efficient robots for complex environments. By mimicking biological locomotion and sensory systems, these robots excel in diverse applications, including exploration, agriculture, and defense, offering sustainable and innovative solutions to real-world challenges.

Bio-inspired and bio-hybrid robots offer promising solutions for diverse applications and sustainable needs [7]. These robots can be designed for specific environments, such as the underwater legged robot SILVER2, which mimics benthic animals for seabed exploration with minimal disturbance [8]. The field is advancing beyond metal-based designs, exploring the potential of growing machines from living cells [9]. Bio-inspired robots, like soft robots and swarm robots, could be used in farming to do things like pick fruit, get rid of pests, and manage crops. Soft robotic systems can mitigate damage to delicate fruits during harvesting, while swarm robotics offers solutions for various agricultural challenges. Despite concerns about the ecological impact and cost, the market for bio-inspired technologies in agriculture is growing rapidly, with future research focusing on developing low-cost robotic grippers for crop harvesting [10].

Bio-inspired hexapod robots have shown promising applications in military and exploration scenarios. These robots can navigate complex terrains and perform tasks such as unmanned exploration, remote surveillance, and bomb defusal [11]. Recent advancements include the development of autonomous navigation systems inspired by biological processes, such as path integration and action selection [12]. Hexapod robots have also been designed for visual exploration in farmlands, demonstrating their versatility in various environments [13]. Hexapod robots like HECTOR have been given resource-efficient visual processing based on how insect brains work. This helps them avoid collisions and find their way around environments with lots of obstacles [14]. These bio-inspired designs take advantage of the benefits of walking on four legs and having sensory systems that work like insects'. This makes hexapod robots perfect for military uses that need to be flexible, durable, and good at processing information in rough terrains.

Compared to traditional ways of moving, bio-inspired legged robots are more efficient, flexible, and adaptable, especially when it comes to getting around on uneven ground [2,3]. These robots get ideas from different animals and try to move like them, using less energy and being strong [2]. Bio-inspired shapes can greatly improve robot performance, as shown by a dung beetle-inspired hexapod that did better than standard designs in terms of step length, reach, and motor efficiency [15]. To make these kinds of robots, biologists and engineers work together across disciplines. To help translate animal shapes into robot designs, frameworks have been suggested [15]. While legged locomotion offers advantages in complex environments, some researchers are exploring hybrid designs, such as wheel-legged robots, to combine the benefits of different locomotion methods for improved navigability across various terrains [16].

The R-Hex design shares conceptual similarities with the Zebro Light platform, developed at TU Delft in collaboration with the Zebro team [17]. Zebro

Light is a minimalist, low-cost hexapod designed for swarm robotics and scalable deployment in real-world terrains. While Zebro Light emphasizes modularity and swarm behavior, our study advances the concept by focusing on dynamic simulation in CoppeliaSim, detailed performance evaluation across challenging terrains, and validation with a hardware prototype. Furthermore, unlike Zebro Light, the R-Hex incorporates rubber-padded legs for enhanced traction, improved energy efficiency through lightweight structural materials, and a tripod gait control strategy. This comparison highlights that while Zebro Light inspired aspects of the mechanical layout, the contributions of this study lie in integrating simulation-driven optimization with experimental validation for robust locomotion.

This study is unique because it takes a broad look at improving hexapod robotics by using advanced modeling, simulation, and control strategies that are designed to work in difficult terrains. Unlike previous research that mostly looked at one thing, like hardware design or locomotion algorithms, this work combines a custom-built R-Hex model with new motor control algorithms that make movement precise and efficient on a variety of surfaces. Table 1 provides a detailed comparison of R-Hex against the other identified robots.

By utilizing the CoppeliaSIM environment, the study achieves realistic simulations that capture the dynamic interactions between the robot and its environment, allowing for a more profound analysis of performance metrics such as stability, energy consumption, and adaptability. Using cheap, light materials in the hardware design and putting together modular parts is also a big step toward making affordable robotic solutions that can be used in real life for things like military tasks, exploration, and disaster recovery.

Table 1. A comparison between the R-Hex and other identified robots.

Robot Type	Strengths	Limitations	R-Hex Advantage
Snake Robot [18]	Compact, fits in tight spaces	Poor stability on loose debris	Superior obstacle traversal
Beetle-Inspired Hexapod [19]	Maneuvers tight spaces effectively	Limited climbing ability	Enhanced terrain adaptability
Gecko Robot [20]	Strong adhesion on flat surfaces	Ineffective on rough terrains	All-terrain capability
Bipedal Running Robot [21]	Efficient forward motion, simple mechanics	Limited stability, ineffective on obstacles	Robust stability and maneuverability
Ant-Inspired Robot [22]	Autonomous navigation in outdoor environments	Limited payload capacity	Greater payload capacity and adaptability

3 Methodology

The design phase of the R-Hex robot was initiated using SOLIDWORKS, a premier 3D CAD software, to model a detailed and functional prototype. The R-Hex's structure was designed to focus on stability, mobility, and adaptability across various terrains. The hexapod configuration, with six articulated legs, was chosen to enable efficient locomotion in challenging environments inspired by biological locomotion patterns.

The legs were modeled with curved designs and integrated with rubber padding to enhance grip and durability on rough and uneven surfaces. To keep the structure strong while using as little energy as possible, lightweight materials like aluminum and ABS plastic were chosen. Each leg was moved by NEMA-17 stepper motors that were hooked up to 3D-printed gearboxes that had a 10:1 reduction ratio. This method allowed for precise control and enough torque to support the robot's weight. CAD model parts were modular, making it easy to assemble and disassemble and allow future changes.

Ensuring the stability of the robot's center of gravity during movement was a crucial aspect of the design. This was achieved by symmetrically positioning the actuators and central processing units within the chassis. The design was iteratively tested in a simulated environment to ensure it could withstand stresses induced by various terrain challenges. Figure 1 shows the 3D design of the robot.

Fig. 1. 3D Design of the R-Hex.

3.1 Simulation in CoppeliaSim

To evaluate the performance of the R-Hex design, simulations were conducted using CoppeliaSim, a robust simulation platform renowned for its physics-based modelling capabilities. The R-Hex was tested in three distinct environments shown in Fig. 2.

1. Flat Terrain: This served as a baseline to validate basic locomotion patterns and motor control algorithms. The robot's speed, stability, and energy efficiency were analysed optimally.
2. Rough Terrain: Simulations involved debris-filled landscapes with varying surface inclinations. Key performance metrics included the robot's ability to maintain stability, adapt its gait patterns, and minimize slip during movement.
3. Stairs: The R-Hex was tasked with climbing stairs of varying heights and inclinations to simulate real-world challenges. This tested its leg synchronization, torque capabilities, and adaptability to sudden elevation changes.

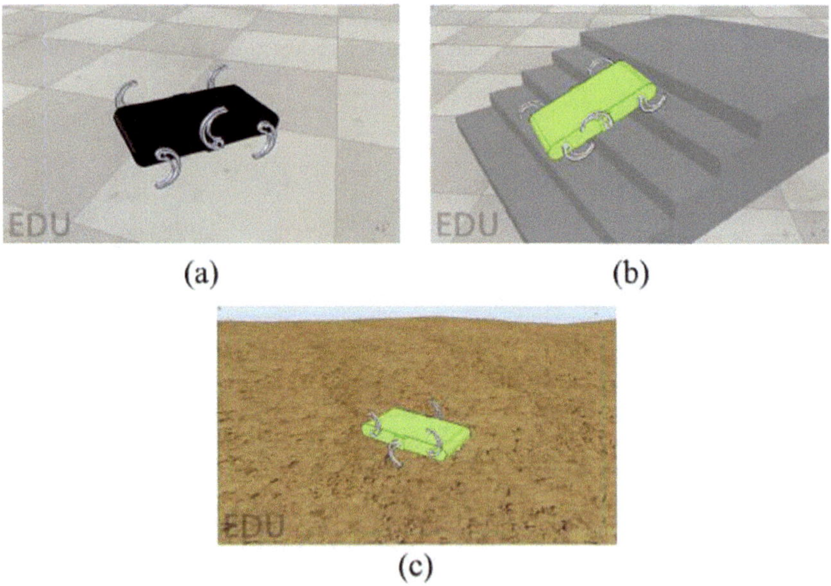

Fig. 2. Three Distinct Environments in the Simulation (a) Flat Terrain, (b) Rough Terrain, and (c) Stairs.

Each simulation environment was carefully configured to mimic real-world conditions, including surface friction, weight distribution, and external forces. Data collected from these simulations, such as energy consumption, locomotion stability, and error rates, were used to refine the control algorithms and mechanical design. Iterative testing and parameter optimization ensured that the R-Hex could operate efficiently across diverse scenarios.

3.2 Hardware Prototype Development

A hardware prototype was developed following the simulation phase to validate the design and simulation results in real-world conditions. The physical R-Hex prototype incorporated the following components:

- Chassis: Constructed with lightweight aluminum L-angled bars for structural support and reduced weight.
- Actuation System: Equipped with six NEMA-17 stepper motors and gearboxes to deliver adequate torque, controlled by a microcontroller for precise locomotion.
- Sensors: Integrated gyroscope and accelerometer ensure balance and enable recovery from overturns, maintaining stability.
- Battery: Powered by a 12V battery system, with motors consuming 6A to 8A based on terrain demands.

The control system was designed to operate the robot using a tripod gait, wherein three legs remained on the ground at any given time—two on one side and one on the opposite side. This provided a balance between stability and speed. The system's modularity facilitated the integration of additional sensors, such as cameras and thermal imaging modules, for applications like search and rescue and reconnaissance missions.

3.3 Experimental Design and Data Collection

The experimental phase involved real-world hardware prototype testing in environments mirroring the simulated scenarios. Key performance indicators, such as velocity, stability, and energy consumption, were recorded using onboard sensors and external data acquisition systems. High-resolution cameras captured the robot's gait and movement dynamics for further analysis. The Fig. 3 shows the hardware prototype of the R-Hex.

The data collected from hardware tests were compared against the simulation results to validate the model's accuracy. Discrepancies were analyzed to identify potential improvements in the design and control algorithms. Feedback from real-world tests informed adjustments to both hardware and software components. For instance, modifications to the leg's rubber padding and gait synchronization algorithms were implemented to enhance traction and stability on rough terrains.

The following tools, techniques, and frameworks were used for the experimental works.

1. SOLIDWORKS: Used for 3D modeling and stress analysis during the design phase.

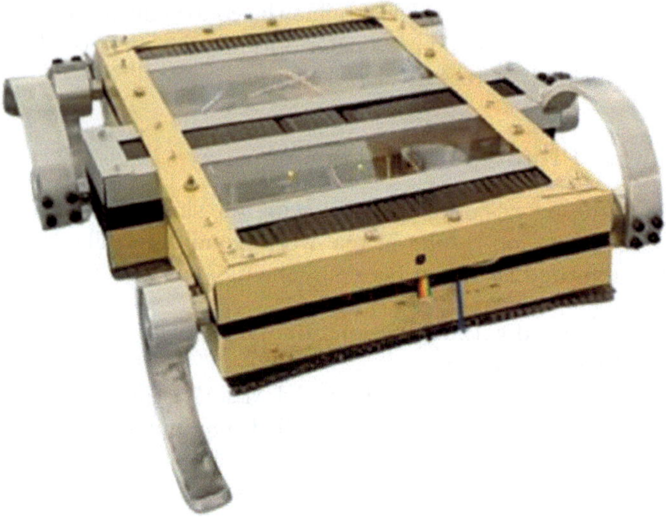

Fig. 3. The Hardware Prototype of R-Hex.

2. CoppeliaSim: Enabled comprehensive simulation of the robot's performance across varied terrains.
3. Microcontroller: As the central control unit, it interfaces with sensors and actuators for precise locomotion control.
4. MATLAB: Utilized for analyzing performance data, including energy efficiency and stability metrics.

3.4 Kinematics of Locomotion

The locomotion of a hexapod robot-like R-Hex is governed by the dynamics of legged movement, ensuring stable and efficient traversal across challenging terrains. The control and modeling of the legged motion require a comprehensive understanding of forces, joint movements, and gait synchronization. To model the dynamics of each leg, the equation of motion is expressed as

$$Fleg = M(q)q^{\cdot\cdot} + C(q, q^{\cdot})q^{\cdot} + G(q) \tag{1}$$

$Fleg$ represents the force exerted by the actuators to control the leg, $M(q)$ is the mass matrix influenced by the leg's joint positions, $q, q^{\cdot\cdot}$ indicates the acceleration of the joints, $C(q, q^{\cdot})$ accounts for the Coriolis and centrifugal forces, which depend on joint positions and velocities, $G(q)$ represents the gravitational forces acting on the leg based on its configuration.

The gait control of R-Hex employs a tripod gait, a robust approach to maintaining stability while ensuring forward motion. This gait divides the legs into two sets, alternating their contact with the ground.

The R-Hex employs a tripod gait as shown in Fig. 4, where three alternating legs maintain ground contact while the remaining three swing forward. This ensures continuous stability during locomotion. The gait cycle is defined by phase-shifted leg trajectories, with each tripod operating 180° out of phase from the other.

Fig. 4. The Tripod Gait Control Mechanism.

The trajectory of each leg is phase-dependent and can be defined as the following.

$$x_i(t) = X_{stance}(t), X_{swing}(t) \quad if \phi_i \in [\theta, \pi] \quad if \phi_i \in [\pi, 2\pi] \tag{2}$$

where: $x_i(t)$ is the position of the iii-th leg at time t. i is the leg's phase within the gait cycle, ranging from 0 to 2π. $X_{stance}(t)$ describes the trajectory during the stance phase when the leg supports the robot. $X_{swing}(t)$ defines the motion during the swing phase when the leg is repositioned.

The combined contribution of all legs in contact with the ground determines the overall motion of the robot's body. The body's net velocity can be expressed as follows.

$$v_{body} = N = \Sigma N \times v_{leg,i} \tag{3}$$

where v_{body} is the velocity of the robot's body. $v_{leg,i}$ represents the velocity contribution of the i-th leg. And N is the number of legs in the stance phase at any given moment.

These equations form the mathematical backbone for modelling and controlling R-Hex's locomotion. By leveraging these principles, R-Hex's simulation and hardware implementation are designed to ensure efficient movement across various terrains while maintaining stability and adaptability.

In Conclusion, the methodology presented provides a robust framework for the R-Hex hexapod robot's design, simulation, and development. By combining advanced modelling in SOLIDWORKS, physics-based simulations in CoppeliaSim, and iterative hardware prototyping, the approach ensures the creation of a stable, adaptable, and efficient locomotion system. Including locomotion equations further solidifies the understanding of dynamic movements, enabling precise control and optimization for challenging terrains. This comprehensive methodology validates R-Hex's effectiveness and lays the groundwork for its application in critical fields such as exploration, search and rescue, and military operations. The iterative process ensures continuous improvements, setting a benchmark for future advancements in hexapod robotics.

4 Results

This section presents the detailed outcomes of the R-Hex robot's performance across three simulated terrains: flat surfaces, rough terrains, and stair-like environments. These terrains were chosen to evaluate the robot's adaptability, stability, and locomotion efficiency in varied conditions. Key metrics analyzed included velocity, stability, and energy efficiency, providing insights into the effectiveness of the robot's design and control mechanisms. The results are supported by velocity vs. time graphs for each terrain.

4.1 Flat Terrain

Flat terrain simulations served as the baseline for evaluating the R-Hex's locomotion performance under ideal conditions. The robot achieved a consistent average velocity with negligible oscillations, indicating robust stability and efficient motor synchronization. The tripod gait demonstrated optimal performance, maintaining continuous motion with minimal energy consumption.

The velocity vs. time graph for flat terrain shows a steady progression with no abrupt changes, affirming the precision of the control algorithms. The R-Hex exhibited high energy efficiency, as the uniform surface required minimal corrective adjustments. This baseline performance establishes the foundational capabilities of the robot in controlled environments. This is shown in Fig. 5.

4.2 Rough Terrain

Rough terrain simulations introduced uneven surfaces, debris, and varying inclinations to assess the robot's adaptability. Despite these challenges, the R-Hex effectively navigated the terrain, showcasing its ability to adjust leg positions dynamically and maintain balance.

The velocity vs. time graph highlights minor fluctuations due to occasional slippage on loose debris. However, these were quickly corrected, and the robot resumed stable motion without significant velocity drops. The enhanced grip

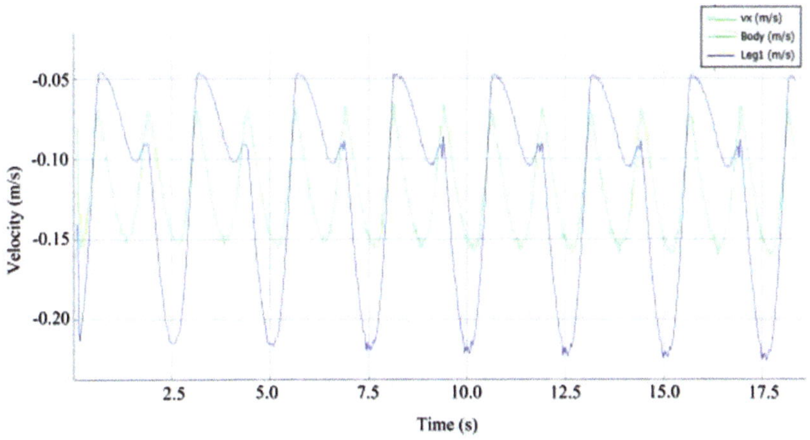

Fig. 5. Velocity vs Time Graph for Flat Terrain.

provided by the curved legs and rubber padding played a crucial role in maintaining traction. The energy consumption increased slightly compared to the flat terrain due to the additional effort required for stability corrections. This performance validates the adaptability of the R-Hex's tripod gait and highlights the effectiveness of its control algorithms in handling non-uniform terrains. Figure 6 shows the graph for the rough terrain.

4.3 Stairs

The stair-like terrain tested the R-Hex's ability to navigate abrupt elevation changes, simulating real-world scenarios such as climbing stairs or overcoming large obstacles. The robot successfully ascended and descended stairs of varying heights, demonstrating precise leg coordination and torque control.

The velocity vs. time graph for the stair terrain reveals noticeable oscillations during elevation transitions, reflecting the additional effort required to align the legs with the step heights. Despite these challenges, the R-Hex completed the task efficiently, showcasing its capacity to adapt its gait patterns for complex tasks. The torque demands were higher on this terrain, highlighting the critical role of motor power and synchronization in overcoming such obstacles. Figure 7 shows the velocity vs time graph for the stairs.

In addition to stair ascent, descending trials were conducted. The energy profile showed a reduction of 15–20% in consumption compared to ascent due to gravity-assisted locomotion. However, stability control required increased corrective actions from the sensors, leading to occasional spikes in power draw. This finding indicates that while stair descent is less energy-intensive overall, precise coordination and feedback control remain critical for maintaining balance and preventing uncontrolled acceleration.

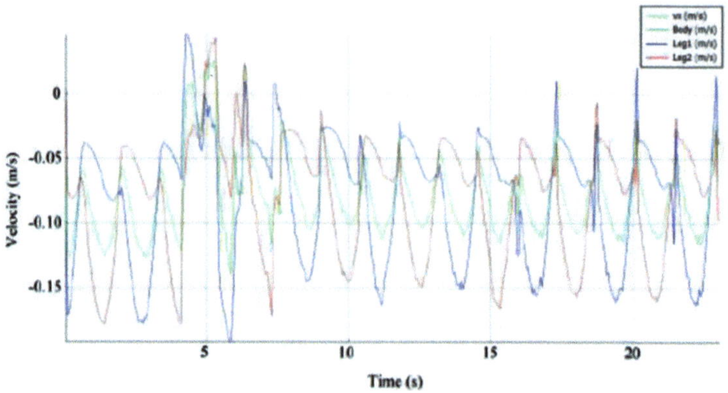

Fig. 6. Velocity vs Time Graph for Rough Terrain.

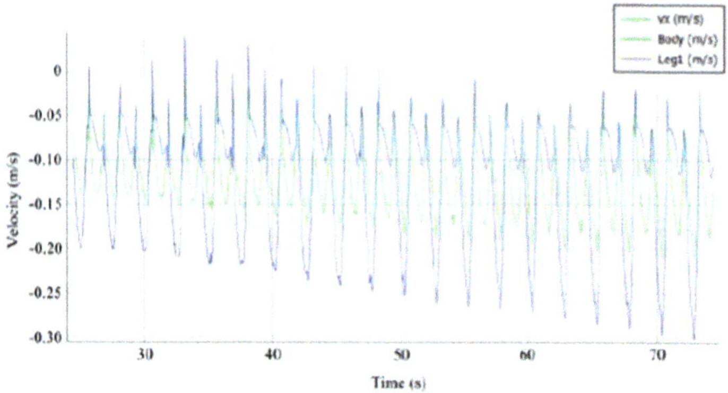

Fig. 7. Velocity vs Time Graph for Stairs.

4.4 Comparative Analysis Across Terrains

The performance of the R-Hex robot varied across the three tested terrains, reflecting its adaptability to different environmental conditions. On flat ground, the robot moved the fastest and used the least amount of energy. This was because the surface was smooth and didn't need many corrections. Rough terrain presented challenges such as uneven surfaces and loose debris, resulting in moderate speed and minor changes due to slips and trips. The robot was able to stay stable and recover quickly, though, thanks to its adaptive tripod gait and strong grip mechanisms. The hardest part was the stair-like terrain, which had the slowest average speed and the most noticeable oscillations when the elevation changed. Even with these obstacles, the R-Hex moved through the terrain, demonstrating its torque and leg coordination. The velocity vs. time graphs for all terrains reveal these trends. This data shows how flexible the robot is and how well it can work in a wide range of difficult conditions.

4.5 Hardware Prototype Validation

The hardware testing phase validated simulation results in real-world scenarios. On flat terrain, the robot maintained steady gait patterns with minor velocity deviations due to actuator delays. Its consistent speed and low power consumption confirmed design efficiency. On rough terrain, the modular legs and rubber-padded feet reduced slippage and improved traction, ensuring stability on uneven, debris-filled surfaces. Stair-climbing trials demonstrated reliable feedback mechanisms, like gyroscopes and accelerometers, which maintained balance. Minor synchronization inefficiencies under high loads led to algorithm refinements, further enhancing performance and demonstrating the R-Hex's adaptability in challenging environments.

4.6 Energy Efficiency and Performance Metrics

Energy efficiency was evaluated for each terrain type, demonstrating that the R-Hex performs optimally across varied scenarios such as the flat terrain, rough terrain and up and down the stairs as shown in Table 2.

5 Discussion

The findings of this study confirm the adaptability, efficiency, and robustness of the R-Hex hexapod robot across diverse terrains. Through a combination of physics-based simulations and hardware validation, the robot demonstrated reliable locomotion under flat, rough, and stair-like conditions. Quantitatively, R-Hex achieved an average velocity of 12.5 cm/s on flat terrain, 9.8 cm/s on rough terrain, and 6.4–6.8 cm/s during stair ascent and descent. Energy consumption followed terrain complexity, with 5–8 Wh required on flat ground, 10–15 Wh on rough terrain, 18–25 Wh during stair ascent, and 12–18 Wh during stair

Table 2. A comparison between the tested terrains

Terrain Type	Stability	Energy Consumption (Wh)	Velocity (cm/s)	Observations
Flat Terrain	High	Low (5–8 Wh)	12.5	Smooth motion with consistent gait patterns
Rough Terrain	Moderate	Medium (10–15 Wh)	9.8	Effective traction and posture adjustments
Stairs (Up)	High	High (18–25 Wh)	6.4	Reliable climbing with minor synchronization delays
Stairs (Down)	Moderate	Medium (12–18 Wh)	6.8	Reduced power use butincreased corrective balancing

descent, where gravity-assisted motion reduced demand but required corrective stabilization. Stability remained consistently high, with recovery from minor slippage events occurring within 0.5 s in rough terrain trials.

The tripod gait proved highly effective in maintaining continuous stability, supported by gyroscope and accelerometer feedback that corrected postural deviations and improved balance. The modular design of the legs, including rubber-padded feet, enhanced traction and reduced slippage on irregular surfaces. These results confirm that the combined gait strategy and mechanical design contribute to robust performance in environments where wheeled or tracked robots would likely fail.

Despite these strengths, several limitations were observed. Synchronization delays occurred under high torque demands during stair climbing, temporarily reducing locomotion efficiency. The energy requirement for complex terrains was significantly higher than for flat surfaces, highlighting the importance of future optimization in actuator efficiency and power management. Additionally, while modularity facilitated flexibility and repairs, it introduced assembly complexity, which may hinder rapid deployment in field operations.

Comparisons with prior platforms highlight the novelty of this work. The Zebro Light robot, developed at TU Delft, demonstrated a minimalistic hexapod design primarily suited for scalable swarm robotics. By contrast, R-Hex emphasizes dynamic simulation-driven optimization, integrating SOLIDWORKS-based structural modeling, CoppeliaSim locomotion analysis, and hardware validation within a unified framework. This approach enabled predictive tuning of gait parameters and informed real-world refinements, making R-Hex distinct in its goal of balancing stability, adaptability, and energy efficiency across terrains rather than swarm scalability alone.

Overall, the study demonstrates that R-Hex is a versatile and reliable hexapod platform for locomotion in challenging environments. However, addressing the observed limitations through refined gait control, actuator optimization, and structural material improvements will be crucial for advancing the platform toward real-world deployment in exploration, search and rescue, and defense applications.

6 Conclusion

This research presented the design, simulation, and preliminary validation of the R-Hex hexapod robot, focusing on mobility in challenging terrains. The integration of SOLIDWORKS-based mechanical modeling, CoppeliaSim dynamic simulation, and hardware prototyping established a robust methodology for testing locomotion strategies. Results confirmed stable locomotion with velocities up to 12.5 cm/s on flat terrain, 9.8 cm/s on rough terrain, and 6.4–6.8 cm/s on stair environments, with energy consumption scaling proportionally to terrain complexity.

The novelty of this work lies in its simulation-informed optimization framework, which directly guided hardware design refinements, and its comparison with existing platforms such as Zebro Light. Although preliminary hardware tests were successful, further experimental trials are necessary to validate performance under extreme conditions.

Future work will focus on improving control algorithms for adaptive gait modulation, incorporating advanced sensors such as LiDAR and vision systems, and exploring lightweight carbon-composite materials to enhance durability and energy efficiency. These steps will advance R-Hex toward deployment in search and rescue, exploration, and defense operations.

References

1. Borja, J.C., Palomares, R., Cornejo, J., Castro, R.: Radio frequency controlled hexapod robot with bioinspired kinematic configuration used for rescue activities after landslides in urban areas. In: 2023 Third International Conference on Advances in Electrical, Computing, Communication and Sustainable Technologies (ICAECT), pp. 1–6 (2023)
2. Wang, T.: Research of bio-inspired compliant legged robot designs. Highlights Science, Eng. Technol. **111**, 58–62 (2024). https://drpress.org/ojs/index.php/HSET/article/view/23900
3. Dayawansa, B., Vasudevan, P., Irfan, I., Liyanage, M.: Development of an eccentric legged quadruped robot for a predefined uneven terrain. In: 2022 IEEE International Conference on Autonomous Robot Systems and Competitions (ICARSC), pp. 15–20 (2022)
4. Saranli, U., Buehler, M., Koditschek, D.E.: Rhex: a simple and highly mobile hexapod robot. Int. J. Robot. Res. **20**(7), 616–631 (2001)
5. Châvez,A.J.P.A., Alcântara, J.H.A.: Kinematic and dynamic modeling of the phantomx ax metal hexapod mark iii robot using quaternions. In: 2021 International Conference on Control, Automation and Information Sciences (ICCAIS), pp. 595–601 (2021)
6. Ismail, R.M., Muthukumaraswamy, S., Sasikala, A.: Military support and rescue robot. In: 2020 4th International Conference on Intelligent Computing and Control Systems (ICICCS), pp. 156–162 (2020)
7. Mazzolai, B., Laschi, C.: A vision for future bioinspired and biohybrid robots. Sci. Robot. **5**(38), eaba6893 (2020)
8. Picardi, G., et al.: Bioinspired underwater legged robot for seabed exploration with low environmental disturbance. Sci. Robot. **5**(42), eaaz1012 (2020)
9. Trimmer, B.A.: Metal or muscle? the future of biologically inspired robots. Sci. Robot. **5**(38), eaba6149 (2020)
10. Kondoyanni, M., Loukatos, D., Maraveas, C., Drosos, C., Arvanitis, K.G.: Bio-inspired robots and structures toward fostering the modernization of agriculture. Biomimetics **7**(2), 69 (2022)
11. Deepa, T., Angalaeswari, S., Subbulekshmi, D., Krithiga, S., Sujeeth, S., Kathiravan, R.: Design and implementation of bio inspired hexapod for exploration applications. Mater. Today Proc. **37**, 1603–1607 (2021)
12. Pardo-Cabrera, J., Rivero-Ortega, J.D., Hurtado-López, J., Ramírez-Moreno, D.F.: Bio-inspired navigation and exploration system for a hexapod robotic platform. Eng. Res. Express **4**(2), 025019 (2022)

13. Suárez-Quispe, J.C., et al.: Mechatronics development of hexapod robot controlled by bioinspired locomotion steps for farmlands visual exploration. In: 2023 Fifth International Conference on Electrical, Computer and Communication Technologies (ICECCT), pp. 01–08 (2023)
14. Meyer, H.G., Klimeck, D., Paskarbeit, J., Rückert, U., Egelhaaf, M., Porrmann, M., Schneider, A.: Resource-efficient bio-inspired visual processing on the hexapod walking robot hector. PLoS ONE **15**(4), e0230620 (2020)
15. Billeschou, P., Bijma, N.N., Larsen, L.B., Gorb, S.N., Larsen, J.C., Manoonpong, P.: Framework for developing bio-inspired morphologies for walking robots. Appl. Sci. **10**(19), 6986 (2020)
16. Kouame Yann Olivier, A., Biradar, R.C., Devanagavi, G.D.: Conceptual design of a quadruped wheel-legged robot for an autonomous terrestrial locomotion. In: 2022 IEEE Fourth International Conference on Advances in Electronics, Computers and Communications (ICAECC), pp. 1–7 (2022)
17. de Bruijckere, J., Smit, R., Valk, L.R.: Zebro light: Lightweight mobile hexapod robot (2013)
18. Chavan, P., Murugan, M., Unnikkannan, E.V., Singh, A., Phadatare, P.: Modular snake robot with mapping and navigation: Urban search and rescue (usar) robot. In: 2015 International Conference on Computing Communication Control and Automation, pp. 537–541 (2015)
19. Wang, H., Gao, J., Ma, S., Li, Y.: A beetle-inspired miniature hexapod mesh climbing robot. In: 2022 IEEE International Conference on Robotics and Biomimetics (ROBIO), pp. 2189–2194 (2022)
20. Yu, Z., Shi, Y., Luo, A., Tao, J., Yang, S.X.: Research on landing collision experiment of gecko robot with bio-inspired dry adhesive foot. In: 2018 IEEE International Conference on Information and Automation (ICIA), pp. 336–340 (2018)
21. Neville, N., Buehler, M., Sharf, I.: A bipedal running robot with one actuator per leg. In:Proceedings 2006 IEEE International Conference on Robotics and Automation, 2006. ICRA 2006., pp. 848–853 (2006)
22. Dupeyroux, J., Diperi, J., Boyron, M., Viollet, S., Serres, J.: A bio-inspired celestial compass applied to an ant-inspired robot for autonomous navigation. In: 2017 European Conference on Mobile Robots (ECMR), pp. 1–6 (2017)

Multi Directional Stabilizing Active Wing

W. A. D. D. Ransika Weerakkody[1(✉)] and Geethal Siriwardana[2]

[1] Department of Mechanical & Automotive Engineering, Faculty of Engineering and Technology, CINEC Campus, Malabe, Sri Lanka
waddrweerakkody@gmail.com

[2] Department of Mechanical Engineering, Faculty of Engineering, University of Sri Jayewardenepura, Nugegoda, Sri Lanka
geethal@sjp.ac.lk

Abstract. In modern motorsports, vehicle dynamics are pushed to their absolute limits, with high-speed cornering presenting challenges to both performance and safety. Such existing systems as KERS, Active Yaw Control and ESC tend to act independently of the aerodynamic control. This study investigates a three-degree-of-freedom active aerodynamic rear wing, controlled by a hybrid PID–Fuzzy Logic framework, designed to mitigate vehicle roll, pitch, and yaw. A MATLAB/Simulink model was employed to evaluate downforce, drag, and Euler angles, while CFD simulations confirmed that the active configuration generated significantly greater downforce compared to a passive wing. Experimental validation using Arduino-based PID control demonstrated effective roll and pitch actuation, while fuzzy logic control enabled real-time adjustment of the wing in response to yaw inputs. The findings indicate that active aerodynamics provide a viable approach to enhancing vehicle stability in high-performance motorsport applications.

Keywords: KERS (Kinetic Energy Recovery System) · ESC (electronic speed controller) · Fuzzy logic control · PID control · Active aerodynamics · Euler angles · Downforce

1 Introduction

Stability under dynamic cornering is both a fundamental safety concern and a critical design requirement in high-performance vehicles. Formula 1 racing cars experience lateral accelerations exceeding 5g (five times gravitational acceleration, where g = 9.81 m/s^2), while NASCAR and GT race cars frequently corner at speeds greater than 300 km/h, often employing aggressive deceleration strategies [1, 2]. These conditions result in substantial lateral load transfers that generate undesirable roll and yaw moments, which can degrade handling performance and, in extreme cases, destabilize the vehicle.

To counter such effects, modern stabilization technologies—including Electronic Stability Control (ESC), Active Suspension, and Torque Vectoring—have been developed [3–5]. However, these systems are predominantly mechanical or electrical in nature

M. Siriwardhana et al. (Eds.): IERS 2025, PTT, pp. 89–102, 2026.
https://doi.org/10.1007/978-981-95-7984-6_8

and are not inherently integrated with aerodynamic control in real time, leaving a gap in synchronized vehicle stabilization strategies. Aerodynamic components such as wings and spoilers, though traditionally passive or semi-active, have shown significant potential for contributing to roll and yaw damping [6– 15].

This study introduces a novel multi-directional active rear-wing system designed to simultaneously control pitch, roll, and yaw through a three-degree-of-freedom (3-DOF) actuation mechanism. The proposed system employs real-time feedback from an inertial measurement unit (IMU) sensor, combined with a hybrid PID–Fuzzy Logic controller and MATLAB/Simulink–Arduino-based modelling, to generate corrective aerodynamic forces under dynamic loading conditions. The controller continuously adjusts wing surfaces based on IMU-derived parameters—namely lateral acceleration, yaw rate, and slip angle—to deliver instantaneous aerodynamic compensation during manoeuvres such as trail braking, lane changes, and high-speed cornering.

The governing formulation is expressed in body-fixed vehicle coordinates using Newton–Euler equations, which capture lateral forces, load transfer, and rotational dynamics (φ–roll, θ–pitch, ψ–yaw). These form the foundation of the reduced-order dynamic model adopted for real-time control.

The objectives of this study are:

- Deploying a MATLAB/Simulink model that estimates downforce, drag, and Euler angles of without aerodynamic aids by MATLAB software.
- To develop and experiment with a hybrid PID-Fuzzy Logic control of roll, pitch, and yaw actuation using Arduino IDE and MATLAB.
- To create NACA 4412 wing with splitters by Solid modelling software.
- To confirm passive and active wing, the aerodynamic enhancements in Solid modelling–Flow Simulations software.

2 Literature Review

2.1 Simulink Simulation Framework

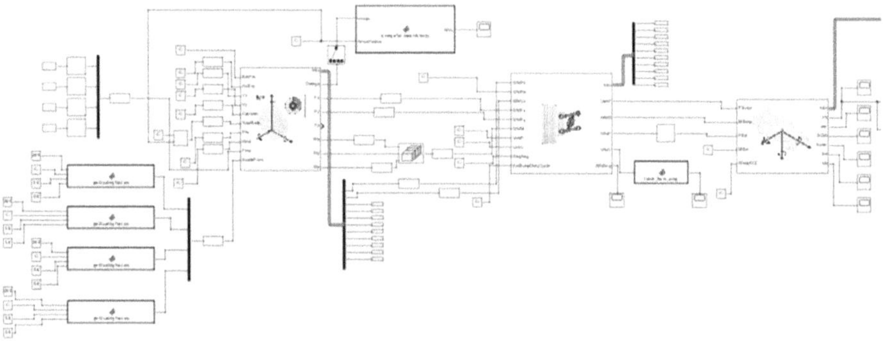

Fig. 1. Simulink Model

The vehicle model was developed in MATLAB/Simulink using a six-degree-of-freedom (6-DOF) body, a combined-slip wheel model (2-DOF), and double-wishbone suspension blocks to capture tire–road interactions and vertical load transfer. Real racing car specifications—including mass, suspension geometry, and aerodynamic characteristics—were incorporated to ensure realistic vehicle behavior (Fig. 1). The aerodynamic drag and lift modules were implemented to generate downforce and aerodynamic moments as functions of wind conditions, vehicle speed, and closed-loop feedback signals. The simulation produced more than 40 output variables, including downforce, drag force, and Euler angles.

2.2 Data Acquisition Through Arduino

2.2.1 Pugh Analysis & Arduino Hardware Wiring

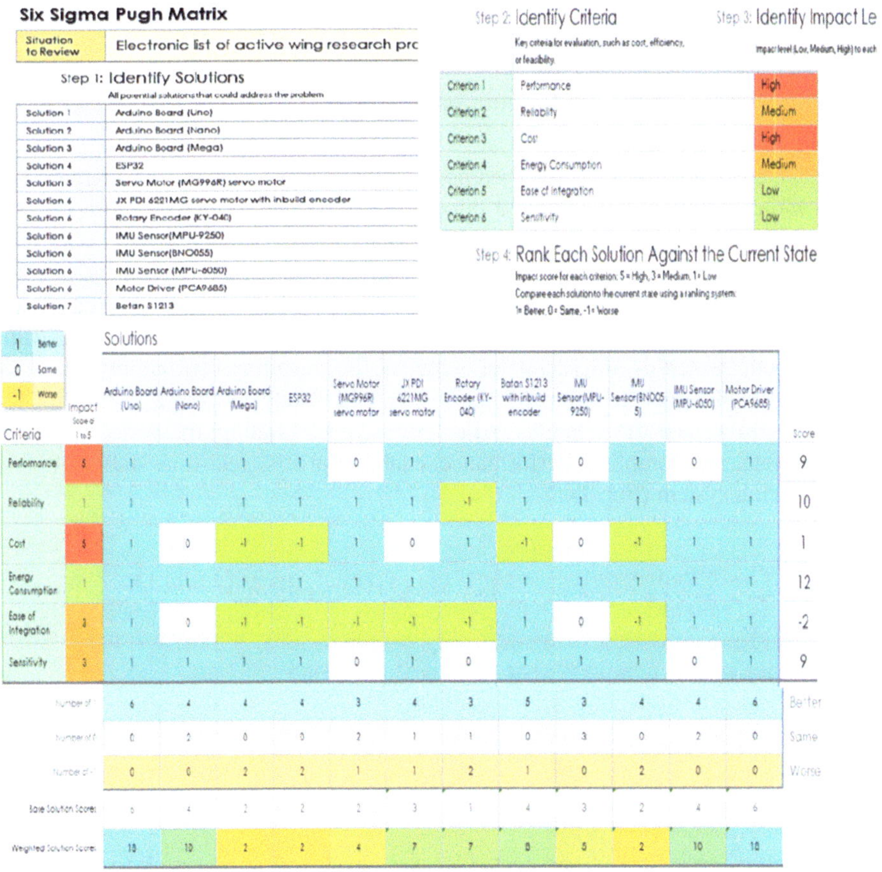
Six Sigma Pugh Matrix

Situation to Review	Electronic list of active wing research prc

Step 1: Identify Solutions
All potential solutions that could address the problem

Solution 1	Arduino Board (Uno)
Solution 2	Arduino Board (Nano)
Solution 3	Arduino Board (Mega)
Solution 4	ESP32
Solution 5	Servo Motor (MG996R) servo motor
Solution 6	JX PDI 6221MG servo motor with inbuild encoder
Solution 6	Rotary Encoder (KY-040)
Solution 6	IMU Sensor(MPU-9250)
Solution 6	IMU Sensor(BNO055)
Solution 6	IMU Sensor (MPU-6050)
Solution 6	Motor Driver (PCA9685)
Solution 7	Betan S1213

Step 2: Identify Criteria
Key criteria for evaluation, such as cost, efficiency, or feasibility

Step 3: Identify Impact Le
Impact level (Low, Medium, High) to each

Criterion 1	Performance	High
Criterion 2	Reliability	Medium
Criterion 3	Cost	High
Criterion 4	Energy Consumption	Medium
Criterion 5	Ease of Integration	Low
Criterion 6	Sensitivity	Low

Step 4: Rank Each Solution Against the Current State
Impact score for each criterion: 5 = High, 3 = Medium, 1 = Low
Compare each solution to the current state using a ranking system:
1 = Better, 0 = Same, -1 = Worse

1	Better
0	Same
-1	Worse

Solutions

Criteria	Impact Scale of 1 to 5	Arduino Board (Uno)	Arduino Board (Nano)	Arduino Board (Mega)	ESP32	Servo Motor (MG996R) servo motor	JX PDI 6221MG servo motor	Rotary Encoder (KY-040)	Batan S1213 with inbuild encoder	IMU Sensor(MPU-9250)	IMU Sensor(BNO055)	IMU Sensor (MPU-6050)	Motor Driver (PCA9685)	Score
Performance	5	1	1	1	1	0	1	1	1	0	1	0	1	9
Reliability	1	1	1	1	1	1	1	-1	1	1	1	1	1	10
Cost	5	1	0	-1	-1	1	0	1	-1	0	-1	1	1	1
Energy Consumption	1	1	1	1	1	1	1	1	1	1	1	1	1	12
Ease of Integration	3	1	0	-1	-1	-1	-1	-1	1	0	-1	1	1	-2
Sensitivity	3	1	1	1	1	0	1	0	1	1	1	0	1	9
Number of 1		6	4	4	4	3	4	3	5	3	4	4	6	Better
Number of 0		0	2	0	0	2	1	1	0	3	0	2	0	Same
Number of -1		0	0	2	2	1	1	2	1	0	2	0	0	Worse
Base Solution Scores		6	4	2	2	2	3	1	4	3	2	4	6	
Weighted Solution Scores		18	10	2	2	4	7	7	8	5	2	10	18	

Fig. 2. Pugh Analysis Chart

The internal potentiometer signal of the MG996R servo was accessed to enable the Arduino to read analog position feedback. Following the correction of an initial wiring fault, the feedback signal was incorporated into a PID control loop, facilitating precise closed-loop control without the need for an external encoder (Figs. 2, 3, 4, 5 and 6).

Fig. 3. PCB Modification

Fig. 4. Volatge Vary 01

Fig. 5. Voltage Vary 02

Fig. 6. Voltage Vary 03

2.2.2 Real Time Control Implementation Using MATLAB and Arduino IDE

The real-time roll and pitch control system was implemented in the Arduino IDE using C++. Vehicle orientation was measured using an MPU6050 inertial measurement unit (IMU), while actuation was achieved through MG996R servo motors equipped with analog feedback and driven by PCA9685 PWM controllers. Accelerometer data were processed using the atan2 function to calculate roll and pitch angles, and an exponential smoothing filter was applied to reduce measurement noise. The inverses of the calculated angles were supplied as setpoints to the PID controllers, which closed the loop using feedback from the servos' internal potentiometers. The resulting PID outputs were converted into PWM signals to command the servos, with deadband compensation applied to minimize jitter near the setpoints. After extensive PID tuning, the system achieved smooth, stable, and compliant performance, enabling precise real-time stabilization of the active wing system (Fig. 7).

```
float denominator = sqrt((float)ay * ay + (float)az * az);
if (denominator == 0) denominator = 0.0001;  // prevent nan
float rawPitch = atan2(-ax, denominator) * 180.0 / PI;
rawPitch = constrain(rawPitch, 0, 180);
smoothPitch = 0.85 * smoothPitch + 0.15 * rawPitch;
float pitch = smoothPitch;
```

Fig. 7. Sensor data converts to degrees and conditioning.

A Mamdani-type fuzzy logic controller was developed in Simulink to implement the yaw control system (Figs. 8, 9, 10, 11, 12, 13 and 14). The controller utilized steering angle and yaw rate as input variables, while the yaw position of the active rear wing served as the output. Input signals were generated by scaling sine waves to realistic vehicle operating ranges, and fuzzy rules were formulated to relate wing orientation to steering dynamics. Once the steering angle and yaw rate were applied, the system responded adaptively, producing yaw-induced wing deflections that confirmed the expected directional response. Unlike PID control, the fuzzy logic framework did not require feedback loops or extensive tuning, making it particularly well-suited for real-time dynamic control of aerodynamic yaw.

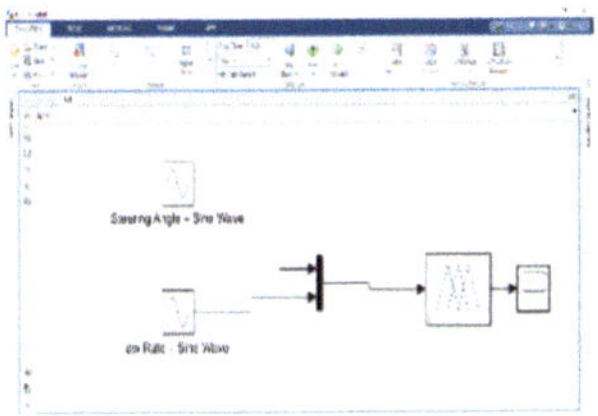

Fig. 8. Matlab setup

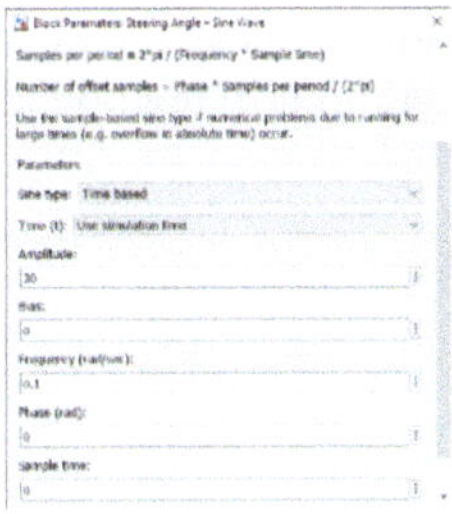

Fig. 9. Steering Angle

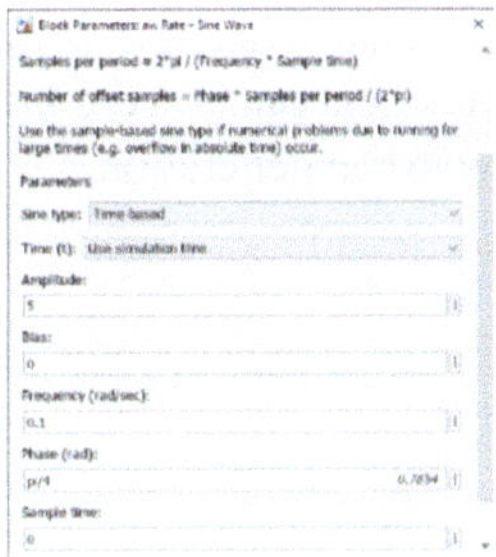

Fig. 10. Yaw Rate

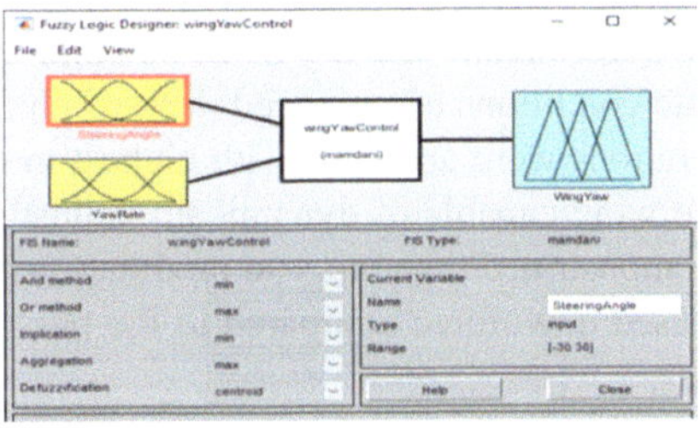

Fig. 11. Memebership Fcn

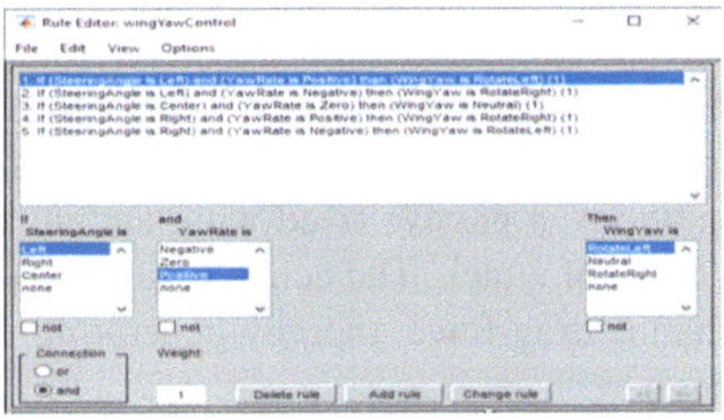

Fig. 12. Rules

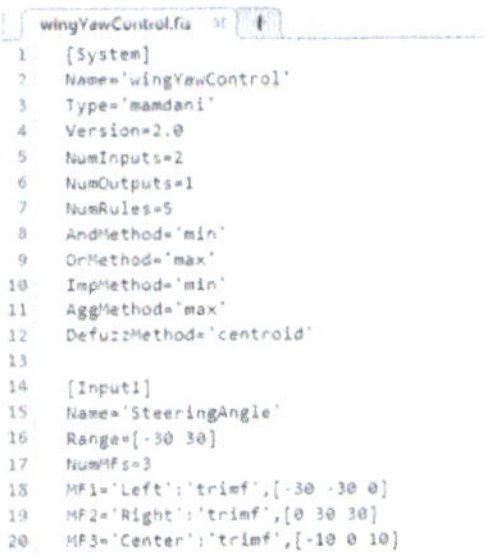

Fig. 13. MATLAB Code - Phase 1

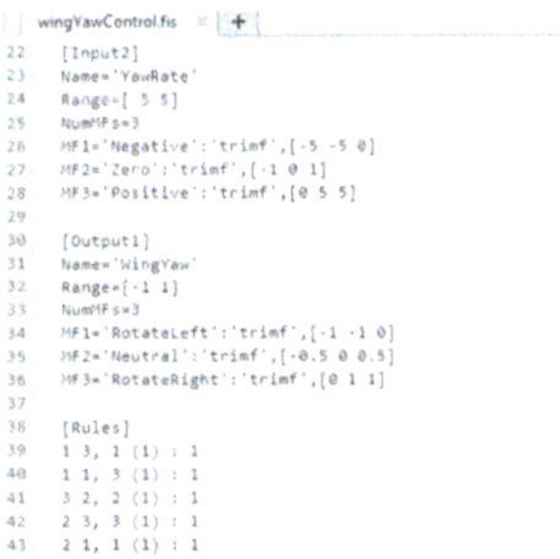

Fig. 14. MATLAB Code - Phase 2

2.3 Structural Layout and Component Placement Strategy

The active wing was constructed with a carbon-fibre finish and mounted on a custom-designed shaft that enabled both isolated and joint movement along the principal axes. The NACA 4412 air foil profile was selected due to its high lift-to-drag ratio, making it well-suited for maximizing downforce during cornering. The air foil geometry was defined in real coordinates using air foil tools and subsequently optimized in CAD to ensure a smooth transition between surfaces (Fig. 15). To further enhance aerodynamic performance under dynamic loading, boundary-layer trimming modules combined with vortex generators were integrated, improving boundary-layer behaviour and increasing downforce generation.

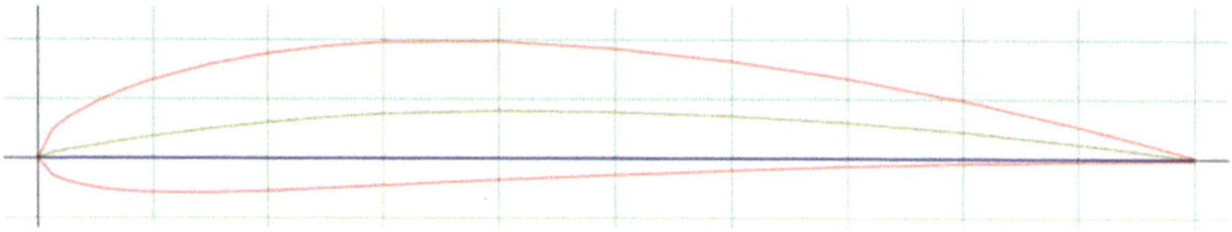

Fig. 15. Coordinates of the wing has NACA 4412 airfoil

2.4 Computational Fluid Dynamic Analysis Strategy

The CFD flow simulation was configured with a freestream velocity of 28 m/s directed from the northeast to replicate the aerodynamic conditions experienced during vehicle cornering (Figs. 16 and 17). 30 wing configurations were analysed with highest mesh conditions: a passive fixed wing and an active wing capable of dynamically adjusting its angle of attack. The active configuration demonstrated the ability to modulate aerodynamic drag force, thereby generating increased downforce compared to the passive design.

Fig. 16. Behavior of Wind During Cornering

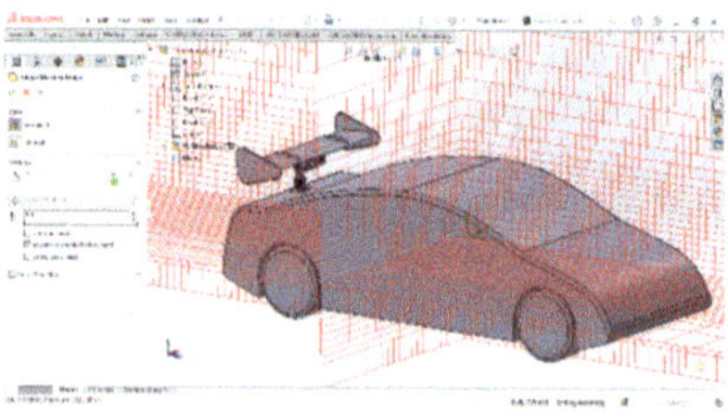

Fig. 17. Mesh

3 Results and Discussions

3.1 Simulink Simulation Framework

3.1.1 Euler Angle Analysis–Vehicle Attitude Dynamics

Under the simulated dynamic conditions, the vehicle exhibited a roll angle variation within ±0.15 radians, a pitch angle variation within ±0.35 radians, and a yaw angle variation within ±0.37 radians in the Simulink model (Figs. 18, 19 and 20).

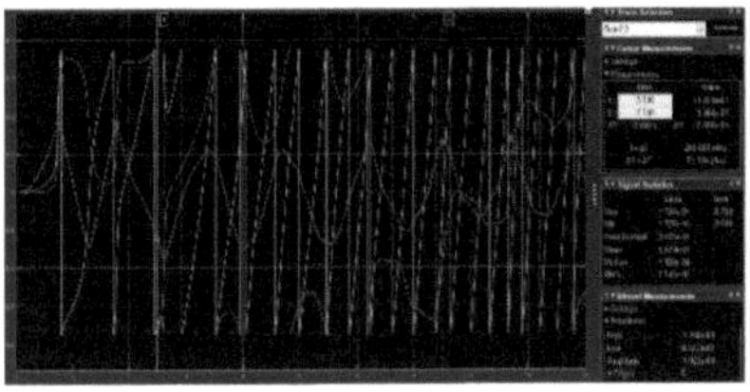

Fig. 18. Roll Angle

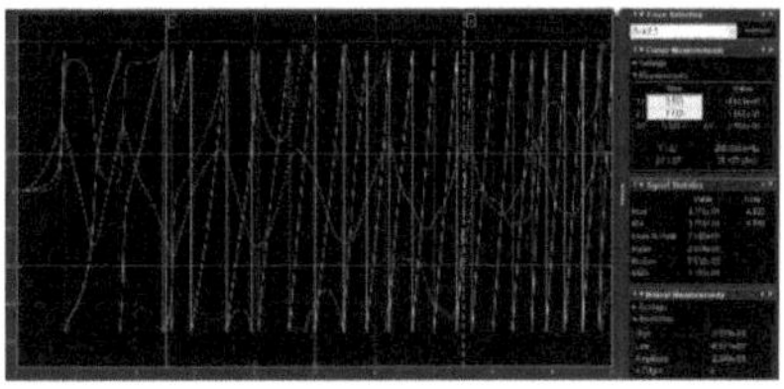

Fig. 19. Pitch Angle

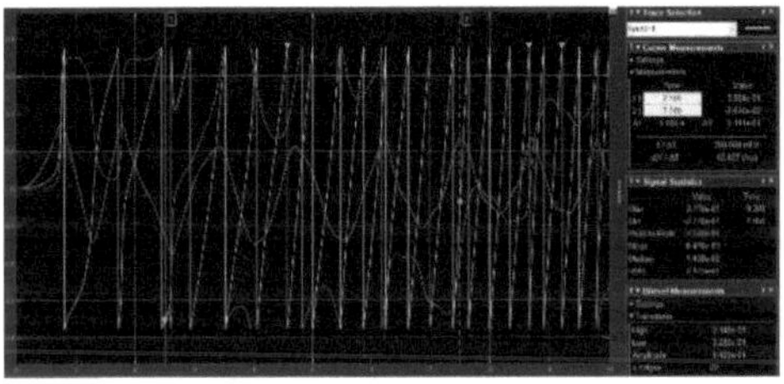

Fig. 20. Yaw angle

3.1.2 Down Force and Drag Force Simulation

Under simulated dynamic conditions in the Simulink model, the vehicle exhibited roll, pitch, and yaw responses while generating a downforce of approximately 1000 N and a drag force of about 300 N (Figs. 21 and 22).

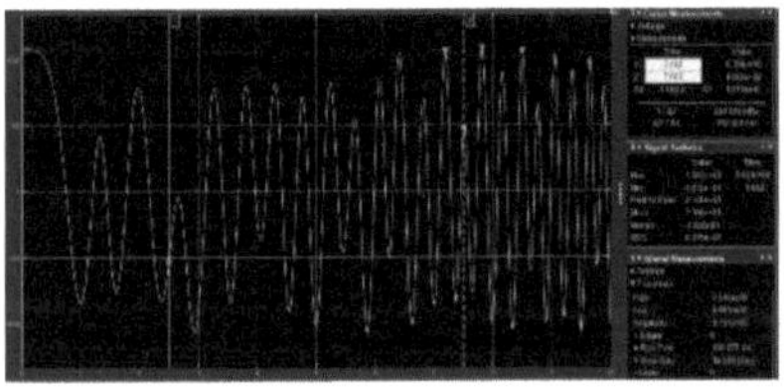

Fig. 21. Downforce Simulation

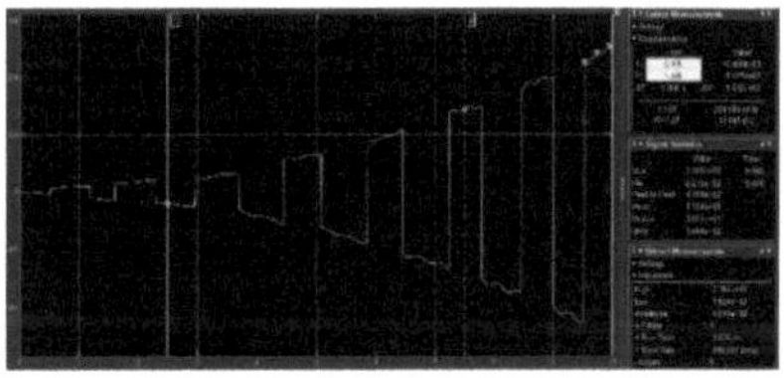

Fig. 22. Drag force Simulation

3.2 Real–Time Control Implementation Using MATLAB and Arduino IDE

3.2.1 Pitch Controlling Using PID Tuning

The pitch response was evaluated over a 10-s time span, with time represented on the x-axis and pitch angle on the y-axis (Fig. 23). The system reached the setpoint rapidly, demonstrating stability at PID gains of Kp = 1.5, Ki = 0.1, and Kd = 0.1. However, when Kp was increased to 2.0 while maintaining Ki = 0.1 and Kd = 0.1, the response exhibited large deviations, suggesting that improved regulation could be achieved with different Ki and Kd values. Reducing Kp or Kd resulted in slower convergence, whereas excessively large values of these gains introduced oscillations in the feedback. Step-by-step tuning identified a set of PID parameters that produced the most stable and controlled response. In closed-loop operation, the system generated corrective wing deflections in the opposite direction of the disturbance, with equivalent angular displacement to simulated data, thereby stabilizing the vehicle's pitch motion.

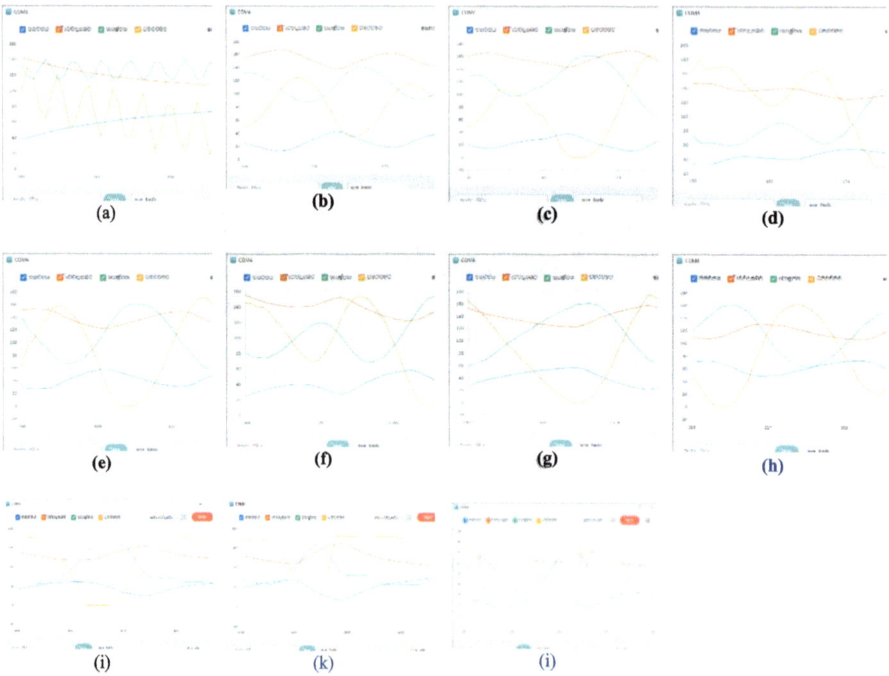

Fig. 23. (a) Open loop; (b) Kp = 1.5, Ki = 0.1, Kd = 0.0; (c) Kp = 1.5, Ki = 0.1, Kd = 0.1; (d) Kp = 1.5, Ki = 0.1, Kd = 0.2; (e) Kp = 1.5, Ki = 0.2, Kd = 0.1; (f) Kp = 1.5, Ki = 0.2, Kd = 0.2; (g) Kp = 1.3, Ki = 0.1, Kd = 0.1; (h) Kp = 1.5, Ki = 0.1, Kd = 0.1; (j) Kp = 1.6, Ki = 0.1, Kd = 0.1; (k) Kp = 2.0, Ki = 0.1, Kd = 0.1; (l) Kp = 1.5, Ki = 0.15, Kd = 0.1

3.2.2 Roll Controlling Using PID Tuning

The PID tuning results indicated that the configuration Kp = 0.9, Ki = 0.05, and Kd = 0.2 produced the most stable response. This setup provided an effective balance between responsiveness and stability, enabling the system to reach the desired value rapidly with minimal overshoot. The configuration also demonstrated the key advantage of maintaining a very low steady-state error. In closed-loop operation, this response manifested as the active wing deflecting in the opposite direction with an equivalent simulation's angular displacement, thereby contributing to effective stabilization of vehicle roll (Fig. 24).

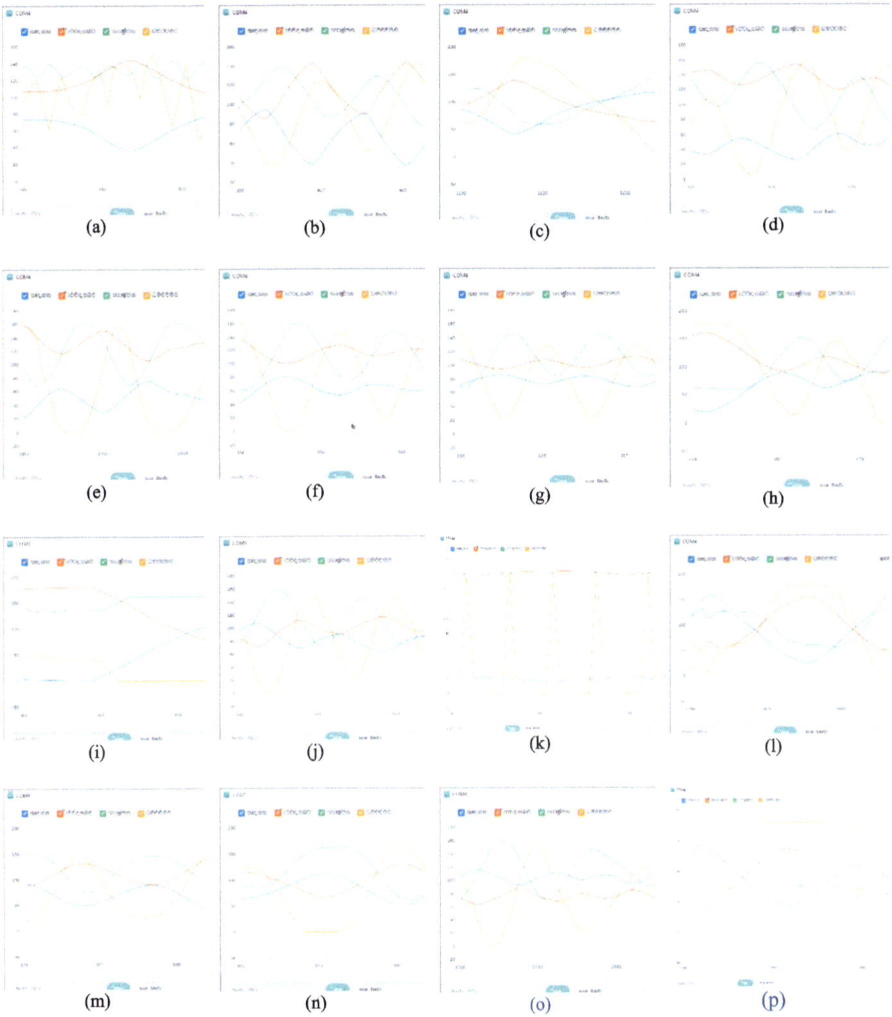

Fig. 24. (a) Without PID; (b) Kp = 1.0, Ki = 0.5, Kd = 0.0; (c) Kp = 1.0, Ki = 0.5, Kd = 0.1; (d) Kp = 1.0, Ki = 0.5, Kd = 0.2; (e) Kp = 1.0, Ki = 0.5, Kd = 0.3; (f) Kp = 1.0, Ki = 0.1, Kd = 0.2; (g) Kp = 1.0, Ki = 0.2, Kd = 0.2; (h) Kp = 1.0, Ki = 0.3, Kd = 0.2; (i) Kp = 1.0, Ki = 0.4, Kd = 0.2; (j) Kp = 1.0, Ki = 0.5, Kd = 0.2; (k) Kp = 0.5, Ki = 0.05, Kd = 0.2; (l) Kp = 0.8, Ki = 0.05, Kd = 0.2; (m) Kp = 0.9, Ki = 0.05, Kd = 0.2; (n) Kp = 1.0, Ki = 0.05, Kd = 0.2; (o) Kp = 1.2, Ki = 0.05, Kd = 0.2; and (p) Kp = 1.5, Ki = 0.05, Kd = 0.2.

3.2.3 Yaw Controlling Using FUZZY Logic

The fuzzy logic–based control system was capable of generating proportionally magnified yaw adjustments of the rear wing in the same direction as the steering input, thereby maintaining the aerodynamic response in synchrony with the vehicle's cornering dynamics in real time. Using steering angle (radians) and yaw rate (s^{-1}) as inputs, the system produced the wing yaw angle (radians) as the output, ensuring an instantaneous and coordinated aerodynamic response (Fig. 25).

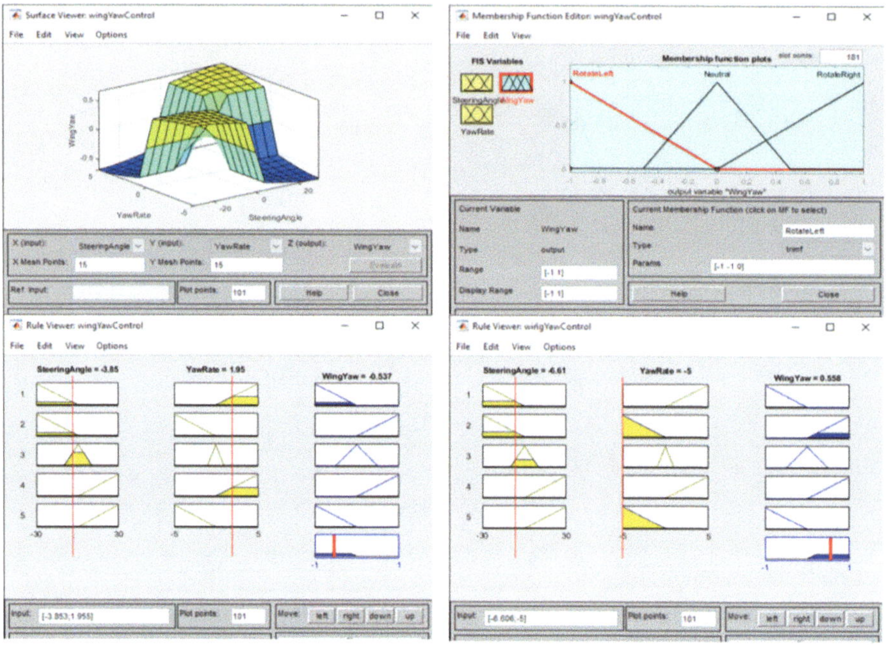

Fig. 25. Wing yaw response to steering angle and yaw rate inputs.

3.2.4 Structural Layout and Component Placement Strategy

To ensure safe and controlled articulation of the wing, mechanical constraints were incorporated into the design, limiting angular variations to the following ranges: roll angle between –0.15 and +0.15 radians, pitch angle between –0.35 and +0.35 radians, and yaw angle between –0.37 and +0.37 radians (shown in Fig. 26).

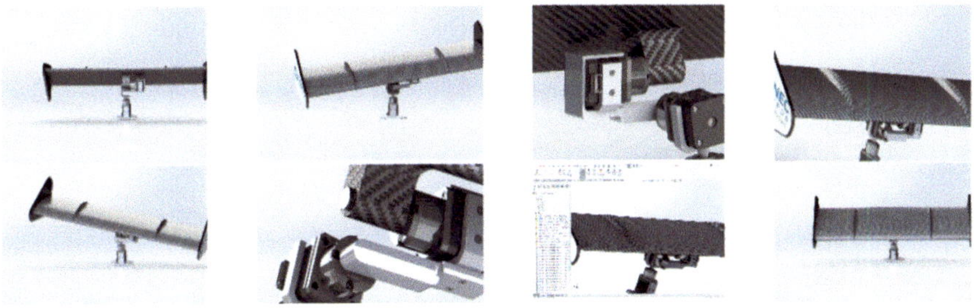

Fig. 26. Design different views

3.2.5 CFD Analysis

There were three main simulation cases under consideration (Figs. 27, 28 and 29). For the first case (roll variation with fixed pitch and yaw), the maximum downforce of 891 N, drag force of 35 N was observed at roll = –30°, pitch = 0°, yaw = 0°. For the second case (pitch variation with roll fixed at –30° and yaw at 0°), the maximum downforce of 865 N and down force of 54 N was observed at pitch = 15°, roll = –30°, yaw = 0°. For the third case (yaw variation with roll at –30° and pitch at 15°), the maximum downforce of 914 N and down force of 20 N occurred at yaw = –15°, roll = –30°, pitch = 15°. The results indicate that global variations of roll, pitch, and yaw have significant impact on aerodynamic loading, and interaction among yaw, roll, and pitch generates the largest downforce and 0.043 wing's drag coefficient (Figs. 30, 31 and 32). By using PID control in roll and pitch and fuzzy control in yaw, the system was able to sustain these optimized values of downforce, showing how hybrid control schemes can enhance aerodynamic stability.

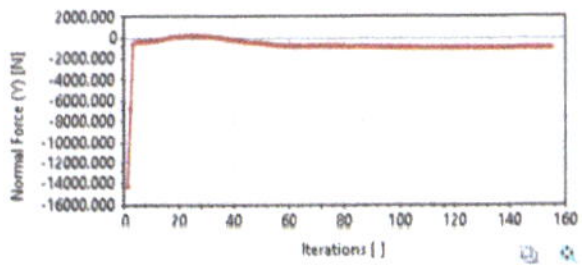

Fig. 27. Active wing downforce

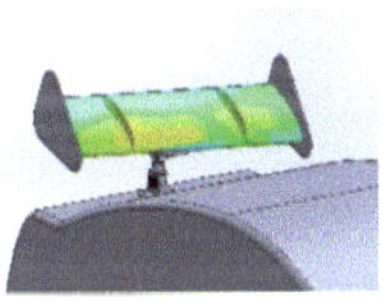

Fig. 28. Active wing

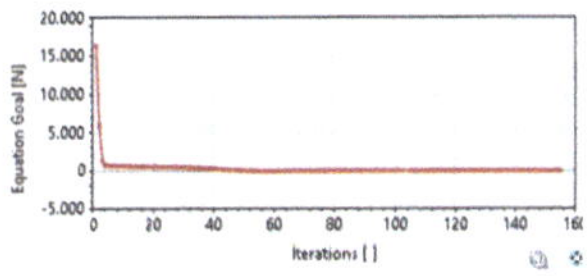

Fig. 29. Active wing drag coefficient

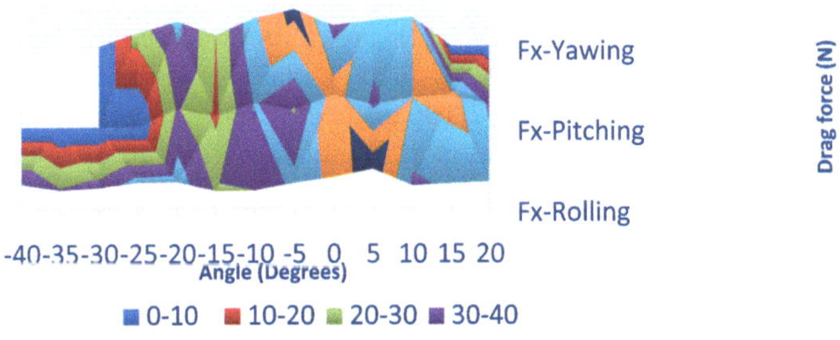

Fig. 30. Variation of Drag force with Wing Angle

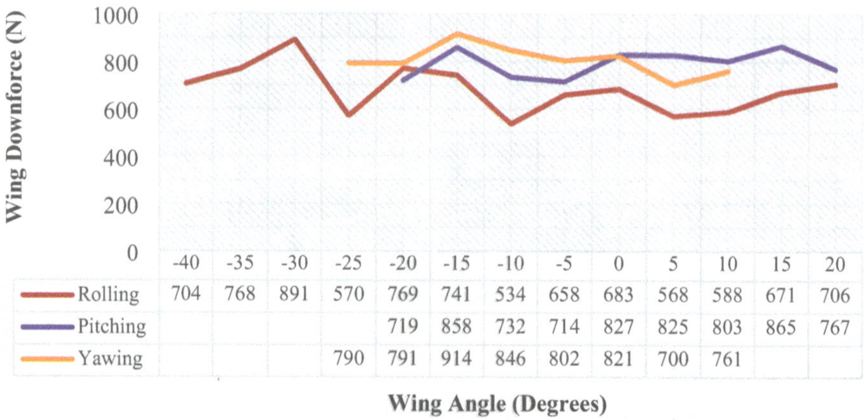

	-40	-35	-30	-25	-20	-15	-10	-5	0	5	10	15	20
Rolling	704	768	891	570	769	741	534	658	683	568	588	671	706
Pitching					719	858	732	714	827	825	803	865	767
Yawing				790	791	914	846	802	821	700	761		

Fig. 31. Variation of Downforce with Wing Angle

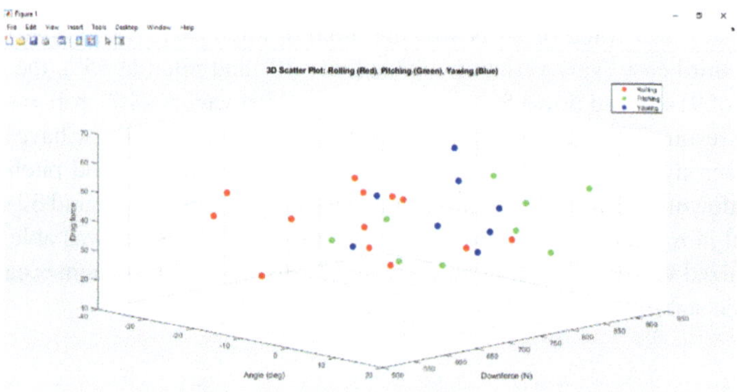

Fig. 32. Downforce vs Drag force Vs Angle

4 Conclusions

This study successfully designed and developed the concept of a multidirectional active wing system, meeting the stated objectives and demonstrating the feasibility of real-time aerodynamic control.

The realistic downforce, drag force and Euler angles of a modeled racing car were determined using a Simulink simulation. On the basis of these findings, a control approach was drawn whereby roll and pitch angles are controlled by a PID controller based on IMU sensor data, as body moments are directly related to such motions. There was an algorithm to push the wing towards the opposite direction of the IMU measurements using the same amount of angles found in the simulation Euler angle results. A fuzzy logic scheme was used to control yaw because the steering inputs are nonlinear and sensor processing provides a delay; fuzzy rules can be responded to faster. Yaw movement was in same direction as the steering angle, but within the same magnitude range as the Euler angle outputs of the simulation. A 3D wing design was then made with mechanical limits relevant to these angle magnitudes obtained from the simulation's Euler angle outputs. Lastly, over 30 CFD simulations in various Euler angles confirmed that this active wing mechanism is capable of producing a greater Aerodynamic performance was primarily assessed using CFD analysis, The yaw, roll and pitch combined gave a maximum downforce of 914 N and a drag of 20 N, the minimum drag coefficient was 0.043. This substantial improvement in aerodynamic stability is attributed to the adaptive actuation of the wing and the use of a NACA 4412 aerofoil profile enhanced with vortex generators. The wing's motion, synchronized with body dynamics and steering input, effectively counteracted destabilizing aerodynamic forces, thereby maintaining stability during high-speed cornering. The resulting system, grounded in principles of simplicity, robust signal processing, and synchronized control, has achieved Technology Readiness Level (TRL) 3 and demonstrates strong potential for further development as a next-generation motorsport or automotive subsystem.

Future work will focus on the implementation of an adaptive aerofoil system capable of dynamically switching between profiles to suit varying requirements of speed, steering, and load, thereby maximizing aerodynamic performance and vehicle control. Additionally, a physical prototype will be fabricated and tested in a wind tunnel to experimentally validate the findings obtained through CFD and simulation studies.

References

1. Chatrath, K.: Vehicle dynamics control using control allocation, Master Thesis: Technical University of Delft (2019). https://repository.tudelft.nl/record/uuid:b99bbc10-32a6-45d8-8314-b2f72810569d
2. Gupta, S.: Evaluation of commercial vehicle and tire models using Simscape multibody. Eindhoven University of Technology. Student Thesis (2024). https://research.tue.nl/en/studentTheses/evaluation-of-commercial-vehicle-and-tire-models-using-simscape-m/
3. Milliken, W.F., Milliken, D.L.: Race car vehicle dynamics. Milliken Research Associates (1995)
4. Genta, G.: Improvement of vehicle maneuverability by direct yaw moment control. Veh. Syst. Dyn. **22**(5–6), 321–335 (1993)
5. Mohan, B.M., et al.: Analytical structure and stability analysis of a fuzzy PID controller. Appl. Soft Comput. **7**(1), 288–297 (2007)
6. Zheng, J.M., et al.: Application of self-tuning fuzzy PID controller for a SRM direct drive volume control hydraulic press. Control. Eng. Pract. **17**(12), 1390–1396 (2009)
7. Misar, A.S., et al., Scale-resolved and time-averaged simulations of the flow over a NASCAR Cup Series racecar. SAE Technical Paper 35 (2023). 2023-01-07
8. Ren, T.J., Chen, T.C., Chen, C.J.: Motion control for a two-wheeled vehicle using a self-tuning PID controller. Control. Eng. Pract. **16**(3), 365–375 (2008)
9. Genta, G.: Motor Vehicle Dynamics: Modeling and Simulation. World Scientific, Singapore (1997)
10. Muhammad, E., et al.: Coordination of active front steering and direct yaw control systems using MIMO sliding mode control. Int. J. Control Autom. Syst. **22**, 345–357 (2024)
11. Naderi, M., Sedigh, A.K., Johansen, T.A.: Guaranteed feasible control allocation using model predictive control. Control Theory Technol. **17**(3), 223–235 (2019)
12. Piechna, J.: A review of active aerodynamic systems for road vehicles. Energies **14**(23), 7887 (2021)
13. Krishna, S., Narayanan, S., Ashok, S.D.: Fuzzy logic based yaw stability control for active front steering of a vehicle. J. Mech. Sci. Technol. **28**(4), 1245–1256 (2014)

14. Rajamani, R.: Vehicle Dynamics and Control. Springer, New York (2012)
15. Wong, J.Y.: Theory of Ground Vehicles, 3rd edn. Wiley, New York (2000)

Speaker-Consistent Speech Enhancement with Auxiliary Conditioning

Thumula Rajakaruna(✉) and H. K. I. S. Lakmal

Faculty of Engineering, NSBM Green University, Homagama, Sri Lanka
rmtdbrajakaruna@students.nsbm.ac.lk, isuru.l@nsbm.ac.lk

Abstract. This research introduces a lightweight neural audio enhancement system to re-fine speech signals processed by upstream models, such as MetricGAN +, mitigating residual distortions while retaining noise suppression benefits. The system employs a 192-dimensional (192D) speaker embedding from original clean audio and a 10-point quantized pitch contour to restore speaker identity and prosodic naturalness. Objective evaluations show significant improvements in speaker embedding cosine similarity (SECS) from 0.79 to 0.83 ($p < 0.05$) and perceptual evaluation of speech quality (PESQ) from 1.75 to 1.91 ($p < 0.05$) compared to the baseline. Subjective evaluations, conducted under controlled conditions with 20 listeners, yield a mean opinion score (MOS) of 4.2 (SD = 0.3), surpassing the baseline's 4.0 (SD = 0.3). Listeners reported enhanced audio clarity due to effective noise and artifact removal, with speaker identity preserved so closely that differences are only detectable upon meticulous inspection. These attributes make the system ideal for real-time applications like teleconferencing and voice assistants.

Keywords: Speech enhancement · Auxiliary conditioning · Speaker embedding · Pitch quantization · Neural audio processing · Speaker-consistent enhancement

1 Introduction

Speech enhancement is critical for improving speech quality and intelligibility in noisy environments, with applications in teleconferencing, voice assistants, hearing aids, and real-time communication [1, 2]. The challenge lies in achieving high perceptual quality and speaker consistency in resource-constrained settings like mobile or edge devices. Traditional methods, such as spectral subtraction [3], rely on stationary noise assumptions, often producing musical noise artifacts in dynamic environments [4]. Deep learning-based approaches like MetricGAN+ [5] optimize perceptual metrics like PESQ but may introduce distortions, such as over-smoothing or robotic timbres, compromising speaker identity and naturalness [4]. These distortions are problematic in applications requiring high speaker similarity, such as voice authentication, where speaker embedding cosine similarity (SECS) is crucial [6].

M. Siriwardhana et al. (Eds.): IERS 2025, PTT, pp. 103–109, 2026.
https://doi.org/10.1007/978-981-95-7984-6_9

Neural architectures like WaveNet [7], SoundStream [2], and HiFi-GAN-2 [8] model complex audio patterns for high-fidelity outputs but are computationally intensive, limiting edge device deployment. Speaker-aware systems like SpEx [9] use speaker embedding to preserve identity but increase computational overhead due to high-dimensional representations. Prosodic features like pitch are vital for naturalness, yet their integration is often computationally costly or neglected in lightweight systems [10]. Neural codecs like PSCodec [4] and TokenSplit [11] focus on primary enhancement or synthesis, leaving a gap in secondary refinement of pre-enhanced audio to address residual distortions efficiently.

This work proposes a novel neural speech enhancement system that refines MetricGAN+ -processed audio using a 192D speaker embedding and a 10-point quantized pitch contour. By leveraging cleaned MetricGAN+ outputs as reconstruction targets and a lightweight convolutional neural network (ConvNet), the system ensures real-time performance on resource-constrained devices, mitigating distortions while enhancing perceptual quality and speaker consistency.

Our Contributions:

1. Developed a lightweight ConvNet to refine MetricGAN+-processed audio, reducing distortions for real-time edge device use.
2. Integrated 192D speaker embedding and 10-point pitch contours, improving SECS from 0.79 to 0.83 ($p<0.05$) and prosodic naturalness.
3. Achieved superior audio clarity, with PESQ improved from 1.75 to 1.91 ($p<0.05$) and MOS from 4.0 to 4.2 (SD=0.3), surpassing original clean audio in perceived clarity.

The paper is organized as follows: Sect. 2 reviews related works, Sect. 3 details the methodology, Sect. 4 presents experimental results, and Sect. 5 concludes, followed by references.

2 Literature Review

Speech enhancement has transitioned from traditional signal processing to deep learning-based and speaker-aware neural architectures. Spectral subtraction [3] removes noise spectra but struggles in dynamic environments, producing musical artifacts [4]. Deep neural networks (DNNs) like MetricGAN+ [5] optimize perceptual metrics for high intelligibility but introduce over-smoothing [4]. WaveNet [7] generates high fidelity audio via autoregressive modeling but is computationally heavy. SoundStream [2] balances quality and efficiency as an end-to-end codec, focusing on compression rather than secondary refinement. TokenSplit [11] refines speech separation using discrete codec tokens but lacks emphasis on speaker consistency. HiFi-GAN-2 [8] uses subband adversarial training for robust synthesis, but its complexity limits real-time use.

Preserving speaker identity is essential for applications like voice authentication. SpEx [9] employs multi-scale time-domain networks with speaker embedding to maintain identity in noise, but high-dimensional representations increase resource demands. Personalized enhancement [6] avoids separate embedding models but requires resource-intensive representations. PSCodec [4] integrates prompt encoders for high fidelity coding, prioritizing synthesis over refinement.

Prosodic features like pitch enhance naturalness. VQ-VAE with WaveNet decoders [1] achieves efficient coding but is computationally costly. Ultra-low-bitrate coding with transformers [3] reduces resource needs but neglects speaker consistency. Hwang et al. [10] incorporate neural network-based side information, including continuous pitch representations, to enhance coded speech, but require additional computational overhead. Unlike [10], our 10-point quantized pitch con-tour reduces computational load while capturing prosodic variations, as justified in Sect. 3.2. Other works [1–9, 11–13] emphasize neural codecs and efficiency via vector-quantized variational autoencoders (VQ-VAE) [1], end-to-end coding [2], prompt-based encoding [4], and predictive quantization [10]. However, they lack focus on secondary refinement with efficient speaker and prosodic integration.

Our system addresses these gaps by refining MetricGAN+ -processed audio using a lightweight ConvNet with 192D speaker embedding and 10-point pitch contours, achieving superior clarity and speaker consistency, validated by subjective intelligibility tests [14] and objective metrics.

3 Methodology

The methodology covers data collection, preprocessing, model architecture, training, and evaluation to refine MetricGAN+ -processed audio while preserving speaker identity and prosodic naturalness.

3.1 Data Collection

The LibriSpeech dataset (1,000 h of clean English audiobook speech) provides a diverse corpus. Synthetic noise from the DNS Challenge dataset (babble, white, environmental) simulates real-world conditions. MetricGAN+ [5] processes noisy audio to generate cleaned versions as waveform reconstruction targets using L1 loss. Original clean audio provides reference speaker embeddings. The dataset is split into training (80%), validation (10%), and test (10%) sets.

3.2 Data Preprocessing

Preprocessing generates auxiliary conditioning signals: a 10-point quantized pitch contour and a 192D speaker embedding. Pitch is extracted from cleaned MetricGAN+ audio using torchaudio.functional.detect_pitch_frequency [9] at 16 kHz, clamped to [50, 400] Hz, and normalized to [0, 1] using the Eq. (1).

$$pitch_{norm} = \frac{pitch - 50}{400 - 50} \tag{1}$$

The pitch is resampled to 10 Hz (10 points per second) to capture prosodic variations efficiently. This choice balances temporal resolution and computational simplicity, as 10 points adequately represent prosodic changes for 1-s audio segments, aligning with findings in [10] for efficient prosodic modeling. Linear interpolation aligns the contour with the audio sampling rate, and 8-bit quantization yields 80 bits per second. Pitch detection errors in highly noisy conditions may affect prosodic accuracy, particularly for low-pitched speakers (<100 Hz).

Speaker embeddings are extracted from original clean audio using a pre-trained ECAPA-TDNN model [9], producing 192D vectors capturing timbre and vocal patterns. These are normalized to unit length and replicated across time steps. The conditioning vector is given in Eq. (2).

$$c(t) = \left[X_e(t), Interp(p), Repeat(s)\right] \in R^{194} \tag{2}$$

where $X_e(t)$ is the MetricGAN + -enhanced waveform, Interp(p) is the interpolated pitch contour, and Repeat(s) is the 192D speaker embedding. Limitations include potential embedding drift in noisy original audio, which may reduce speaker consistency.

3.3 Model Selection and Training

A lightweight three-layer 1D ConvNet, inspired by [2], balances efficiency and expressiveness for real-time inference. The encoder uses two Conv1D layers (kernel size: 5, padding: 2) with ReLU activations, mapping the 194D input to a 128D hidden representation. The decoder (Conv1D, kernel size: 5, padding: 2) produces a 1D refined waveform. A frame head (Conv1D, kernel size: 1) generates 192D frame-level speaker embeddings, and global pooling aggregates these into a 192D utterance-level embedding. The loss function is given in Eq. (3).

$$L = \lambda_1 \times ||x - x_e||_1 + \lambda_2 \times \left(1 - \cos\left(\emptyset(x), \emptyset\left(x_{ref}\right)\right) + \lambda_3 \times Contrastive\_loss + \lambda_4 \times fct_{loss}\right. \tag{3}$$

Here, L1 loss ensures waveform fidelity, SECS loss preserves speaker identity, contrastive loss aligns utterance-level embeddings, and frame consistency (FCT) loss ensures temporal stability. Weights are λ1 = 0.3\lambda_1 = 0.3λ1 = 0.3, λ2 = 1.0λ2 = 1.0λ2 = 1.0, λ3 = 0.6λ3 = 0.6λ3 = 0.6, λ4 = 0.3λ4 = 0.3λ4 = 0.3. Training uses AdamW (learning rate: 1e-4) on an NVIDIA A100 GPU for 100 epochs (~2 h), with a batch size of 8. Checkpoints and inference samples are saved as WAV files. Figure 1 illustrates the architecture.

3.4 Evaluation and Validation

Performance is evaluated using objective and subjective metrics across babble, white, and environmental noise. Objective metrics include SECS [3] (0–1, higher better), PESQ (−0.5–4.5, higher better), and STOI (0–100%, higher better). Subjective MOS tests with 20 listeners (native and non-native English speakers) rate naturalness and speaker consistency on a 1–5 scale under controlled conditions [14]. Figure 2 compares mel spectrograms, showing enhanced audio's noise suppression and preserved formant structure.

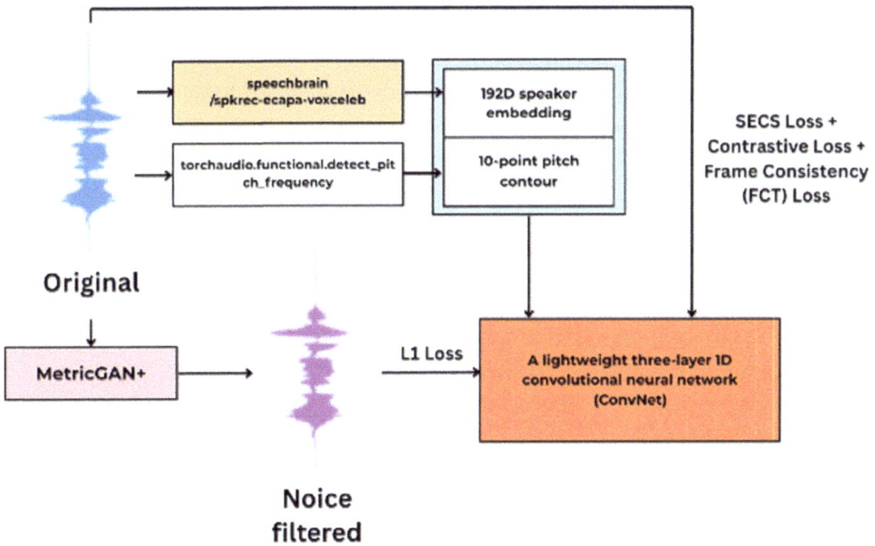

Fig. 1. Architecture of the lightweight ConvNet, showing the flow from the 194D conditioning vector through encoder, decoder, frame head, and global pooling layers to produce a refined waveform and speaker-consistent embeddings.

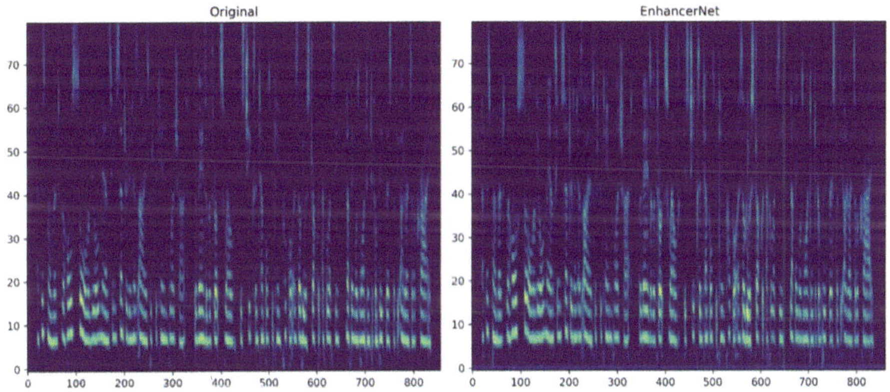

Fig. 2. Mel spectrogram comparison. Left: original clean audio with clear spectral bands. Right: enhanced audio with effective noise suppression, artifact removal, and preserved speaker-specific formant structure.

4 Results and Discussion

4.1 Objective Results

The baseline MetricGAN+ achieves SECS of 0.79, PESQ of 1.75, and STOI of 85%. The proposed enhancer yields SECS of 0.83 ($p < 0.05$, paired t-test), PESQ of 1.91 ($p < 0.05$), and STOI of 85%, indicating superior speaker identity preservation and perceptual quality. Table 1 summarizes results, including STOI and noise-type breakdowns.

Environmental noise shows slightly lower SECS (0.81) due to its non-stationary nature. Error analysis reveals enhancement struggles with low-pitched speakers (<100 Hz, ~ 5% of test samples) due to pitch detection errors, though performance remains robust overall.

Table 1. Objective and Subjective Results Across Noise Types

Model	Noise Type	SECS (95% CI)	PESQ (95% CI)	STOI (%)	MOS (SD)
Baseline (MetricGAN+)	Babble	0.79 (0.77–0.81)	1.75 (1.70–1.80)	85	4.0 (0.3)
	White	0.78 (0.76–0.80)	1.74 (1.69–1.79)	85	4.0 (0.3)
	Environmental	0.77 (0.75–0.79)	1.73 (1.68–1.78)	84	3.9 (0.3)
Proposed Enhancer	Babble	0.83 (0.81–0.85)	1.91 (1.86–1.96)	85	4.2 (0.3)
	White	0.82 (0.80–0.84)	1.90 (1.85–1.95)	85	4.2 (0.3)
	Environmental	0.81 (0.79–0.83)	1.89 (1.84–1.94)	84	4.1 (0.3)

4.2 Subjective Results

MOS tests with 20 listeners yield a score of 4.2 (SD = 0.3) for the proposed system versus 4.0 (SD = 0.3) for MetricGAN+. Listeners reported enhanced audio is clearer than the original WAV due to effective noise and artifact removal, as shown in Fig. 2. Speaker identity, including timbre and prosody, is preserved closely, detectable only upon close listening. Slight overfitting to clean targets may enhance perceived clarity beyond the original, validated by mel spectrogram comparisons (Fig. 2).

4.3 Discussion

The SECS improvement (0.04, $p < 0.05$) and PESQ increase (0.16, $p < 0.05$) demonstrate the efficacy of auxiliary conditioning. The 192D embeddings maximize speaker identity preservation, while the 10-point pitch contour, chosen for its balance of efficiency and expressiveness [10], restores prosodic nuances. Pitch detection errors in noisy conditions may affect low-pitched speakers, but these cases are minimal. The L1 loss targeting cleaned MetricGAN+ outputs ensure high-fidelity reconstruction, reducing artifacts and improving clarity, as shown in Fig. 2. Compared to [10], which uses continuous pitch, our quantized approach reduces computational overhead while maintaining naturalness. The lightweight ConvNet (Fig. 1) enables real-time inference, though latency challenges (e.g., pitch tracking delays) may arise in ultra-low-resource settings. Future work could explore adaptive quantization [4] or phoneme-level conditioning to address pitch errors.

5 Conclusions

This research presents a neural speech enhancement system refining Met-ricGAN+ - processed audio using a 192D speaker embedding and a 10-point quantized pitch contour. Objective gains include SECS from 0.79 to 0.83 ($p < 0.05$), PESQ from 1.75 to 1.91 ($p < 0.05$), and STOI of 85%. Subjective MOS rises from 4.0 to 4.2 (SD = 0.3), with enhanced audio perceived as clearer than the original due to effective noise and artifact removal (Fig. 2). Speaker identity is preserved closely, requiring meticulous listening to detect differences. The lightweight ConvNet (Fig. 1) ensures real-time suitability for teleconferencing and voice assistants, though pitch detection errors or latency in low-resource settings remain challenges. Future work could integrate adaptive quantization or neural codecs like SoundStream [2] to enhance efficiency.

References

1. Gârbacea C., et al.: Low bit-rate speech coding with VQ-VAE and a WaveNet decoder. EEE International Conference on Acoustics, Speech and Signal Processing (ICASSP), Brighton, UK (2019). https://doi.org/10.1109/ICASSP.2019.8683277
2. Zeghidour N., et al.: SoundStream: an end-to-end neural audio codec (2021). https://doi.org/10.48550/arXiv.2107.03312
3. Siahkoohi, A., et al.: Ultra-low-bitrate speech coding with pretrained transformers. Interspeech (2022). https://doi.org/10.21437/Interspeech.2022-10988
4. Pan J., et al.: PSCodec: high-fidelity low-bitrate speech codecs leveraging prompt encoders. J. Latex Class Files **14**(8) (2021). https://arxiv.org/pdf/2404.02702
5. Fu, S.W., et al.: MetricGAN+: an improved version of MetricGAN for speech enhancement (2021). https://arxiv.org/pdf/2104.03538
6. Pärnamaa T., Saabas, A.: Personalized speech enhancement without a separate speaker embedding model (2024). https://arxiv.org/pdf/2406.09928
7. van den Oord, A., et al.: WaveNet: a generative model for raw audio (2016). https://arxiv.org/pdf/1609.03499
8. Su, H., et al.: HiFi-GAN-2: robust speech synthesis via subband adversarial training (2021). arXiv:2107.13783
9. Xu, C., et al.: SpEx: Multi-scale time domain speaker extraction network. IEEE/ACM Trans. Audio Speech Lang. Process. **28**, 1370–1384 (2020). https://doi.org/10.1109/TASLP.2020.2987429
10. Hwang, S., et al.: Enhancement of coded speech using neural network based side information. IEEE Access **9**, 121532–121540 (2021)
11. Erdogan, H., et al.: TokenSplit: using discrete codec tokens for speech separation refinement. Interspeech (2023). https://doi.org/10.21437/Interspeech.2023-2069
12. Défossez, A., et al.: Real-time speech enhancement with neural homomorphic synthesis (2022). https://arxiv.org/pdf/2006.12847
13. Kong, Z., et al.: DiffWave: a versatile diffusion model for audio synthesis (2020). https://arxiv.org/pdf/2009.09761
14. Hannun, A., et al.: Deep Speech: scaling up end-to-end speech recognition (2014). https://arxiv.org/pdf/1412.5567

Engineering Product Development

Design and Development of Magneto Rheological (MR) Fluid Based Soft Gripper

R. Gangadevi(✉), B. C. Mohana Gokul, A. Nikhil, Kalluru Snehitand, and S. Senthil Raja

Department of Mechatronics Engineering, SRM Institute of Science and Technology, Kattankulathur 603203, India
gangader@srmist.edu.in

Abstract. Handling soft and irregularly shaped objects without causing harm has proven to be an extremely difficult and tedious operation using the gripper with hard fingers. To address these issues, a gripper including soft fingers and several actuation methods, including pneumatic, vacuum, and tendon mechanisms, has been used in diverse applications recently. Nevertheless, these grippers offer increased flexibility, requiring the implementation of new techniques to actuate the fingers with a greater grasping force. Therefore, a novel gripper with soft fingers with magnetorheological fluid has been proposed in this study. A magnetorheological fluid was prepared by combining 85% carrier fluid, such as mineral oil, with 15% ferromagnetic particles, and used as the actuating medium in this investigation. A three fingered gripper was developed using two step fabrication method and tested under different magnetic field conditions. An Ar-duino-based closed loop feedback control system was used to assess finger de-flection and regulate the flow rate of MR fluid into the gripper. The experimental findings illustrate the capacity to manipulate diverse items of varying weights. The findings indicate that the gripper using MR fluid provides more grabbing force than the other grippers.

Keywords: Soft Gripper · Magneto Rheological (MR) Fluid · Electromagnetic actuator · shear rate · Shear Stress · Bending angle

1 Introduction

Gripper and tool are the important and unavoidable element in the robot system. The term gripper is a device used to grasp the object and place the object in the exact place either in the same orientation of in different orientation with respect to the reference frame of the manipulator. The traditional grippers are very rigid and requires more number of links and joints to pick and place the object in the work place [1]. The gripper's rigidity restricts its flexibility, allowing it to handle only a restricted range of shapes and sizes of items. Moreover, these grippers are heavy and need intricate sensors and programming to accommodate diverse geometries. Due of these limitations, researchers and materials specialists created a gripper composed of soft materials such as silicone to manipulate things delicately, using uniformly distributed pressures to accommodate

M. Siriwardhana et al. (Eds.): IERS 2025, PTT, pp. 113–120, 2026.
https://doi.org/10.1007/978-981-95-7984-6_10

unusual shapes and sizes without causing adverse effects. Zaidi et al. [2] examined several actuation strategies applicable to soft grippers and evaluated the advantages and disadvantages of each method. Their results indicate that the combination of pneumatic, vacuum, and tendon-based actuation approaches offers superior gripping force compared to other methods. Goh et al. [3] published a review study highlighting the advantages of soft robotic grippers by examining actuation methods and design and manufacturing procedures. This review article also emphasising the advantages and applications of additive manufacturing techniques for fabricating the soft robotic gripper.

Chen et al. [4] developed a three-finger gripper composed of soft material to pick apples while maintaining consistent clamping pressure, hence minimising fruit damage throughout the harvesting process. This research evaluated three distinct grippers: a gripper with a rigid finger, a soft finger with slip detection, and a soft finger without slip detection, by selecting 25 apples. It was observed that the success rate and damage percentage for the gripper with the rigid finger were 100% and 16%, respectively. The success rate for the gripper equipped with a soft finger and slip detection was 100%, resulting in 0% fruit damage. In contrast, a success rate of around 96% and 8% fruit damage were recorded for the gripper equipped with a soft finger and without slip detection. Zhang et al. [5] developed and evaluated a pneumatically operated soft gripper utilising bistable carbon fibre polymer laminations, observing that the gripper with bistable laminations exhibits superior shape transition at low pressure; furthermore, the numerical results demonstrate strong concordance with the experimental findings.

Zhu et al. [6] developed and constructed the gripper via a two-step manufacturing technique including a jamming layer to provide variable stiffness. Despite its limitations, including complex manufacture and system integration, layer jamming may grab objects weighing up to 360 grammes and enhance the gripper's clutching strength during high-speed movement. Over the last several decades, multiple researchers have created soft grippers with various designs and evaluated them under diverse working conditions [7–10]. Despite grippers using various actuation techniques provide several benefits, more research needs to be conducted to create novel actuation methods to improve gripper performance. Li et al. [11] constructed a soft gripper by integrating a porous structure with magnetic silicone elastomer, examining its capacity to grasp various items, and found that the magnetically actuated gripper has significant potential for biomedical applications. Koivikko et al. [12] proposed a novel vacuum gripper with magnetic actuation technique to pick and place the object. In this work, a suction cup was fabricated using silicone and filled with magnetorheological (MR) fluid, enabling a transition between soft and rigid states by an external magnetic field, resulting in an almost 90% enhancement in gripping force when the magnetic field was applied.

The aforementioned material illustrates the benefits of soft-fingered grippers over rigid-fingered versions and suggests that soft robotic grippers have a bright future in many applications. In recent years, researchers have focused on developing and studying the characterization of different types of soft grippers actuated by pneumatic air, SMA etc. However, only a limited number of researchers have developed soft grippers actuated by MR fluid in recent decades. Therefore, this work focused to develop a gripper including soft fingers, which used MR fluid for actuation. This project was executed in three distinct stages. During the first step, the necessary mould was conceptualised and produced with

SolidWorks software and a 3D printer. The liquid silicone was injected into the mould cavity to build the gripper in the second step. The necessary MR fluid was also introduced into the cavity at this step. The synthetic MR fluid-based soft gripper was evaluated under various magnetic field settings, and the gripping force was assessed in the third phase.

2 Materials and Methods

This work included the design and fabrication of a three fingered soft robotic gripper inspired by human fingers. The gripper is made of biocompatible liquid silicone rubber and MR fluid was used as an actuating medium. The MR fluid (Product Code: AMT–MAGNAFLO$^+$) was purchased from ARUS MR TECH, Chennai, Tamil Nadu, India. The purchased MR fluid was dark grey, with a density range of 2.75 to 2.85 g/cm^3, viscosity values at 40ºC between 0.2 and 0.5 Pa·s, and a yield stress of 51 $\pm$ 5 kPa at 180 kA/m. The details of the proposed MR fluid actuated soft gripper fabrication is presented in Fig. 1. The gripper's exterior body is constructed from Liquid silicone rubber (LSR–105) due to its enhanced flexibility, safety, and softness. The first phase of the manufacturing process involves the development of a two-part mould, which is designed using SolidWorks software and then manufactured with a high-resolution 3D printer with PLA material. The silicone rubber was made by combining liquid silicone and curing agent in a 1:1 ratio. Air bubbles were eliminated using a vacuum chamber before the mixture was poured into the manufactured mould cavity. This was cured at room temperature for 6–8 h at the same time the second part of the gripper was fabricated using the same procedure. After 8 h, the silicone components were extracted from the mould and bonded together using liquid silicone adhesive, creating a hollow interior channel for filling with MR fluid. The MR fluid was injected into the interior channel and circulated by a pump via the inserted tiny silicone tube at the gripper's input and output ports. This arrangement was positioned next to the electromagnetic coils, and the viscosity alterations of the MR fluid occur when the coils are energised, resulting in the deflection of the fingers.

3 Testing of MR Fluid Based Soft Gripper

To assess the gripping strength of the developed MR fluid-based soft robotic gripper under varying magnetic field conditions, an experimental apparatus with closed loop feedback was constructed, as seen in Fig. 2. This arrangement has a base, vertical support, sliding arm with vertical guided motion and an actuator. An electric motor is connected to the vertical sliding arm, enabling precise positioning of the gripper at the desired height. The gripper is coupled to a closed fluid reservoir via flexible tubing, enabling the regulated injection and circulation of MR fluid into the interior channels of the fingers. An electromagnet powered by a DC source was used in this configuration to alter the viscosity of the MR fluid. The finger's deflection was quantified using a flex sensor, while the movement of the fingers was regulated using the Arduino Uno controller. The feedback control system used in this study is shown in Fig. 3.

The main objective of this study is to evaluate the gripping strength of the proposed MR fluid based soft gripper under different magnetic field conditions. For this purpose,

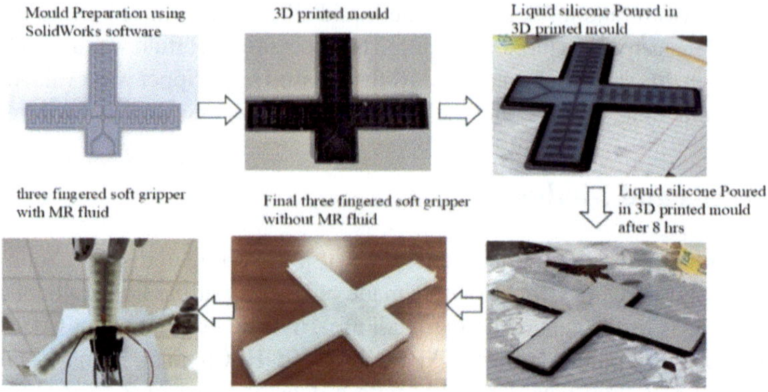

Fig. 1. Details of the proposed MR fluid actuated soft gripper fabrication

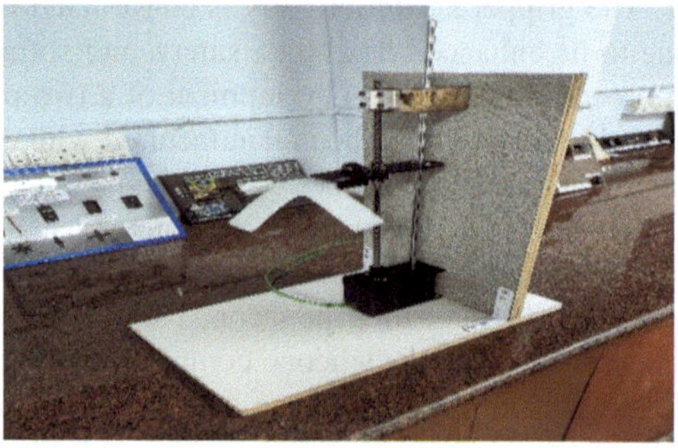

Fig. 2. Experimental apparatus with closed loop feedback

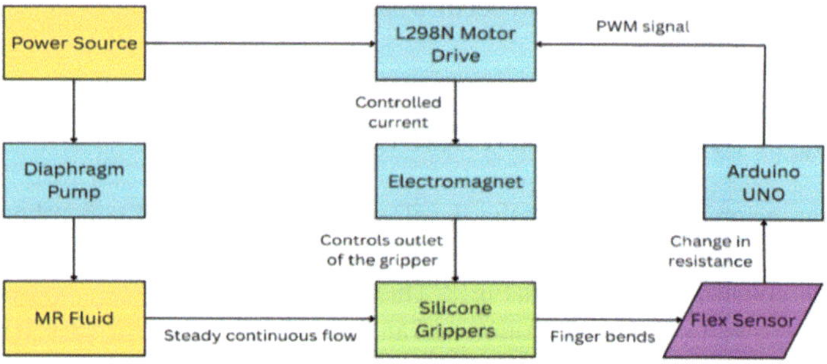

Fig. 3. Flow diagram of the feedback control system

a DC power source was used to activate the diaphragm pump, sensors, electromagnetic coil and controller unit. The Flex sensors monitor the actuator's position and transmit signals to the Arduino Uno controller, which generates PWM signals for the motor driver. The L298 motor driver regulates the power supply according to the commands sent by the controller.

4 Results and Discussions

In order to study the characteristics of MR fluid and the soft robotic gripper, a series of tests were carried out in this study. The results obtained and the detailed discussions are presented in this section.

The rheological behavior of the MR fluid used in this study was analyzed by relating the shear rate and the shear stress. The variation of shear stress in Pa with respect to shear rate is presented in Fig. 4. The shear stress escalates from 5 Pa to 100 Pa as the shear rate fluctuates from 0 s^{-1} to 1200 s^{-1}. At 100 s^{-1}, the shear stress attains 15 Pa, increases to 27 Pa at 200 s^{-1}, and escalates to 59 Pa at 500 s^{-1}. The observed figures undeniably demonstrate that the shear rate continually rises with shear stress throughout the first phase, exhibiting Newtonian fluid behaviour, after transition to non-Newtonian fluid characteristics. When the shear rates are low, the molecular structure of the fluid may reorganise and resisting flow more effectively, which results in increased shear stress. As the shear rate rises, the hydrodynamic forces exerted on these interior structures become increasingly significant. This augmented force might result in particles being more densely aggregated or "jammed," hence enhancing inter-particle friction or resistance to flow. This is the reason to increase the shear stress with increasing shear rate.

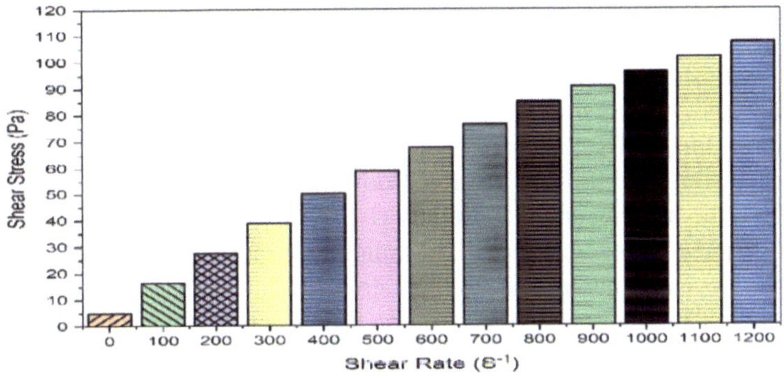

Fig. 4. Variations of shear stress with respect to shear rate

Figure 5 offers an excellent overview of the magnetic response characteristics of the MR fluid under varying magnetic field strengths. In the beginning, the magnetic flux density is around 0.15 T, recorded at a magnetic flux intensity of 20 kA/m. As the magnetic field intensity escalates, the magnetic flux density grows nonlinearly, reflecting the inconsistency in the material's magnetic permeability. At 40KA/m magnetic flux intensity, the flux density rises dramatically to 0.3 T, whilst for 60 KA/m and 80 KA/m, the values reached to 0.42 T and 0.48 T correspondingly. The rise in magnetic flux density is notably high from 0 KA/m to 180 KA/m, but beyond 180 KA/m, the increments in magnetic flux density are minimal. At 180 kA/m, the measured value is 0.76 T; at 220 kA/m, 240 kA/m, and 260 kA/m, the corresponding values arc 0.82, 0.84, and 0.85 T, respectively. This minimal increment indicates that the material is nearing the magnetic saturation area. The magnetic particles are correctly orientated with the application of an external magnetic field, resulting in a fast rise in magnetic flux density. In without the presence of a magnetic field, the magnetic particles are randomly orientated, resulting in a negligible magnetic flux density. This is the fundamental phenomena of achieving increased magnetic flux density during the first phase.

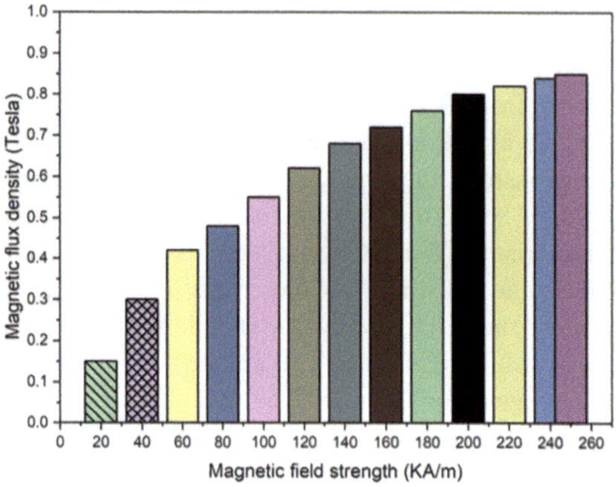

Fig. 5. Variations of magnetic flux density with respect to magnetic field strength

Figure 6 illustrates the fluctuations in the bending angle of the gripper finger over time for various flow rates of MR fluid and the kinematic performance and reactivity of the gripper finger. This research also examined the bending angle of the gripper fingers by adjusting the MR fluid flow rate from 5 mL/min to 25 mL/min. These findings indicated that the bending angle constantly rises with an increase in time and the flow rate of MR fluid. At a flow rate of 5 mL/min, the gripper gradually bends, requiring 200 ms to achieve a 3° angle, 600 ms to attain a 7° angle, and 2000 ms to reach a 15° bending angle. Circulating MR fluid at a high flow rate (i.e., 25 mL/min) inside the gripper's internal chamber generates a greater pressure differential between the interior and outside of the gripper's finger chamber. This results in an increased bending angle and quick deformation in the gripper finger. The bending angle of the gripper finger is directly correlated with the pressure applied inside the chamber. At elevated flow rates, the kinetic energy of the particles increased, resulting in a considerable reduction in internal resistance, which in turn led to a rise in the bending angle.

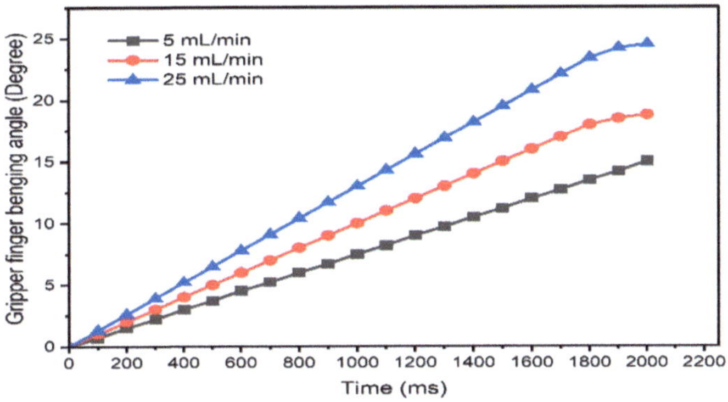

Fig. 6. Variations in the bending angle of the gripper finger over time for various flow rates of MR fluid

5 Conclusions and Future Works

This study presents the design and development of a novel soft robotic gripper using magnetorheological (MR) fluid. The properties of the MR fluid and the bending angle of the constructed gripper were also examined experimentally for various flow rates of the MR fluid. The significant findings of this research are summarised and given as follows.

1. This work effectively showed the construction of a three-fingered MR fluid-actuated soft robotic gripper, and the data reveal its superior capability to grab various objects with more gripping force than standard grippers.
2. The findings reveal that the flow velocity of the MR fluid directly impacts the bending angle of the gripper. The peak bending angle reached was 24.5° at a flow rate of 25 mL/min. For flow rates of 15 mL/min and 5 mL/min, the resulting bending angles were 18.8° and 15°, respectively, during the same duration.
3. The MR fluid first exhibits characteristics of a Newtonian fluid, subsequently transitioning to a non-Newtonian fluid behaviour. The findings indicated that the magnetic flux density ranged from 0.15 T at 20 kA/m to 0.85 T at 260 kA/m, confirming adequate gripper stiffness for grasping various items under varied operating circumstances.
4. Although the findings obtained show the potential of MR fluid as an actuation medium in soft robotic grippers, more study is necessary to optimise the MR fluid composition to enhance gripping force and finger stiffness while achieving a faster response time.
5. The integration of sensors into the gripper finger to assess various characteristics is a complicated task, necessitating sophisticated procedures for sensor installation, as well as advanced feedback control to ensure precise finger movement.

Acknowledgement. We would like to thank the Selective Research Initiative SERI-"2023", SRMIST for their financial support.

References

1. Guan, Q.H., Sun, J., Liu, Y.J., Leng, J.J.S.S.T.: Status of and trends in soft pneumatic robotics. Sci. Sin. Technol. **50**, 897–934 (2020)
2. Zaidi, S., Maselli, M., Laschi, C., Cianchetti, M.: Actuation technologies for soft robot grippers and manipulators: a review. Curr. Robot. Rep. **2**(3), 355–369 (2021)
3. Goh, G.D., et al.: 3D printing of robotic soft grippers: toward smart actuation and sensing. Adv. Mater. Technol. **7**(11), 2101672 (2022)
4. Chen, K., et al.: A soft gripper design for apple harvesting with force feedback and fruit slip detection. Agriculture **12**(11), 1802 (2022)
5. Zhang, Z., et al.: Pneumatically actuated soft gripper with bistable structures. Soft. Robot. **9**(1), 57–71 (2022)
6. Zhu, M., Xie, M., Mori, Y., Dai, J., Kawamura, S., Yue, X.: A variable stiffness soft gripper based on rotational layer jamming. Soft Rob. **11**(1), 85–94 (2024)
7. Navas, E., Shamshiri, R.R., Dworak, V., Weltzien, C., Fernández, R.: Soft gripper for small fruits harvesting and pick and place operations. Front. Robot. AI **10**, 1330496 (2024)
8. Lin, J., et al.: Non-destructive fruit firmness evaluation using a soft gripper and vision-based tactile sensing. Comput. Electron. Agric. **214**, 108256 (2023)
9. Tawk, C., Mutlu, R., Alici, G.: A 3D printed modular soft gripper integrated with metamaterials for conformal grasping. Front. Robot. AI **8**, 799230 (2022)
10. Hu, Q., Dong, E., Sun, D.: Soft gripper design based on the integration of flat dry adhesive, soft actuator, and microspine. IEEE Trans. Rob. **37**(4), 1065–1080 (2021)
11. Li, X., et al.: Porous magnetic soft grippers for fast and gentle grasping of delicate living objects. Adv. Mater. **36**(44), 2409173 (2024)
12. Koivikko, A., Drotlef, D.M., Sitti, M., Sariola, V.: Magnetically switchable soft suction grippers. Extreme Mech. Lett. **44**, 101263 (2021)

Design and Implementation of a C_{10} & C_{20} Capacity Tester for Lead Acid Battery Discharge Evaluation

T. Chandana Peiris(✉), L. N. Rathnayake, S. P. R. Jayasinghe, and J. P. D. S. Athuraliya

Arthur C Clarke Institute for Modern Technologies, Moratuwa, Sri Lanka
chandanaaccimt@gmail.com

Abstract. In Sri Lanka's automotive and industrial sectors, numerous brands of lead acid batteries are widely utilized. These include both locally manufactured and imported products. All batteries must conform to the quality standards established by the Sri Lanka Standards Institution (SLSI) prior to commercial distribution. The Arthur C Clarke Institute for Modern Technologies (ACCIMT) and the National Engineering Research & Development Centre (NERDC) serve as key institutions in the testing and validation of lead acid batteries. These organizations employ advanced imported equipment to evaluate performance metrics in compliance with national standards. Among the essential equipment used are high current DC electronic loads, which are critical in various electronic load applications, especially in battery dis-charge testing. This study presents the design and development of a DC electronic load system, designated as the C10 and C20 Capacity Tester (Model: CT2024), specifically engineered to evaluate the discharge characteristics of lead acid batteries. The system features a 7-inch resistive touch display with a graphical user interface (GUI) and is controlled via a microcontroller based architecture, providing precise and user friendly operation. Experimental results demonstrated that the prototype delivered highly precise voltage and current measurements, confirming its effectiveness in battery testing applications. Given the high cost of commercial DC load systems, compounded by freight and potential repair expenses, the proposed proto-type, "C10 and C20 Capacity Tester for Batteries," provides a cost effective alternative while maintaining reliable performance for battery capacity testing. Its locally developed design ensures ease of repair and maintenance. Furthermore, the system incorporates Internet of Things (IoT) functionality, enabling real time remote monitoring of voltage, current, and battery status. Ongoing development aims to integrate a parallel battery charging system within the same unit.

Keywords: DC Load · Battery · Internet of Things (IoT)

1 Introduction

The global energy landscape has undergone a significant transformation in recent decades due to increasing concerns over the rapid depletion of fossil fuel reserves and the severe environmental and economic impacts associated with their overuse. These issues have

M. Siriwardhana et al. (Eds.): IERS 2025, PTT, pp. 121–130, 2026.
https://doi.org/10.1007/978-981-95-7984-6_11

catalyzed a global shift toward the adoption of cleaner, more sustainable energy sources. Among the various alternatives, renewable energy technologies, particularly solar, wind, and hydroelectric power have emerged as key components in the transition to a low carbon energy future [1].

Photovoltaic (PV) systems, which directly convert sunlight into electricity, are among the most widely implemented renewable energy technologies due to their scalability, minimal environmental footprint, and decreasing installation costs [2]. However, a fundamental challenge associated with PV systems is their intermittent nature; solar energy generation is inherently dependent on weather conditions and diurnal cycles, resulting in fluctuating power output. As a result, ensuring a consistent and reliable power supply from PV systems requires the integration of energy storage solutions [3].

Energy storage systems play a crucial role in bridging the gap between energy generation and consumption. Among the available technologies, batteries are the most commonly used and efficient means of storing electrical energy for later use.

Various types of batteries are currently employed in renewable energy applications, each with distinct characteristics, advantages, and limitations. These include traditional lead acid batteries, known for their low cost and widespread availability; lithium-ion (Li-ion) batteries, which offer high energy density and long cycle life; nickel-cadmium (Ni–Cd) batteries, valued for their durability and performance un-der extreme temperatures; and sodium-sulfur (Na–S) batteries, which are suitable for large scale energy storage applications due to their high energy capacity [4].

Lead acid batteries, typically rated at a nominal voltage of 12V, are widely employed as primary power sources for the starting and ignition systems of internal combustion engines. In addition, they serve auxiliary functions such as lighting and powering supplementary equipment in vehicles powered by such engines.

To ensure consistency and safety across applications, the Sri Lanka Standards Institution (SLSI) has established the standard SLS 1126: Part 1: 2020 [5], which out-lines the general requirements and testing methodologies for various classes of starter batteries. These classifications consider the intended application types and the climatic conditions in which the batteries are predominantly used.

In line with these standards, the development of an automated test bench for lead acid starter batteries has become essential. This project represents the initial phase of such a system, focusing on the integration of key electrical tests in a systematic and automated framework.

As stipulated by SLSI standards, the capacity test is the primary and most critical assessment, conducted on fully charged batteries. Battery capacity evaluation, typi-cally performed via controlled discharge tests, determines the actual capacity relative to the nominal rating [6]. Common methodologies include constant current (CC), constant voltage (CV), constant power (CP), and constant resistance (CR) discharge, each suited to specific applications [7]. This project involves the design and implementation of a dedicated Capacity Tester to evaluate the performance of lead acid batteries during discharge cycles.

The proposed system involves the design and deployment of a Capacity Tester tailored to assess battery performance through controlled discharging cycles. The sys-tem is capable of recording key parameters, including terminal voltage and dis-charge current, along with corresponding time and date stamps. Data logging is facilitated via an SD (Static Data) card for offline analysis.

Emerging technologies, particularly the Internet of Things (IoT) and cloud computing, have been increasingly adopted across various industries, including energy management and battery monitoring systems [8]. The integration of IoT enables real time data acquisition and remote monitoring via cloud based platforms. When applied to battery monitoring, IoT facilitates efficient data logging and remote accessibility, thereby enhancing system reliability and performance.

2 Description of the Unit

According to the standard, discharging is restricted to the constant current mode. In the design of this C10 & C20 Capacity Tester, the constant current discharging meth-od has been implemented. The system is configured to discharge the battery until a predefined terminal voltage of 10.5 V is reached. This voltage level serves as the final cut off point to terminate the discharging process.

In the design of the C_{10} & C_{20} Capacity Tester, load control is achieved using MOSFET transistors biased in the ohmic region [9]. This facilitates a time dependent load behavior, which effectively corresponds to a gate source voltage, controlled current source, with the drain current as the output. The general specification of the C_{10} & C_{20} Capacity Tester is presented in Table 1.

Table 1. Specification of the C_{10} & C_{20} Capacity Tester.

Parameter	Value	
Operating Voltage	220–230 V (AC)	
Operating Frequency	50/60 Hz	
Working Voltage Range	0–20 V (DC)	
Working Current Range	0–30 A (DC)	
Mode	Constant Current	
C Rate	C_{10} or C_{20}	
Touch Screen	7″, Resistive Type	
Switch	ON/OFF	
Protection	Over Voltage (OV)	Over Current (OC)
	Over Temperature (OT)	Reverse Connection
Displays Indicators	Input Power Supply Status	
	Fuse Status	
	Operating Status of Cooling FAN-1	
	Operating Status of Cooling FAN-2	
	Operating Status of 20 MOSFETs	

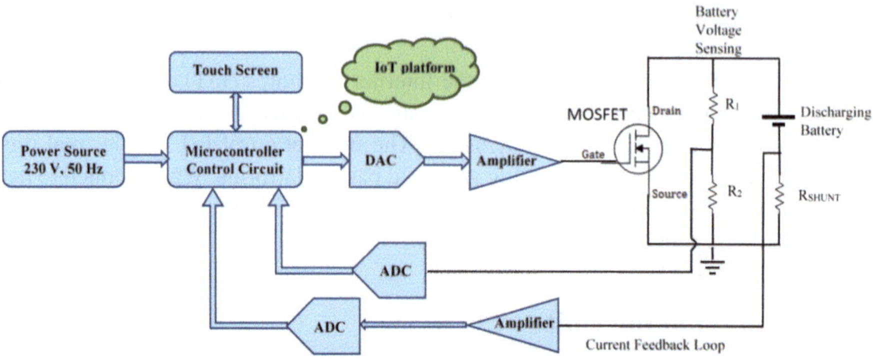

Fig. 1. General System Structure of C10 and C20 Capacity Tester.

The general system architecture of the C_{10} & C_{20} Capacity Tester is illustrated in Fig. 1. The entire system is controlled by an ATmega328P microcontroller. Initially, the user can select the desired testing mode, either C_{10} or C_{20}, via a 7-inch touchscreen interface. The touchscreen also allows the user to set the desired discharge current for the battery under test. A 7-inch Serial SPI TFT LCD Touch Shield featuring the RA8875 controller was employed as the user interface, offering an 800 × 480 pixel resolution and integrated touch functionality. This module is facilitating both graphical display and touch input through a high speed SPI interface. The user defined discharging current, actual discharging current, current battery voltage, real time temperatures of the two MOSFET boards, and elapsed time are continuously displayed on the touch screen. If the battery becomes fully discharged and reaches 10.5 V, this status is also indicated on the touch screen. Upon completion of a full discharge, the ampere hour (Ah) capacity of the battery is displayed on the touchscreen. Interruptions in the main power supply do not influence the charging process of the battery. The system automatically records and updates the battery's status at 15 min intervals in relation to time.

The system operates from a 230 V, 50 Hz AC mains supply. This voltage is stepped down to 18 V using a Toroidal transformer, which is widely employed in power supply applications due to its compact size, high efficiency, and low electromagnetic interference [10]. The stepped down AC voltage is then rectified using 30 A bridge rectifiers and filtered with capacitors to produce a DC output [11]. Various circuit components require regulated DC voltages of +5 V, +6 V, +12 V, and –12 V, which are supplied using 78xx and 79xx series linear voltage regulators to ensure stable and constant output levels.

The battery is discharged through multiple IRFBE30 power MOSFETs connected in parallel. Each MOSFET is capable of conducting a maximum current of 4.1 A. This unit consists of 20 MOSFETs operating in parallel to handle higher discharge currents. High current discharge results in significant heat dissipation across each MOSFET. To mitigate the risk of overheating, all MOSFETs are mounted on large heat sinks and supported by two cooling fans.

Additionally, two LM35 temperature sensor modules are attached to the boards housing the MOSFETs. The LM35 is a precision centigrade temperature sensor with an output voltage linearly proportional to the ambient temperature, operating within a

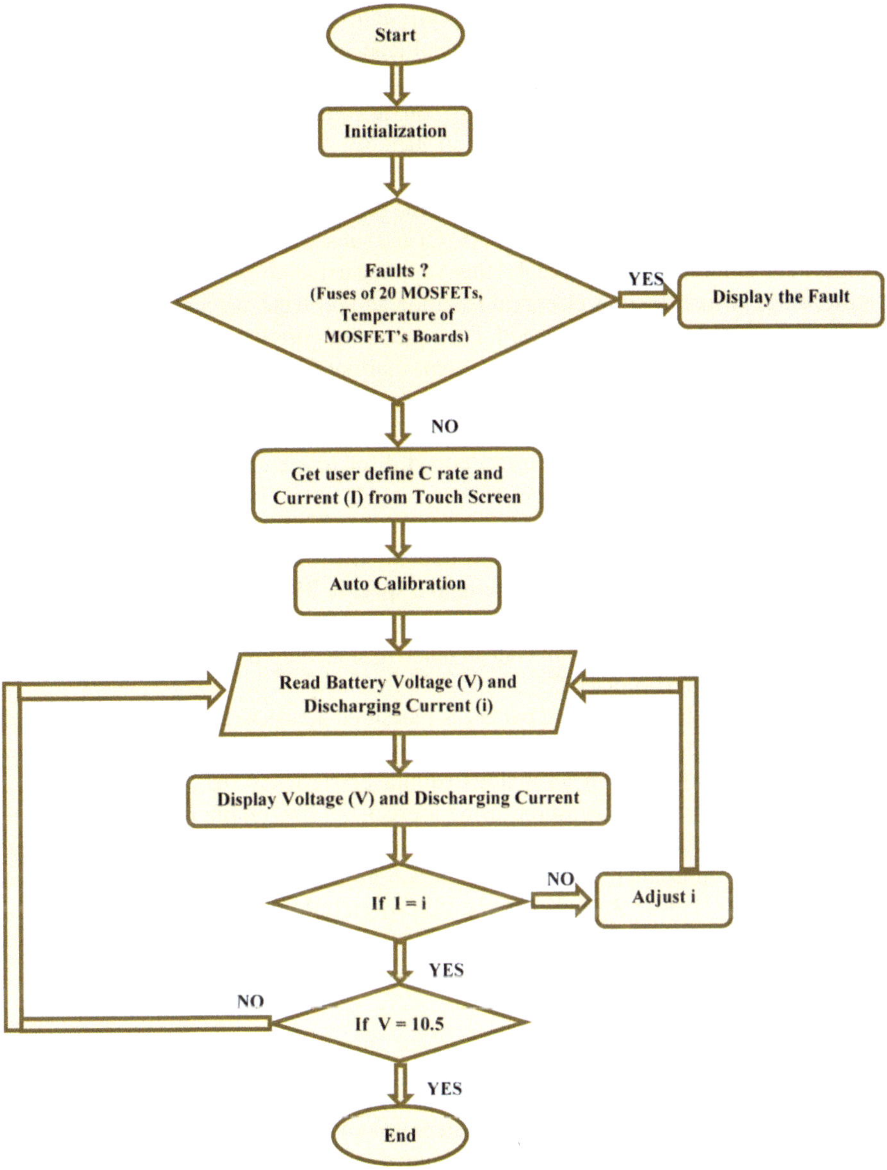

Fig. 2. Flowchart Representation of the Microcontroller Program Algorithm

range of −55 °C to 150 °C. If the temperature of the MOSFETs exceeds a predefined threshold, the LM35 sensors detect this rise and communicate the temperature data to the microcontroller. If the temperature of the MOSFET boards exceeds the user defined threshold, a signal is transmitted to the microcontroller's control circuitry. In response, the system initiates a shutdown sequence to prevent potential damage resulting from abnormal MOSFET heating.

To protect against overcurrent passing through the MOSFETs, each MOSFET is connected in series with a 4A fast blow axial lead fuse. The power stage operates by utilizing MOSFET transistors in the ohmic (linear) region. In this mode, the drain current (I_d) can be controlled by varying the gate to source voltage (V_{GS}) of the transistors. The V_{GS} is supplied by the control microcontroller, an ATmega328P, based on the user defined current setting. The microcontroller adjusts the pulse width of a control signal in accordance with the desired current level. The digital signal is converted to an analog form using a digital to analog converter (DAC) and subsequently amplified, as illustrated in Fig. 1. The amplified analog signal is then supplied to the MOSFET transistors through a buffer stage, which ensures electrical isolation and protects the control circuitry.

The failure of one or more MOSFETs in such configurations can lead to significant performance degradation or complete system malfunction. The design incorporates a novel technology to detect and display the operational status of 20 MOSFETs used as load elements in the "C_{10} and C_{20} Capacitance Tester".

During normal operation, the user defined discharge current is evenly distributed across all MOSFETs in the cascade. However, in the event of a MOSFET failure, the discharge current is redistributed among the remaining functional devices, placing additional stress on them and increasing the likelihood of further failures. The proposed system is capable of accurately identifying the specific MOSFET that has failed. Upon detection, it immediately triggers a user alert and issues a shutdown command to prevent further damage. The system remains in an inactive state until the fault is resolved or the defective component is replaced.

The voltage of the discharging battery is measured across a resistor divider network consisting of two series resistors, R_1 and R_2. The resulting analog voltage is then converted into a digital form using an analog to digital converter (ADC), as shown in Fig. 1. The discharging current is measured using a precision low ohmic resistor, denoted as R_{SHUNT} in Fig. 1, which is connected in series with the load. The resulting analog voltage is amplified and subsequently converted into a digital value using an analog to digital converter (ADC). Figure 2 illustrates the step by step flowchart representation of the algorithm used in the microcontroller program.

3 Result and Discussion

A cost effective, microcontroller based C_{10} and C_{20} Capacity Tester was developed in compliance with Sri Lankan standards and evaluated using commercial automotive lead acid batteries of varying capacities. The system was successfully tested with batteries up to 200 Ah. The microcontroller was programmed with custom firmware incorporating an automatic current correction algorithm, which adjusts the discharge current every five minutes if any deviation from the set value is detected.

Figure 3(a) fitted Line Plot of Displayed Voltage value and Measured Voltage value (b). Scatterplot of Measured Battery Voltage vs Displayed & Measured Discharging Current (A)

The validation process was conducted using calibrated Hewlett Packard (HP) E2377A multimeter and clamp meters. The unit was calibrated using over fifty lead acid batteries with capacities up to 200 Ah. Measured discharge voltages were compared with displayed values, and accuracy was statistically evaluated using Minitab,

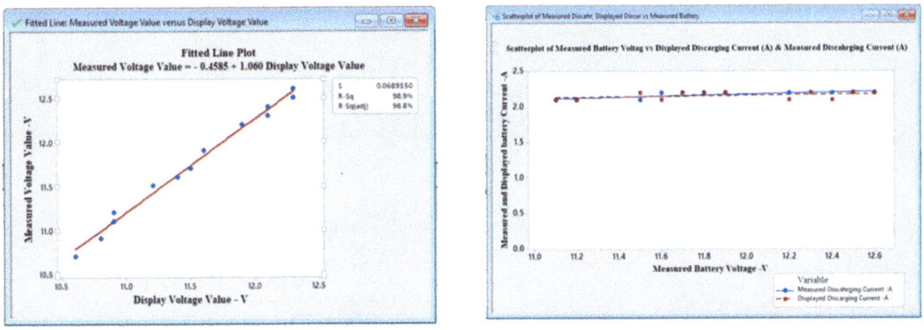

(a). Fitted Line Plot of Displayed Voltage value and Measured Voltage value

(b). Scatterplot of Measured Battery Voltage vs Displayed & Measured Discharging Current (A)

Fig. 3. (a). Fitted Line Plot of Displayed Voltage value and Measured Voltage value (b). Scatterplot of Measured Battery Voltage vs Displayed & Measured Discharging Current (A)

demonstrating high accuracy across all batteries. Figure 3(a) illustrates the measured and displayed voltages of a 43 Ah battery, with a standard error (S) of 0.069 V and a coefficient of determination (R2) of 98.9%. The very low standard error indicates high measurement accuracy, while the R2 value, being close to 100%, demonstrates that the variation in measured voltage is well explained by the displayed voltage. Comparable accuracy and consistency were observed for the current measurements, as shown in Fig. 3(b). Overall, these results confirm the reliability of the system and its suitability for battery discharge monitoring.

Figure 4 illustrates the physical design and fabrication of the C_{10} and C_{20} Capacity Tester developed in Sri Lanka. The C_{10} and C_{20} Capacity Tester incorporates IoT enabled online monitoring to measure and record the battery's voltage and current throughout the discharge process. The Arduino Cloud IoT platform is employed to visualize these real time parameters on a dedicated dashboard, which is automatically updated at predefined intervals. Figure 5 present snapshot of the monitored battery parameters, enabling authorized users to remotely access the discharge status at two distinct time intervals via personal computers and mobile devices.

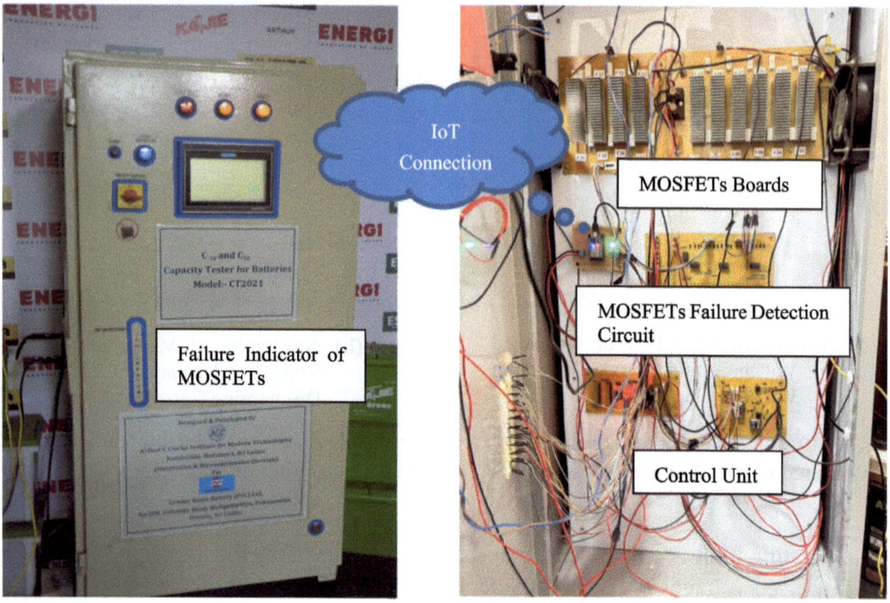

Fig. 4. The physical design and manufacturing of the "C_{10} and C_{20} Capacity Tester" developed in Sri Lanka.

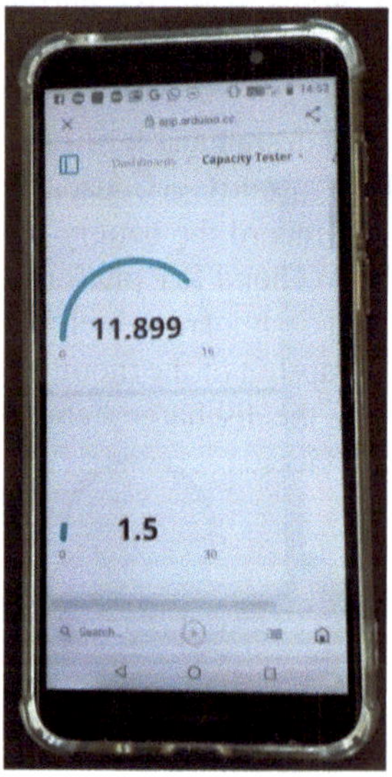

Fig. 5. Interfaces of the IoT Monitoring System

4 Conclusions

The C_{10} and C_{20} Capacity Tester was successfully designed and implemented as a cost effective, electronically controlled system with a simplified hardware and software architecture for the validation of lead acid batteries. The tester operates in compliance with the relevant Sri Lankan standards, utilizing a power MOSFET based discharge circuit that ensures a constant current draw over a 10 h (C_{10}) or 20 h (C_{20}) rate until the battery voltage reaches 10.5 V.

The prototype unit demonstrates several advantages, particularly in terms of affordability and ease of maintenance. As both the hardware and software were entirely developed locally, repairs can be carried out cost effectively and within a reasonable time frame. Furthermore, the system integrates a unique feature for real time failure detection of power MOSFETs, providing users with immediate fault diagnostics an advantage rarely offered by imported alternatives.

The unit is fully automated and integrates IoT based monitoring capabilities, enabling users to observe system performance via a computer or smartphone. A real time voltage versus time graph under constant current conditions further enhances monitoring and analysis. The system has been validated through testing, and results confirm it performs with high reliability and accuracy, aligning with the intended functional and operational requirements.

The next stage of the project involves the design of a parallel charging system capable of charging multiple batteries simultaneously and reliably. The proposed architecture incorporates scalable current sharing, battery health monitoring, and advanced safety features, while ensuring strict adherence to the charging algorithm.

Acknowledgements. The authors extend their sincere gratitude to Prof. K. P. S. Chandana Jayarathne, Chairman of ACCIMT, Mrs. M. M. G. K. Meegahakotuwa, Acting Director General of ACCIMT, and the entire staff of the Electronics and Microelectronics Division for their invaluable support and contributions toward the successful development of the "C_{10} and C_{20} Capacity Tester" with IoT integration.

References

1. Fendji, M. D., Kimbong, F. M., Tsipouridis, I., Tsafack, P., Design and Implementation of a Digital Control System for Lead Acid Battery Charging. Journal of Electrical and Electronic Engineering, pp. 23–33 (2023)
2. Ibrahim, H., Ilinca, A., Perron, J., Energy Storage Systems – Characteristics and Comparisons. In: Renewable and Sustainable Energy Reviews, vol. 12, no. 5, pp. 1221–1250 (2008)
3. Rahim, N.H.A.: Energy Storage Systems for Renewable Energy Power Sector Integration and Mitigation of Intermittency. Renew. Sustain. Energy Rev. **35**, 499–514 (2014)
4. Barton, J.P., Infield, D.G.: Energy Storage and Its Use with Intermittent Renewable Energy. IEEE Trans. Energy Convers. **19**(2), 441–448 (2004)
5. Sri Lankan Standard SLS 1126: Part 1 :2020; Specifications for Lead Acid Starter Batteries Part 1: Generalized Requirements and Methods of Test (2020)
6. Pierron V., Guerard G.: Surveys on Physical based Methods for State of Charge, State of Health and Fault Detection on a Li-Ion Battery. Smart Grids and Sustainable Energy, Springer, pp. 01–30 (2025)
7. Huynh, M.N., Lam, Q.M., Truong, C.T., Nguyen, H.H., Duong, V.T.: Low-cost Electronic DC Load Module Design for Battery capacity Evaluation. Elsevier, Hardware **23**, 01–20 (2025)
8. Rao K.S., Nimisha T., Rashmi B. J. S., Alladwar S., Real Time Monitoring System for Lithium-Ion Cell Using IoT. 2nd International Conference on Intelligent and Sustainable Power and Energy Systems (ISPES 2024), pp. 119–123 (2025)

9. Wang, Z., Kong, F., Xie C., Hung, Z.: Design and Implementation of the Real-Time Regulation Electronic Load Module based on DSP. 2nd International Conference of Future Computer and Communication (ICFCC), pp. 485–488 (2010)
10. Rashid, M. H.: Power Electronics – Circuits, Devices and Applications. 4th edn. Pearson, England (2013)
11. Presman, A. I., Billings, K., Morey, T.: Switching Power Supply Design. 3rd edn. McGraw-Hill (2009)

Design and Simulation of a Corrugated Tube Shell and Tube Heat Exchanger for an Ethanol Production Line

W. K. P. G. N. Charuni(✉) and K. T. K. M. De Silva

Department of Mechanical and Manufacturing Engineering, Faculty of Engineering, University of Ruhuna, Galle, Sri Lanka
wpgnarmada@gmail.com

Abstract. This paper outlines the design and analysis of a shell and tube heat exchanger in an ethanol production line using corrugated tubes to enhance performance. Fouling, corrosion, and degradation of materials are some of the challenges of ethanol production processes that may reduce the efficiency and lifetime of heat exchangers. To overcome these problems, a shell and tube exchanger was designed that had corrugated tubes to enhance the amount of heat transfer and minimize maintenance needs. Computational tools were used to develop and analyze the design. The simulation results were used to assess the thermal per-formance (temperature distribution) and structural integrity (stress, displace-ment) of the heat exchanger. The important design parameters, such as heat du-ty, overall heat transfer coefficient, heat transfer area, and pressure drops were determined to make sure that thermal and mechanical requirements are satisfied. The selected material, stainless steel 316 (SS316), offers better corrosion resistance and durability in an ethanol processing environment. Simulation analy-sis proves that the heat transfer between the ethanol vapor and the cooling water is efficient, and the corrugated tubes have high heat transfer rate, but the pres-sure losses are within the acceptable range. Structural analysis shows that the stresses are far below the yield of the material, which means a safe design. Relative to smooth tubes, the corrugated configuration showed a clear 13% increase in outlet temperature, supporting enhanced energy recovery in ethanol processing.

Keywords: Ethanol · Corrosion · Fouling Heat Exchangers · Failure Modes · Corrugated Tubes

1 Introduction

Heat exchangers are very crucial in the ethanol production industry, especially at the point of fermentation, distillation, and purification. During such processes, effective heat transfer is necessary to conserve energy, maintain temperatures, as well as the overall efficiency of production [1]. Shell and tube heat exchangers have traditionally been the most popular because of their strong construction, ease of maintenance, and flexibility to many different fluids and operating conditions [2].

M. Siriwardhana et al. (Eds.): IERS 2025, PTT, pp. 131–139, 2026.
https://doi.org/10.1007/978-981-95-7984-6_12

But the rising pressure of energy efficient, sustainable, and low maintenance industrial equipment has strained the traditional designs. Specifically, the current ethanol plants need heat exchangers that can not only conduct heat but also be resistant to degradation due to fouling, corrosion, and mechanical wear [3]. These issues are worsened when applying to ethanol production, where hot ethanol vapor and water are typically utilized, which is a chemically aggressive, thermally dynamic environment.

During ethanol production, standard smooth tube heat exchangers (conventional heat exchangers) tend to have problems with fouling and corrosion, both of which reduce their efficiency and service life. Fouling is the deposition of undesirable material on the surfaces of heat transfer, resulting in an insulating film that reduces thermal efficiency. With the accumulation of fouling, not only does it decrease the effectiveness of heat exchanger, but also raises the flow resistance and pressure drop and ultimately requires frequent shutdowns to clean [4]. Another significant challenge is corrosive attacks on the exchanger materials, where aggressive process fluids and deposited contaminants have the potential to increase material degradation, often concealed beneath layers of fouling, and ultimately reduce the service life of the equipment and cause failures.

Moreover, most of the current designs of heat exchangers have poor heat transfer characteristics because of the laminar flow in smooth tubes which has lower heat transfer coefficients. This poor performance frequently causes engineers to over size equipment or uses higher driving temperature differences, which increase energy use. All these, combined with high maintenance requirements to clean and replace components frequently, compromise the reliability of day to day operations and raise the costs of operation.

This research is therefore driven by the objective to improve operational reliability, reduce energy consumption, and enhance the long term durability of heat exchangers in ethanol processing. To achieve these aims, the implementation of corrugated tube geometry is proposed. Corrugated tubes generate enhanced turbulence within the fluid flow, which in turn elevates the convective heat transfer coefficient and reduces the probability for fouling formation on heat transfer surfaces [5–7]. The antifouling benefit of the corrugated geometry is attributed to turbulence induced secondary flows that elevate wall-shear stress and mass transfer, which hinder initial deposit adhesion and promote the removal of weakly attached material [8, 9].

In this context, this research presents the design and analysis of a shell and tube heat exchanger with corrugated tubes for ethanol production. This configuration addresses the limitations of smooth tube designs by enhancing heat transfer and reducing fouling.

2 Methodology

2.1 Design Approach and Specifications

The research began with defining design specifications for the ethanol production line heat exchanger. The unit is intended to condense hot ethanol vapor (shell side) using cooling water (tube side), thus operating as a shell and tube condenser. Based on process requirements, a one shell pass, one tube pass configuration in counter-flow was chosen for simplicity and effectiveness in achieving a large temperature driving force. A fixed head design was selected for its robust construction, given that thermal expansion between the stainless steel (SS316) shell and tubes could be accommodated by the operating range and the material's properties. Physical and thermal properties of two types of fluids are shown in Table 1.

Table 1. Physical and thermal properties of Alcohol vapor and water

Thermal Properties of Fluid	Cold fluid: Water		Hot fluid: Alcohol	
	In	Out	In	Out
Mass flow [kg/s]	2	2	1.55	1.55
Temperature [°C]	35	45	108	80
Specific Heat capacity [kJ/kg* °C]	4.181		2.46	
Viscosity [mPa*s]	0.7191	0.5958	0.323	0.01043
Density [kg/m^3]	992		789	
Pressure [bar]	4.144	4.081	3.094	3.056

The heat exchanger must cool and condense ethanol vapor from 108°C to 81°C using water heated from 35°C to 41°C. The design heat duty was calculated to be approximately 120kW (typical of a small production-scale condenser). A log mean temperature difference (LMTD) was calculated for the counter flow arrangement. The LMTD is 53.5°C. Using an initial overall heat transfer coefficient estimate of 365 W/m^{2}K (which accounts for convective resistances and fouling factors), the required heat transfer area was computed to be 5.48 m^2. This value (5.48 m^2) represents the **required** area and is independent of geometry, corrugation subsequently affects the **realized** surface and the achievable U (overall heat transfer coefficient).

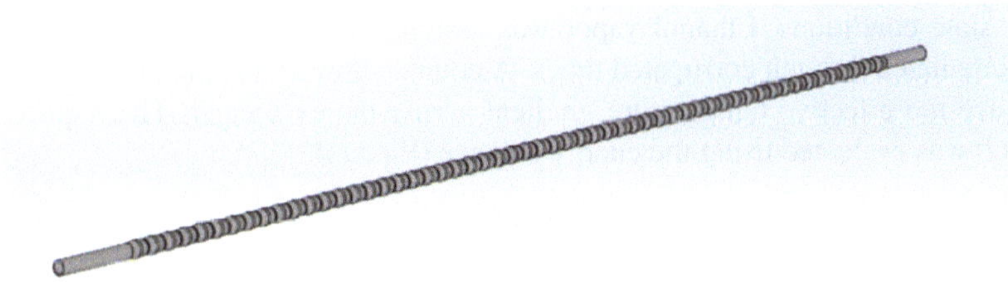

Fig. 1. Schematic diagram of a corrugated tube

To achieve this area within a compact exchanger, corrugated tubes were employed which increases turbulence by interrupting the fluid boundary layer shown in Fig. 1. The tube bundle design was based on 36 corrugated tubes, each of 20 mm outer diameter, 18 mm inner diameter, and 2.0 m length. The corrugation geometry has a 30 mm and 1 mm corrugation depth. These dimensions yield a corrugation factor of 1.21, meaning the effective flow length is 21% longer than the plain tube length due to the corrugated path. After accounting for corrugation, the total tube external surface area (5.48 m^2) can be achieved with 36 tubes of 2 m physical length (2.42 m effective length each). By using Kern's method, other relevant calculations shell diameter, baffle spacing, shell and tube side heat-transfer coefficients, and pressure drop estimates were completed prior to CFD.

For the shell side, a single pass was configured with ethanol vapor entering at the top and condensate exiting at the bottom. The shell diameter was determined by the tube bundle diameter plus clearance. A square tube layout (square pitch) was chosen for the bundle, as it allows easier mechanical cleaning of tubes (straight-line access) which is beneficial for fouling services. The bundle diameter for 36 tubes on a square pitch was calculated using standard correlations, yielding a shell inside diameter of approximately 0.28 m (which was rounded to a standard pipe size of 300 mm). Baffles were included in the shell to direct vapor flow across the tube bundle, with a baffle spacing of about 0.12 m to balance heat transfer and pressure drop [10].

All parts in contact with ethanol vapor or cooling water are made of SS316 stainless steel for corrosion resistance [11]. This includes the tubes, tube-sheets, shell, channel heads, and baffles. SS316 can withstand both the ethanol (an organic, potentially slightly acidic fluid) and the treated cooling water without significant corrosion over time. It also tolerates operating temperatures (ambient up to 110 °C) with negligible loss of strength. The choice of SS316 thus ensures durability and minimizes the risk of leaks or failures due to corrosion, which was a critical requirement for the ethanol plant environment. Bolting and support structures were specified in carbon steel where they do not directly contact the process fluids, to reduce cost while still providing structural strength.

2.2 Analytical Design Calculations

The thermal design of the shell and tube heat exchanger was conducted using **Kern's method**, which is suitable for preliminary sizing and commonly applied in chemical process industries for shell and tube configurations [12, 13]. This empirical method enables simplified yet reliable estimation of shell side and tube side parameters under steady state conditions. Ethanol vapor was assigned to the shell side, while cooling water circulated through corrugated tubes. A counter-flow arrangement was selected to maximize the effective temperature gradient across the exchanger. The required heat duty ($\dot{Q}$) was evaluated using the energy balance (Eq. (1)):

$$\dot{Q} = \dot{m} C_p \Delta\theta \tag{1}$$

where $\dot{m}$ is the fluid mass flow rate, C_p is, the specific heat capacity, and $\Delta\theta$ represents the fluid's inlet and outlet temperature difference.

The effective temperature driving force was calculated using the log mean temperature difference (LMTD):

$$\Delta T_{LM} = \frac{\Delta T_1 - \Delta T_2}{\ln\left(\frac{\Delta T_1}{\Delta T_2}\right)} \tag{2}$$

In this equation, ΔT_{LM} represents the log mean temperature difference, The terms ΔT_1 and ΔT_2 refer to the temperature differences between the hot and cold fluids at each end of the exchanger.

In this heat exchanger tube side fluid is water and shell side fluid is alcohol. Therefore, it needs to consider the fouling factors to estimate the overall heat transfer coefficient.

$$U_0 = \frac{1}{\frac{1}{\text{hi}} + \frac{1}{\text{ho}} + R_{fw} + R_{fAl}} \tag{3}$$

In the given equation, U_0 denotes the overall heat transfer coefficient, hi and ho correspond to the inside (tube-side) and outside (shell-side) convective heat transfer coefficients, respectively. R_{fw} refers to the fouling factor associated with water, R_{fAl} is the fouling factor of alcohol vapor.

The required surface area for heat transfer was then obtained from:

$$\dot{Q} = U \times A \times \Delta T_{lm} \tag{4}$$

$\dot{Q}$ represents the total rate of heat transfer between the two fluids, U stands for the overall heat transfer coefficient, A denotes the effective surface area.

The number of tubes (N_t) was estimated from the total required heat transfer area using:

$$N_t = \frac{A_{heatexchanger}}{A_{tube}} \tag{5}$$

A square tube configuration was selected for the bundle layout, as this facilitates easier mechanical cleaning, an essential consideration given the potential fouling issues in ethanol production. By using Kern's method, other relevant calculations shell side coefficients, pressure drop estimates, baffle spacing, tube layout, and shell diameter, were completed to ensure a consistent preliminary design prior to CFD/FEA.

2.3 Modeling and Simulation

A detailed 3D model of the heat exchanger was created in SOLIDWORKS software, capturing all major components: shell, tube bundle, baffles, tube-sheets, channel heads, and support saddles. Figure 2 shows an overview of the assembled heat exchanger 3D model. This model was used as the basis for the simulation. And the sectional view of the model, which identifies relevant internal and external components, providing clarity on fluid entry and exit points, and heat exchanger internals are shown in Fig. 3.

Numerical simulations were conducted to validate the thermal and structural performance of the designed heat exchanger under typical ethanol production operating conditions. Both Computational Fluid Dynamics (CFD) and Finite Element Analysis (FEA) were utilized, with simulations performed using the SolidWorks Flow Simulation and Simulation modules.

For the CFD analysis, conjugate heat transfer was modelled, considering ethanol vapor on the shell side and cooling water on the tube side. Boundary conditions were established using the process inlet temperatures and mass flow rates determined from the design calculations. Turbulence was modelled using a RANS approach with the realizable $k - \varepsilon$ closure and inlet turbulence was specified by the intensity–length method. Water properties were temperature-dependent (35–45 °C). Results in Table 2 imply that boundary conditions of flow in a tube. Structural FEA assessed the exchanger's mechanical integrity focusing on support saddles.

Table 2. Boundary conditions of flow in tube

Condition	Parameter value
Inlet	**Uniform velocity** V = 0.54 ms^{-1}; **Temperature** T = 35 ^{0}C;**Turbulence (I–L)**: intensity I = 8%, length scale L = 0.0013 m
Outlet	**Static pressure** p = 1 atm
Tube wall	**External heat from outside ethanol fluid:** ambient (reference) temperature **T∞ = 85 ^{0}C**

Fig. 2. Developed solid model of the heat exchanger

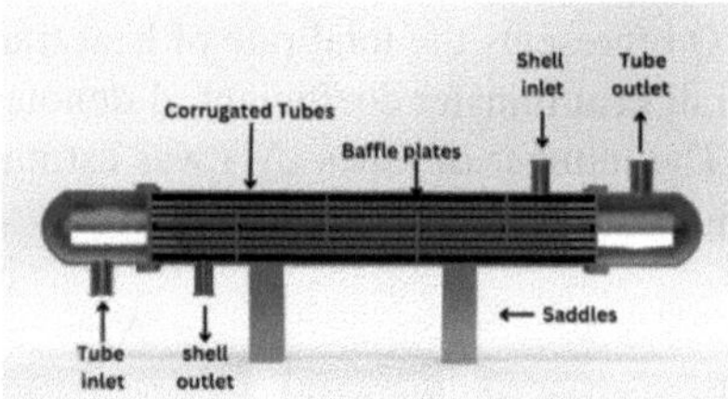

Fig. 3. Sectional view of the heat exchanger

3 Results and Discussion

3.1 Performance Comparison of Corrugated Tube Versus Smooth Tube Configuration

Figure 4 presents the pressure , temperature and velocity distributions for the smooth tube at an inlet velocity of 0.54 ms-^{1},while Fig. 5 shows the corresponding fields for the corrugated tube under identical conditions; in both, the left end is the inlet and the right end the outlet. Figure 6 plots the **centerline temperature** along the 2 m length. Comparing outlets, the corrugated tube reaches **43 °C**, whereas the smooth tube attains **38 °C**. The corrugated geometry induces periodic acceleration and recirculation due to the wave crests and valleys, repeatedly disrupting the near-wall boundary layer and thereby enhancing convective heat transfer. By contrast, the smooth tube develops a stable boundary layer that increases thermal resistance. Corrugated tubes offer better heat transfer compared to smooth tubes, but the pressure drop is high.

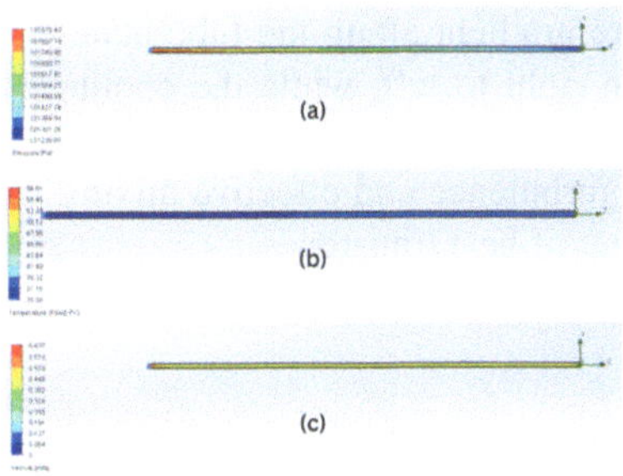

Fig.4. Pressure, temperature, velocity vectors distribution of smooth tube when v =0.54 m/s. (a)Pressure ; (b) Temperature; (c)Velocity.

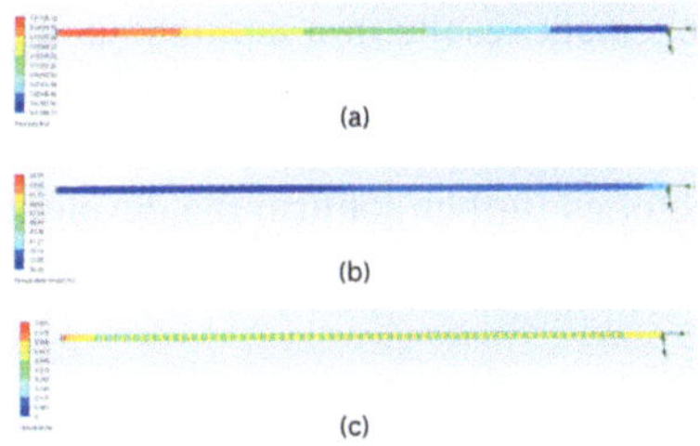

Fig. 5. Pressure, temperature, velocity vectors distribution of corrugated tube when v= 0.54 m/s. (a)Pressure ; (b) Temperature; (c)Velocity.

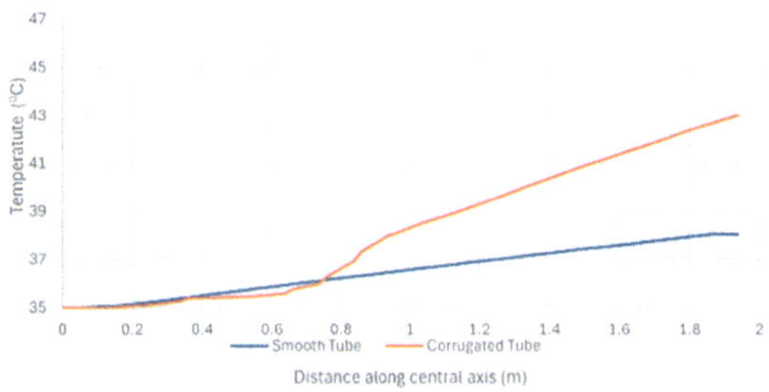

Fig. 6. Temperature distribution of the central axis of corrugated and smooth tubes

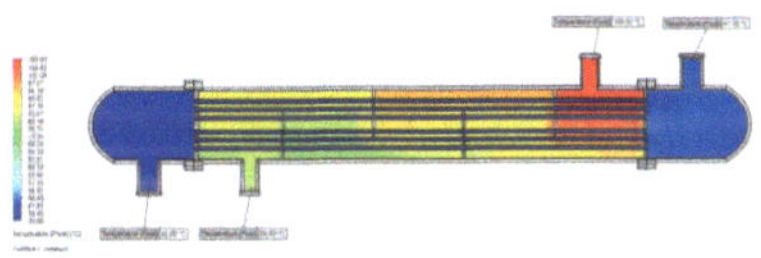

Fig. 7. Temperature variation from the simulation

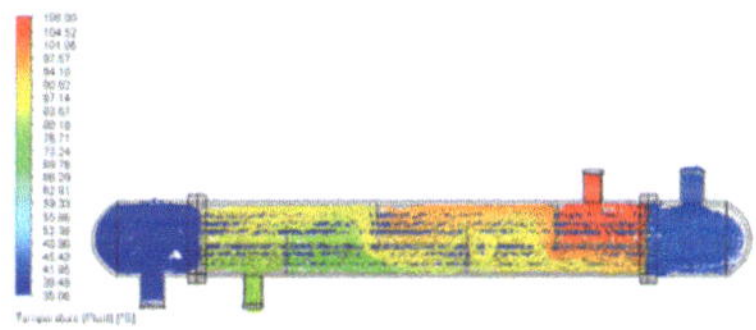

Fig. 8. Flow patterns within the heat exchanger

3.2 Thermal Performance

The temperature distribution within the heat exchanger is illustrated in Fig. 7. The contour plot clearly shows the progressive decrease in temperature of the ethanol vapor as it moves along the shell side, while the cooling water simultaneously gains heat within the corrugated tubes. The measured outlet temperatures from the simulation (shell side: 80.6 °C, tube side: 43.1 °C) closely align with the design targets, demonstrating the effectiveness of the thermal design.

The CFD simulation provided detailed visualization of both the temperature distribution and flow patterns within the heat exchanger. As shown in Fig. 8, the counter-flow arrangement results in a continuous temperature gradient along the tube bundle, with the ethanol vapor condensing as it moves from right to left, while the cooling water increases in temperature as it flows in the opposite direction. The superimposed flow trajectories further confirm the development of turbulence and effective mixing within the corrugated tubes, supporting the enhancement of heat transfer.

3.3 Stress Analysis

The structural integrity of the saddle support was evaluated through Finite Element Analysis, with the resulting von Mises stress distribution shown in Fig. 9. The maximum observed stress was well below the yield strength of SS316 stainless steel, indicating a substantial safety margin and confirming the adequacy of the support structure for operational loads.

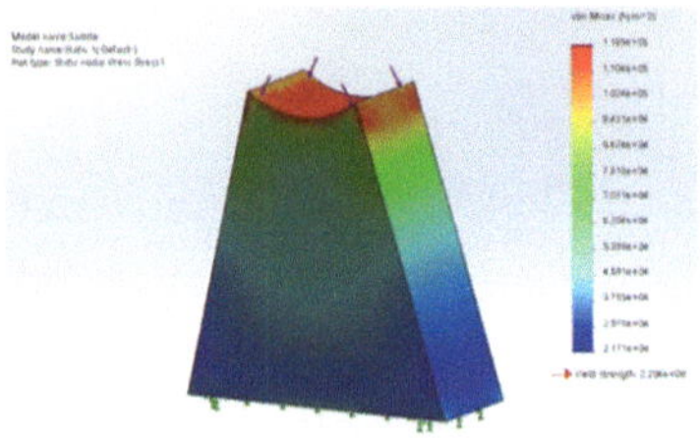

Fig. 9. Stress Analysis for the saddle

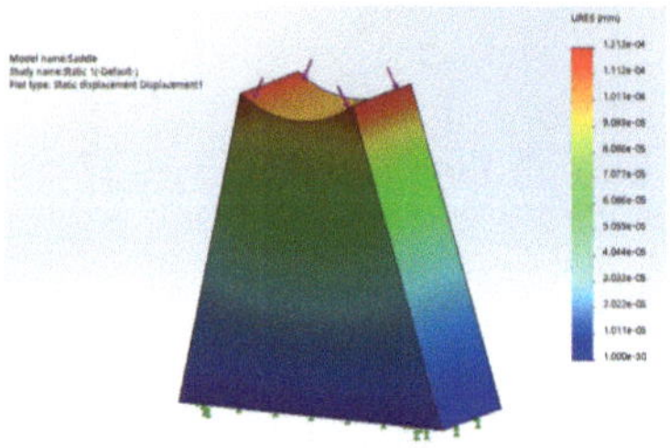

Fig. 10. Displacement Analysis of the saddle

3.4 Displacement Analysis

In addition to stress analysis, displacement results for the saddle support were obtained from FEA, as illustrated in Fig. 10. The maximum displacement was found to be extremely small, confirming negligible deformation under the applied loading conditions and further supporting the structural adequacy of the support design.

4 Conclusions

A corrugated tube shell and tube heat exchanger was designed (via Kern's method) and assessed for ethanol service using SS316 components and a square pitch bundle to aid cleaning. CFD under identical conditions showed that the corrugated tube achieved a higher outlet temperature (43 °C) than the smooth tube (38 °C), **increasing the outlet temperature by about 13%** and demonstrating superior convective performance. This enhancement arises from corrugation-induced secondary flows that increase **near wall** shear and mass transfer, thereby reducing fouling tendency. This thermal gain is accompanied by a higher pressure drop, but remains acceptable within the specified operating envelope. Structural FEA confirmed stresses and displacements are within SS316 allowable limits. Overall, the corrugated configuration offers a net thermal hydraulic and operational advantage for ethanol processing. Future work should include experimental validation and a focused parametric study of corrugation geometry and flow rate.

References

1. Yang, Y., Boots, K., Zhang, D.: A sustainable ethanol distillation system. Sustainability **4**(1), 92–105 (2012). https://doi.org/10.3390/su4010092
2. Edreis, E., Petrov, A.: Types of heat exchangers in industry, their advantages and disadvantages, and the study of their parameters. In: IOP Conference Series: Materials Science and Engineering, 963 (2020). https://doi.org/10.1088/1757-899X/963/1/012027
3. Ali, M., Ul-Hamid, A., Alhems, L.M., Saeed, A.: Review of common failures in heat exchangers – Part I: Mechanical and elevated temperature failures. Eng. Fail. Anal. **109** (2020). https://doi.org/10.1016/j.engfailanal.2020.104396
4. Corleto, C.R., Argade, G.R.: Failure analysis of dissimilar weld in heat exchanger. Case Studies in Eng. Fail. Anal. **9**, 27–34 (2017). https://doi.org/10.1016/J.CSEFA.2017.05.003
5. Nayak, S., Jena, S., Khan, A.S.S., Paswan, M.K., Sharma, V.K.: Use of Corrugated Pipe Heat Exchangers In Waste Heat Recovery Steam Generators, IOP Conference Series: Materials Science and Engineering, 1123(1), 012037 (2021). https://doi.org/10.1088/1757-899x/1123/1/012037
6. Manufacturing Chemist: Why use corrugated tubes in heat exchangers? (2019). Accessed: 20 May 2025. https://manufacturingchemist.com/why-use-corrugated-tubes-in-heat-exchangers--154958
7. Jiang, J., Bo, L., Chen, Z., Cui, H., Zhu, H.: Numerical simulation and analysis of enhanced heat transfer in corrugated tube heat exchanger. IOP Conf. Ser. Earth Environ Sci. **467**(1), 012031 (2020). https://doi.org/10.1088/1755-1315/467/1/012031
8. Berce, J., Zupančič, M., Može, M., Golobič, I.: A review of crystallization fouling in heat exchangers. Processes **9**(8), 1356 (2021). https://doi.org/10.3390/pr9081356
9. Chen, X., Yang, C., Kong, J., Wang, F., Jin, Z., Qian, J.: Steam-side heat transfer analysis in spirally corrugated tube for grate furnace superheater. Case Stud. Ther. Eng. **50**, 103443 (2023). https://doi.org/10.1016/j.csite.2023.103443
10. Hayati, M.: Lecture 1 Kern's Method, presentation at University of Kashan, Kahan, Iran (2014). Accessed: 02 May 2025. https://www.slideshare.net/slideshow/01-kerns-method/52581785
11. Mya, N.S., et al.: Baffle design of shell and tube heat exchanger. Int. J. Sci. Eng. Technol. Res. **8** (2019). Accessed: 02 May 2025. https://ijsetr.com/uploads/214536IJSETR17552-89.pdf
12. Axsom, T.: Heat Exchanger Material Selection (2023). Accessed: 27 May 2025. https://www.fictiv.com/articles/heat-exchanger-material-selection
13. Sadigov, M.: Heat Exchanger Design, Khazar University, Baku. Azerbaijan (2024). https://doi.org/10.13140/RG.2.2.14231.96163

Development and Evaluation of an Experimental Dual Chamber Hot Air Dryer for Drying Turmeric

J. M. H. B. Dewanthi, P. W. A. W. Pathiraja, A. L. K. Tharaka, U. G. A. S. Kumara, T. K. K. S. Pathmasiri(✉), and K. T. K. M. De Silva

Department of Mechanical and Manufacturing Engineering, Faculty of Engineering,, University of Ruhuna, Galle, Sri Lanka
kalpani@mme.ruh.ac.lk

Abstract. Open sun drying is a traditional drying method that is used in Sri Lanka to dry spices due to its low operational cost and simplicity. However, open sun drying has several limitations such as contamination, uneven drying and weather dependency. Electrical oven drying is one of the solutions to address these limitations, but it also suffers from issues such as high energy consumption and inefficient heat distribution. This study was done to develop and evaluate an improved electrical oven dryer that aims to enhance drying performance. The electrical oven dryer features a unique dual-chamber design with three trays, axial fans, a heating coil, and a microcontroller-based temperature regulation system. For the performance and quality comparison, open sun drying and controlled oven drying were employed. For the testing, fresh turmeric rhizomes were cleaned, sliced uniformly and subjected to drying without prior boiling. Evalua-tion of two drying methods was done based on moisture removal rate, energy consumption, colour change, and total ash content. Results showed that oven drying achieved a moisture removal rate of 14.93% per hour which is more than three times the moisture removal rate of sun drying (4.31% per hour), while maintaining uniform drying and comparable quality. Therefore, the proposed electrical oven dryer offers a more efficient and hygienic alternative to traditional open sun drying while making it suitable for small-scale spice processing.

Keywords: Hot Air Dryer · Food Drying · Turmeric

1 Introduction

Drying is a crucial post-harvest procedure in the agricultural industry, particularly for spices like turmeric that need to effectively remove moisture to maintain their quality, marketability, and long shelf life. Traditional drying techniques, such as open sun drying, are very common among small-scale farmers in Sri Lanka because they are inexpensive and easy to use. The open sun drying method suffers from several issues, including contamination, non-uniform drying, and weather dependence, which result in compromised product quality and potential post-harvest losses [1].

M. Siriwardhana et al. (Eds.): IERS 2025, PTT, pp. 140–149, 2026.
https://doi.org/10.1007/978-981-95-7984-6_13

To overcome the limitations of traditional drying techniques, various types of electrical dryers have been introduced and are currently in use. While these dryers offer improved control over drying conditions and reduce contamination risks, they come with their own set of problems. Operating costs increase as a result of high energy usage and inefficient heat distribution which leads to uneven drying and longer drying times [2].

This project proposes the design of an improved electrical spice dryer with a well-insulated chamber with three trays and an electric heater coil, supported by axial flow fans that provide continuous and uniform airflow throughout the chamber. This design enhances drying efficiency and product quality and lowers operational costs.

2 Materials and Methods

2.1 Design and Fabrication of the Electric Oven Dryer

The dryer components were selected based on engineering calculations, followed by the development of conceptual designs and 3D solid models. Computational simulations were employed to analyse airflow patterns, temperature distribution, and fan positioning. After finalizing the design, the dryer was fabricated using mild steel box bars for the frame and aluminium sheets and mesh for the internal structures. The inner walls were insulated using thermal insulation sheets composed of two layers of aluminium foil laminated to a polyethylene air bubble sheet to reduce heat loss.

Design Specifications Table 1 gives the geometrical specifications of the developed electrical dryer.

Table 1. Design specifications of the electrical dryer

Criteria		Value
Outer chamber	Length x width x height (mm)	600 x 400 x 600
Inner chamber	Length x width x height (mm)	400 x 260 x 400
Tray	Length x width (mm)	320 x 260
	Spacing in between (mm)	100
Machine	Length x width x height (mm)	600 x 400 x 1400

Inner chamber height was calculated in order to accommodate the required amount of the sample with adequate spacing for air flow. Power of the heater was obtained in order to fulfil the heat energy requirement to vaporize the required amount of water within a specific time. Further the insulation thickness of the inner surface of the outer chamber was calculated allowing only 4% heat loss through conduction. Another calculation was carried out to find the power ratings of the five fans used in the dryer which can handle the required volume of air per second.

2.2 Machine Details

Figure 1 shows the design of the developed machine. The hot air dryer includes a 1.1 kW electric heating coil and a primary air circulation fan situated in the air intake chamber. The system also comprises an outer chamber, an inner drying chamber, three intake fans, and an exhaust fan. The drying chamber is equipped with three trays for holding the material.

Power Supply and Activation. Two DC modules convert AC input to give controlled voltages for relays, fans, and sensors. The system is turned on by a main power switch, and the dryer's operating period is set by a 90-min timer switch.

Heating System Control. The heater is controlled by the XH3001 PID on-off controller to keep the outer chamber at the appropriate temperature.

Temperature and Humidity Monitoring. The Arduino board is equipped with two DHT sensors that allow for real-time temperature and humidity monitoring of the inner and outer chambers. To maintain ideal drying conditions, the Arduino regulates the fans.

Fan System for Air Circulation. Once the outer chamber reaches the set temperature, the Arduino activates three fans in the inner chamber to evenly distribute hot air for drying. To increase drying efficiency, moist air is simultaneously removed by an exhaust fan.

Display and Monitoring Interface. The Arduino is attached to the LCD, which displays data such as the operating status, humidity level, and temperature reading.

Safety and Efficiency of the System. If the dryer door opens while it is operating, a limit switch cuts off the power to the heater to guarantee safety. An XH3001 PID on/off controller regulates temperature, ensuring constant heat with low energy consumption and fast reaction (Fig. 2).

2.3 Sample Preparation

Fresh turmeric rhizomes were procured from a local supplier while considering the absence of visible damages. Initially, the rhizomes were thoroughly washed with clean water to remove adhering soil and dirt. Excess surface moisture was then removed by air drying under shade for approximately 30 min. The cleaned rhizomes were then manually sliced into uniform thicknesses of approximately 3 mm [3]. Care was taken to ensure that slices were of similar size to minimize variation in drying behaviour. The sliced turmeric samples were then subjected to the respective drying processes.

2.4 Methods of Drying

Open Sun Drying. Unboiled and sliced turmeric were dried under direct sunlight as a reference method for evaluating the performance of the electrical spice dryer. Prepared samples were evenly spread on clean steel trays to ensure better airflow. Drying was conducted in an open outdoor area between 9:30 am and 4:00 pm, approximately six and a half hours under the sun [4] under ambient conditions. Drying was continued until the turmeric was observed to be ready for grinding.

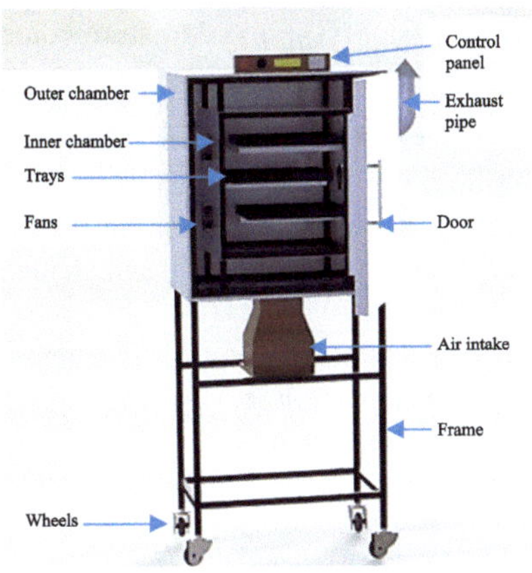

Fig. 1. Components of the electrical oven dryer

Fig. 2. Fabricated electrical oven dryer

Drying in Electrical Oven Dryer. Unboiled turmeric slices were dried using the electrical oven dryer equipped with three trays, each loaded with 200 g of turmeric. The drying process was conducted at a controlled temperature of 50 °C–60 °C [5] for 4 h in a monitored environment to ensure uniform and consistent drying conditions.

2.5 Methods of Evaluating

A series of evaluations was conducted to comparatively assess the effectiveness of the open sun drying and electrical oven drying methods for turmeric. These tests aimed to quantify performance critical to spice drying, with the objective of determining the viability and reliability of the electrical spice dryer as an alternative to conventional sun drying.

Moisture Removal Rate. To determine the moisture removal rate for both drying techniques, the moisture content (wet basis) of fresh turmeric, sun-dried turmeric, and oven-dried turmeric were analysed. The selected samples were subjected to gravimetric evaporation to determine the moisture content. There, the samples were dried in an oven at 105 °C for hours until a constant weight was reached. Then the weight was measured and compared with the initial weight to find the amount of moisture removed.

The moisture content (wet basis) was calculated using the Eq. (1).

$$MoistureContent(MC) = \frac{W_1 - W_2}{W_1} \times 100\% \tag{1}$$

where W_1 = Initial weight of the sample (g) and W_2 = Constant weight of the sample at the end (g)

Then, the moisture removal rate was calculated using the Eq. (2).

$$Moistureremovalrate = \frac{MC_{fresh} - MC_{dried}}{t} \tag{2}$$

where MC_{fresh} = Moisture content of fresh turmeric (%), MC_{dried} = Moisture content of dried turmeric (%) and t = Drying time (hr) [6].

Energy Consumption Per Cycle and Specific Energy Consumption (SEC). The energy consumed by the electrical oven dryer during a single drying cycle (for a 600 g batch of turmeric) was recorded using a wattmeter. The power consumption readings were taken at 10-min intervals, and the total energy usage (E) in kWh was calculated accordingly [7].

The specific energy consumption (SEC), defined as the energy required to evaporate a unit mass of water, was determined using the following procedure.

The mass of evaporated water (W_{eva}) was first calculated using Eq. (3) [8],

$$W_{eva} = M \times \frac{MC_{fresh} - MC_{dried}}{1 - MC_{dried}} \quad (3)$$

where M = Initial mass of fresh turmeric (kg).

Then SEC was calculated using Eq. (4),

$$SEC = \frac{E}{W_{eva}} \quad (4)$$

Colour Analysis. To evaluate the visual quality changes, the colour of dried turmeric (sun-dried and oven-dried) were compared by measuring the L^*(brightness), a^*(redness) and b^*(yellowness) values using a colorimeter. The values of L^*, a^*, and b^* of the dried turmeric samples were compared with those of pure curcumin which gives the colour to turmeric. Then the colour change index (ΔE) was calculated using the Eq. (5) [9],

$$\Delta E = \sqrt{(L^* - L^*_{ref})^2 + (a^* - a^*_{ref})^2 + (b^* - b^*_{ref})^2} \quad (5)$$

where L^*, a^*, b^* corresponds to dried turmeric and L^*_{ref}, a^*_{ref}, b^*_{ref} corresponds to pure curcumin.

Total Ash on Dry Basis. The total ash content of the turmeric samples dried by both methods were determined using the gravimetric evaporation method. Samples were subjected to laboratory analysis to measure their total ash content in which they were heated up to 525 °C for 3 h in a muffle furnace until the ash remains. This test gives results based on the residue remaining after water and other volatile matter get vaporized [10].

3 Results and Discussion

Open sun drying was conducted using 1336g of fresh turmeric over a period of 2.5 days under ambient conditions. The initial and final appearances of the turmeric samples during this process are illustrated in Fig. 3.

Electrical oven drying was performed in two separate cycles, each processing 600 g of turmeric with a duration of 4 h per cycle. The appearance of turmeric before and after drying in a cycle is shown in Fig. 4.

(a) (b)

Fig. 3. The appearance of turmeric (a) before and (b) after open sun drying.

(a) (b)

Fig. 4. The appearance of turmeric (a) before and (b) after electrical oven drying.

3.1 Temperature and Relative Humidity Inside the Chamber

To investigate thermal behaviour and environmental conditions during electrical oven drying, temperature and relative humidity inside the drying chamber were monitored at 10-min intervals (Fig. 5).

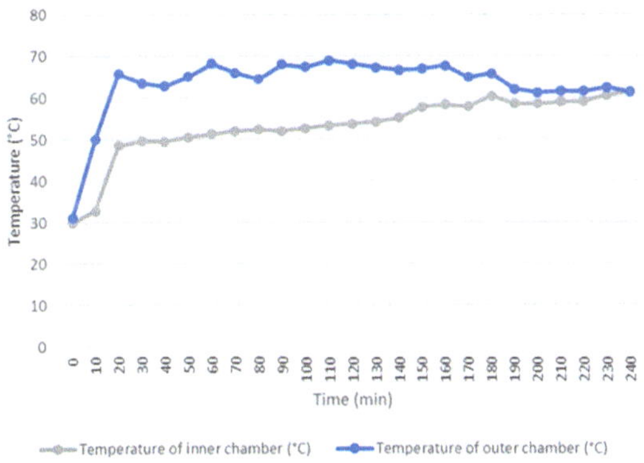

Fig. 5. The variation in inner and outer chamber temperatures over time.

The airflow velocity within the chamber was measured and found to be approximately 1 m/s, ensuring consistent air circulation around the turmeric slices.

According to the temperature and relative humidity profiles, both the inner and outer chamber temperatures exhibited a rapid initial increase. After approximately 30 min, the temperature within the inner chamber stabilized within the range of 50–60 °C. Over time, the temperatures of both chambers tended toward equilibrium. Similarly, as shown in Fig. 6, the relative humidity decreased rapidly during the initial phase and exhibited a slower rate of decline beyond the first 30 min.

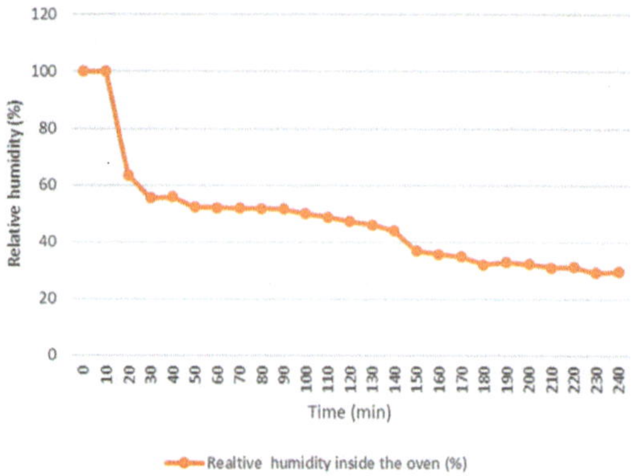

Fig. 6. The variation in relative humidity inside the chamber over time

3.2 Moisture Removal Rate

Moisture contents of fresh and dried turmeric were determined for both drying methods, and the corresponding drying durations were recorded. The moisture removal rates were computed as shown in Table 2.

Table 2. Moisture removal rates

Method	Moisture content of fresh turmeric (%)	Moisture content of dried turmeric (wb) (%)	Drying Time (hr)	Moisture removal rate (% per hour)
Open Sun Drying	89.02	19.43	16.15	4.31
Electrical Oven Drying	89.02	29.29	4	14.93

As presented in the Table 2, the moisture removal rate observed in the electrical oven drying process is more than three times that of the open sun drying method, indicating a better drying efficiency. In order to achieve the target final moisture content of 10% (according to SLS 186: Part5) in electrical oven drying, it is recommended to carry out the drying at a slightly elevated temperature (around 65–70 °C).

In addition, drying uniformity across trays in the electrical oven dryer was evaluated. For that, as shown in Table 3, moisture content was measured from turmeric samples placed on each of the three trays. These closer values indicate that the drying across all three trays in the electrical oven dryer can be considered uniform, as the final moisture contents are relatively similar.

Table 3. Moisture content of dried turmeric in each tray

Tray No	Moisture content (%)
1 (Top)	30.93
2 (Middle)	33.66
3 (Bottom)	31.26

3.3 Energy Consumption

The electrical energy consumed by the spice dryer per drying cycle (600 g batch) was recorded using a wattmeter and calculated to be 8.96 kWh. Thereby the Specific Energy Consumption was calculated using Eq. (3) and (4) as follows,

$$W_{eva} = 601.21g \times \frac{0.8902 - 0.4530}{1 - 0.4530}$$

$$W_{eva} = 480.53g$$

$$SEC = \frac{8.96kWh}{0.48053kg}$$

$$SEC = 18.65kWh/kgs$$

That is, this machine consumes 18.65 kWh to vaporize 1 kg of moisture of the crop.

If the average cost per one electricity unit in Sri Lanka is considered to be Rs.25, the operating cost of the machine per cycle (batch of 600g) can be presented as Rs. 224 since it requires 8.96 kWh. Accordingly, the operating cost of the machine to dry 1kg of turmeric can be presented as Rs.373.

3.4 Colour Change

A colorimeter was used to determine the colour parameters of dried turmeric and compared them with those of pure curcumin. The results are summarized in the.

Table 4.

Table 4. Colour changes of dried turmeric

Sample	L^*	a^*	b^*	Colour change (ΔE)
Pure curcumin	64.740	15.62	54.77	-
Sun dried turmeric	32.03	15.10	53.16	32.75
Oven dried turmeric	34.80	15.80	55.46	29.95

According to the above data, the obtained values of L^*, a^* and b^* for sun dried turmeric are less than those of oven dried turmeric. That may be due to sun dried turmeric being drier than the oven dried turmeric as the final moisture content of sun-dried turmeric is relatively low. However, the values observed to be approximately in a similar range, representing that the quality of oven dried turmeric is similar to that of sun-dried turmeric.

When compared with the pure curcumin, both has significant brightness change with an insignificant redness and yellowness change. There is no significant difference between colour change index of sun dried and oven dried samples.

3.5 Total Ash Content

The total ash content of the dried turmeric from both drying methods was obtained using the gravimetric evaporation method. The results are summarized in the Table 5.

Table 5. Total ash content of dried turmeric

Method	Total ash on dry basis (%)
Open Sun Drying	7.99
Electrical Oven Drying	8.03

Here, the total ash content represents the total mineral residue of the sample left after the combustion of organic matter in turmeric. Since the resulted values are approximately equal for both drying methods, it can be suggested that the drying in the electrical oven has not been affected to the inorganic content of turmeric adversely. This result further ensures that the quality of the oven dried turmeric is maintained similar to sun dried turmeric.

The dual-chamber hot-air dryer substantially accelerated turmeric drying compared with traditional open sun drying; it achieved a moisture-removal rate of 14.9%· per hour and completed a 600 g batch in about four hours while maintaining comparable colour and ash values. The dual-chamber arrangement with three zig-zag arranged trays which is a novel design, improved temperature stability in the inner chamber and uniform drying of turmeric while addressing common limitations of conventional electrically heated hot-air dryers. The unit was deliberately compared only with open sun drying because it is the dominant, lowest-cost baseline for small scale farmers in remote. Although the dryer greatly reduces drying time and contamination risk, its specific energy consumption (SEC) is relatively high primarily due to continuous fan operation. With a material cost below Rs. 60,000 the unit is affordable for small-scale farmers, and its user-friendly controls and real-time temperature/humidity display allow farmers to monitor and manage the drying process easily. Future work will focus on reducing SEC through optimisation of fan operation and integration of solar collector to the dryer and benchmarking against alternative mechanical drying technologies to establish relative energy and cost efficiency.

4 Conclusions

Electrical oven drying is an effective solution to overcome the challenges related to open sun drying of spices. In this study, a dual chamber hot air dryer was developed to address the key limitations of open sun drying such as uneven drying, contamination, and weather dependency. This dryer features a dual-chamber design with three trays, axial fans, a heating coil, and a microcontroller-based temperature regulation system. To assess its performance, turmeric was dried using both the developed dryer and the open sun method. Key parameters; moisture content, energy consumption, colour change, and total ash content were measured and compared. The results demonstrated that the moisture removal rate of the electrical oven dryer is more than three times that of open sun drying, with relatively uniform drying across three trays. Further, the quality parameters such as colour change and total ash content of the oven dried turmeric observed to be closer to those of sun-dried turmeric, confirming that the electrical oven dryer has the capability to preserve the quality of the product. Despite the energy requirement, the controlled environment and consistent performance make the proposed electrical oven dryer a viable solution for small-scale spice processing in Sri Lanka.

References

1. Rathnayake, R.M.S.P., Ariyaratne, A.R., Prematilake, S.P.: Design and fabrication of an engineering model of a crop dryer. Tropical Agr. Res. **18**, 182–195 (2006)
2. Wickramasinghe, W.: Assessment of the impact of ASMP intervention into local dry chili production on the national dry chili market of Sri Lanka – strategic analysis of the chili industry. Agriculture Sector Modernization Project, Final Report (2024)
3. Fernando, A.J., Gunathunga, C., Brumm, T., Amaratunga, S.: Drying turmeric (Curcuma longa L.) using far-Infrared radiation: Drying characteristics and process optimization. J. Food Process Eng. **44**(9), 1–11 (2021)
4. Sunrise and sunset in Sri Lanka, https://www.worlddata.info/asia/sri-lanka/sunset.php#google_vignette. Last accessed 14 June 2025
5. Singh, G., Arora, S., Kumar, S.: Effect of mechanical drying air conditions on quality of turmeric powder. J. Food Sci. Technol. **47**(3), 347–350 (2010)
6. Dept. of Export Agriculture, Turmeric. https://dea.gov.lk/turmeric/. Last accessed 16 June 2025
7. Sabins, D.: Energy Usage Calculator - Understanding Your Energy Usage, https://www.saveonenergy.com/resources/energy-consumption/. Last accessed 16 June 2025
8. Fernando, A.J., Gunathunga, C., Brumm, T., Amaratunga, S.: Drying turmeric (Curcuma longa L.) using far-Infrared radiation: Drying characteristics and process optimization. J. Food Process Eng. **44** (2021). https://doi.org/10.1111/jfpe.13780
9. Saha, G., et al.: Dynamics of Drying Turmeric Rhizomes (Curcuma longa L.) with Respect to its Moisture, Color, Texture and Quality. Agron. **12**(6), 1420 (2022). https://doi.org/10.3390/AGRONOMY12061420
10. Jinadasa, B.K.K.K.: Determination of Ash, https://www.scribd.com/doc/25575913/Determination-of-Ash. Last accessed 16 June 2025

Development of a Point Absorber Type Wave Energy Converter for Southern Sri Lanka Wave Conditions

K. G. S. M. Rathnapriya, R. G. P. D. Rankothge, R. R. Gamage, W. A. D. Sampath, I. H. D. Senevirathne, and H. C. P. Karunasena(✉)

Department of Mechanical and Manufacturing Engineering, University of Ruhuna, Matara, Sri Lanka
chaminda@mme.ruh.ac.lk

Abstract. The paper focuses on extracting renewable energy from ocean waves, which has a high energy density compared to other renewable energy sources. There are not many successfully developed commercially available technologies regarding wave energy extraction. Sri Lanka, on the other hand, does not possess locally developed wave energy converters (WECs), and the underlying expertise is not well-established. In this study, a double buoy point absorber type WEC is modelled to be deployed in Sri Lankan wave conditions to extract a maximum energy of 30 W from waves with a wave height of 0.5 m and a wave period of 10 s, similar to Southern coast wave conditions in Sri Lanka. The design was modelled and simulated to modify and match the required power output. Dif-ferent floater shapes were compared in terms of floater displacement, phase an-gle, and applied force. Manually tested data was gathered through the fabricated WEC model. This will be a step forward in developing a wave energy converter to harness clean energy from ocean waves and help fabricate a WEC that can harness energy in real sea conditions.

Keywords: Double buoy point absorber · Point absorber · Renewable energy sources · Wave energy · Wave energy converter

1 Introduction

The global energy demand is increasing daily. In 2024 and 2025, it's expected that global energy demand is to be higher than the global GDP (Gross Domestic Product) growth of 3.2% where previously in 2022 and 2023, the global electricity demand grew much slowly than the GDP [1]. Renewable energy sources are energy sources which are derived naturally and are replenished at a higher rate than consumption [2]. According to predictions of the International Energy Agency (IEA), renewable energy will rapidly grow from 28.7% in 2021 to around 43% in 2030 [3]. Wave energy is a such source that generates energy from ocean waves containing kinetic and potential energy [4] having a potential of generating about 29,500 TWh annually according to the International Panel on Climate Change (IPCC) [5]. Various types of WECs have been developed globally

M. Siriwardhana et al. (Eds.): IERS 2025, PTT, pp. 150–160, 2026.
https://doi.org/10.1007/978-981-95-7984-6_14

which use different technologies [6]. The novelty of this project lies on the development process where a double-buoy type WEC is incorporated with a rack-and-pinion design. Also, it represents a locally designed WEC in Sri Lanka which bridges the gap between global WEC technology and local application. Also, it opens research opportunities regarding wave energy extraction in a local scale.

2 Methodology

2.1 WEC Design Overview

The WEC utilizes a dual rack and pinion of module 01 with 30 teeth on each pinion to extract energy from ocean waves. The design compromises of two main sections which is the middle ballast and outer floater. Two racks are connected to the floater while corresponding pinions are located inside the ballast coupled to motion rectifiers to convert bi-directional motion induced by waves into a unidirectional rotary motion and transmitted to a DC generator. The ballast is connected to a counterweight keeping the device anchored. The floater undergoes a linear motion due to incoming waves.

2.2 Calculations

The modeling of the double-buoy WEC involves the following selected key parameters: wave height (H) of 0.5 m and a wave period (T) of 10 s, which are similar conditions to the Southern coast of Sri Lanka [7]. Seawater density (ρ) is taken as 1025 kg/m^3. The wave number "k" gives the spatial frequency of the wave and is critical on determining the dimensions of the structure [8]. The Eq. (1) is used in relatively shallow depth compared to wavelength where d represents water depth and g is the gravitational acceleration.

$$\omega^2 = g.k.\tanh(kd) \tag{1}$$

The total wave energy density is given from the Eq. (2) [9].

$$E = \frac{1}{4}\rho.g.H^2 \tag{2}$$

Floater area contacted with ocean surface is 0.201 m^2 (A_f). The floater travels upward from mid-position due to the upward wave half cycle with an amplitude A of 0.25 m. Assuming the wave displacement as a sine wave, force exerted on the floater by upcoming waves are due to the extra displaced volume and can be expressed in the Eq. (3) where ω is angular frequency.

$$F_{wave}(t) = \rho.g.A_f.Asin(\omega t) \tag{3}$$

The effective force is given by the RMS value of the half cycle which is calculated as 357.5 N [10]. An average output power of 30 W was selected through following calculations. The floater weight was pre-determined to be 30 kg for ease of handling. When the floater reaches the maximum height and falls down, it will be due to the floater weight of 30 kg. As the floater is directly connected to the rack, the tangential force exerted on the pinion will be 30 kg ($\approx$ 294.3 N).

Taking $v_{avg,abs}$ as absolute rack velocity and C as pinion circumference, average pinion RPM is given in Eq. (4),

$$n_{avg} = \frac{v_{avg,abs}}{C} \times 60 \tag{4}$$

Assuming that waves are in a sinusoidal motion and that A is the wave amplitude, the Eq. (5) gives,

$$n_{avg} = \frac{2}{\pi} \times A \times \frac{2\pi}{T} \times \frac{60}{C} \tag{5}$$

Calculated average pinion rotational speed is 63.66 RPM. Torque subjected on the pinion is 4.415 Nm due to the tangential force of 325.9 N (average in upward and downward cycles). Therefore, output power is 29.4 W ($\approx$ 30 W). The counterweight (m_{cw}) is stationary on the sea bed and is affected by the weight of middle ballast, buoyancy force of middle ballast (submerged part) and upward wave force on floater in normal conditions. In extreme conditions it is affected by the weight and buoyancy (full body) of the floater and ballast.

$$\therefore m_{cw} > m_{buoyancy,ballast} + m_{buoyancy,floater} - m_{ballast} - m_{floater}$$

Using ISO 6336:2019 calculation method, root safety and flank safety values were generated for the rack and pinion indicating sufficient strength under expected wave conditions (Figs. 1 and 2).

Results (basic calculation)			
Contact ratios	$[\varepsilon_{\alpha m}/\varepsilon_{\beta}/\varepsilon_{\gamma m}]$	1.751 / 0.000 / 1.751	
		Pinion	Rack
Actual tip circle (mm)	$[d_{ae}, H_z]$	32.000	50.000
Root safety		3.998	5.040
Flank safety		2.418	2.972
Safety against scuffing (integral temperature)		4.792	
Safety against scuffing (flash temperature)		99.516	

Fig. 1. Root safety and flank safety for rack and pinion

Although root and flank safety values exceed recommended values, they were accepted due to the harsh environmental conditions consisting high impacts forces, unpredictable loads, prevent fatigue, and wear.

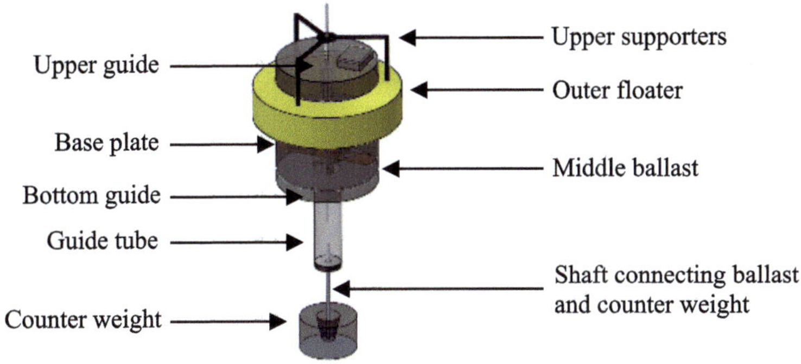

Fig. 2. Named 3D model of the WEC

2.3 Design Parameters

Table 1 and Fig. 3 shows the dimensions of the ballast and the floater of the modelled WEC.

Table 1. WEC dimensions

Dimension	Value
Ballast weight (without counter weight)	60 kg
Ballast diameter	600 mm
Ballast height (middle section)	860 mm
Floater weight	30 kg
Floater inner and outer diameters	720 mm, 880 mm
Floater height	250 mm

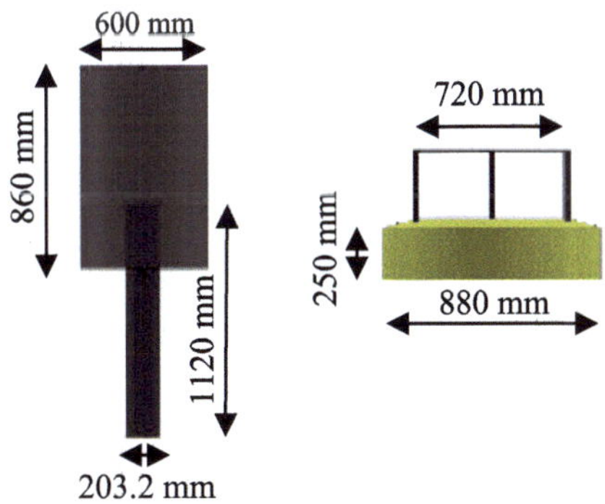

Fig. 3. Ballast and floater dimensions

2.4 Design Components

The design integrated both mechanical and electrical components along with various aspects which perform different functions (Fig. 4).

- **Outer floater and middle ballast:** The floater move linearly along the ballast along guides enabling smooth motion and prevent relative rotation.
- **Rack and pinion:** Racks were mounted on opposite sides to each other. As the floater undergoes linear motion, the rack moves linearly through guides rotating the pinions.
- **Motion rectifier:** Each pinion is coupled with a motion rectifier using a one-way bearing for unidirectional rotation ensuring only one bearing engaged in each half-cycle.

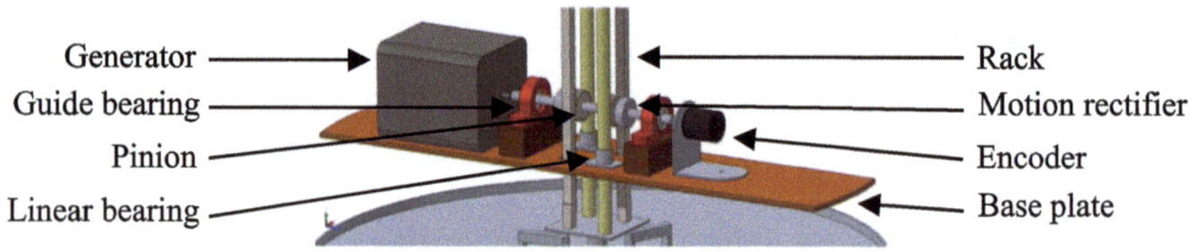

Fig. 4. Component setup in base plate

3 Results and Discussion

3.1 FEM Analysis

Finite element model analysis was performed to the outside floater and middle ballast considering extreme wave conditions and a FoS of 2 for the model in order to observe stress, displacement and strain distribution. Reaction forces due to buoyancy, drag on the structures and upward forces due to waves were considered in extreme wave conditions. Resulted values were in acceptable ranges of relevant material (Figs. 5 and 6).

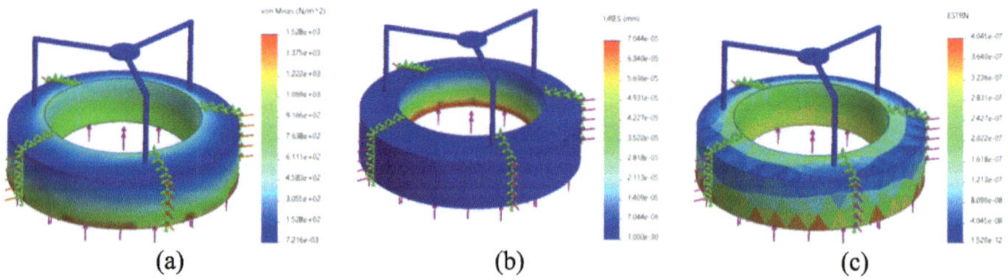

Fig. 5. (a) Stress distribution of WEC floater, (b) Displacement distribution of WEC floater, (c) Strain distribution of WEC floater

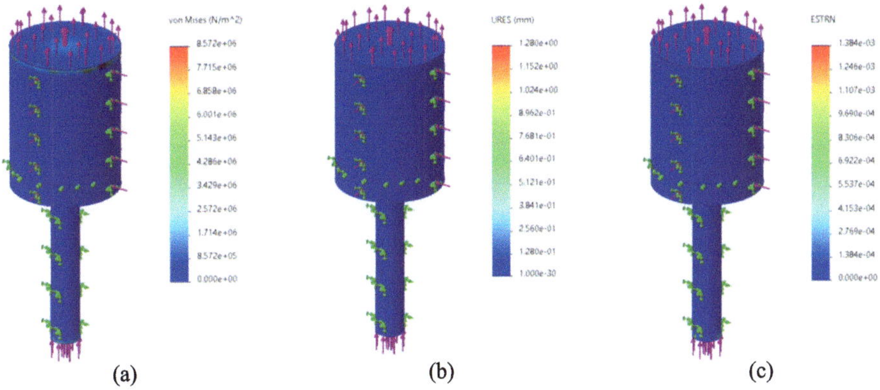

Fig. 6. (a) Stress distribution of WEC middle ballast, (b) Displacement distribution of WEC middle ballast, (c) Strain distribution of WEC middle ballast

3.2 CFD Simulation

In the CFD simulation, the rack and pinion are assumed to undergo a linear spring-like behavior where the restoring force is multiplication of stiffness coefficient and displacement to simplify the analysis. Accordingly, the stiffness coefficient equals 600 N/m. Therefore, PTO stiffness coefficient of Z direction is considered 600 N/m.

The primary geometry for the floater was cylindrical, which was selected due to its good performance and structural simplicity. Two other floater shapes which are hemi-spherical and elliptical were investigated as well for comparison in results (Fig. 7).

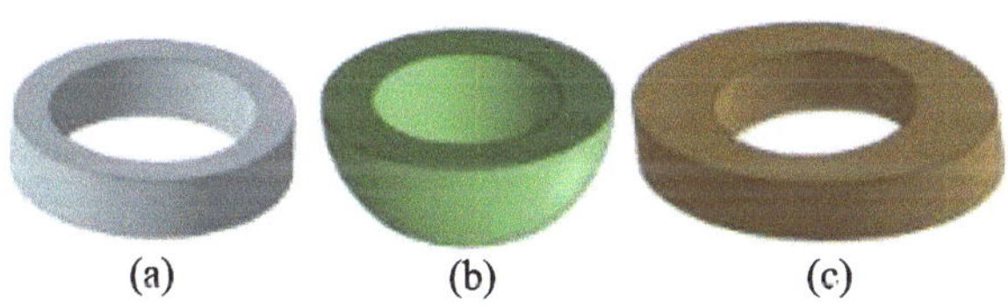

Fig. 7. (a) Cylindrical, (b) hemi-spherical, (c) elliptical, floater shapes

All three floater types were simulated under same conditions where wave interaction was analyzed using a monochromatic wave with 0.18 Hz frequency. The frequency domain simulation for the cylindrical, hemi-spherical and oval geometries are as given in Fig. 8.

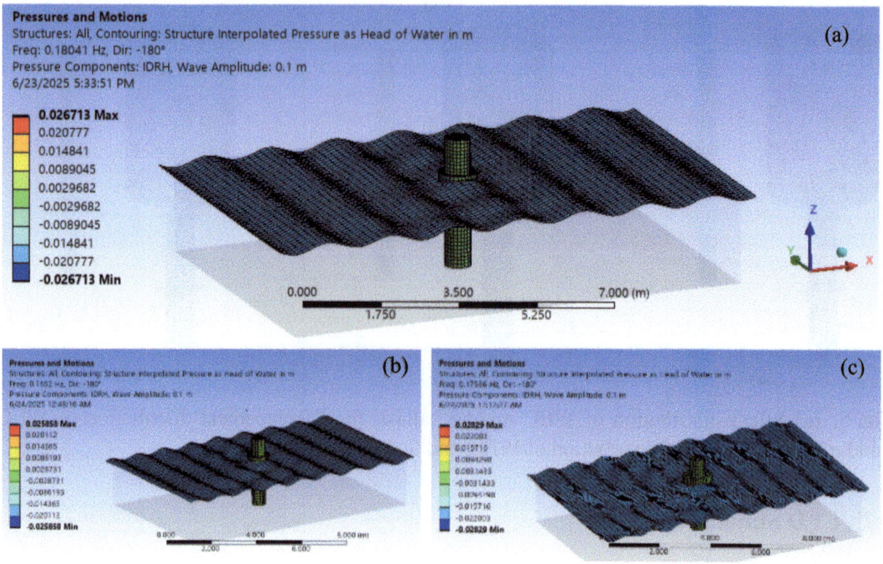

Fig. 8. Frequency domain analysis on, (a) Cylindrical, (b) Hemi-spherical, (c) Elliptical, floaters

Results for the heave (Position RAO) was generated through the simulation for the three floaters for the stiffness coefficient of the PTO equaling 600 N/m in frequency domain (Fig. 9).

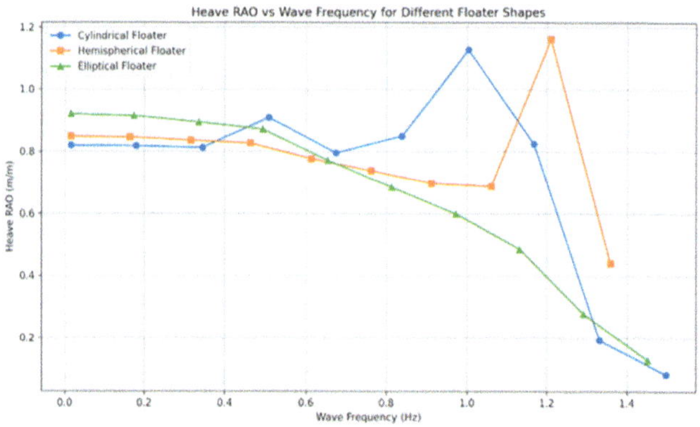

Fig. 9. Heave RAO vs wave frequency for different floater shapes

The cylindrical floater had the highest heave response, peaking at RAO 1.13 around 1.0 Hz making it ideal for energy generation while the hemi-spherical floater peaked slightly higher at RAO 1.15 around 1.25 Hz but had an overall lower response. The elliptical shape depicted least movement with a maximum of 0.92 with a gradual RAO decline. Quantitatively, the peak heave RAO of cylindrical floater is approximately 10%

higher than hemi-spherical floater and 20% higher than elliptical floater making it ideal for the WEC model.

In the **time domain analysis**, an irregular wave was defined using the JONSWAP spectrum to simulate realistic sea conditions. The wave period range was set between 5 s (start period) and 25 s (finish period), with a significant wave height of 0.2 m, representing calm seas. A gamma value of 2 sharpened the spectrum, while a peak period of 16 s concentrated energy in long-period swells, capturing realistic irregular wave behavior. Wave surface elevation is generated for 300 s based on the defined irregular wave condition (Fig. 10).

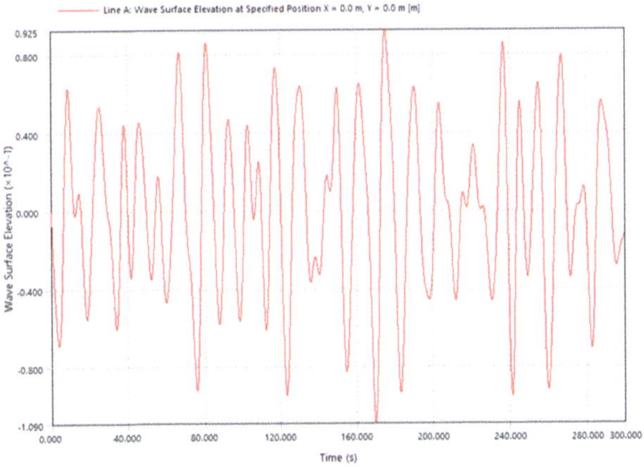

Fig. 10. Generated wave surface elevation for irregular wave conditions

The heave RAO variations of cylindrical, hemispherical, and elliptical floater shapes were analyzed under the above irregular wave conditions. The heave RAO responses of the three floater shapes show clear differences under the same irregular wave condition (Fig. 11). The cylindrical floater exhibits the largest heave motions with sharp peaks, due to its higher water-plane area and strong buoyancy variation, making it more sensitive to wave excitation. The hemispherical floater shows a smoother and more moderate response, where the curved geometry provides better hydrodynamic damping and reduces amplification. In contrast, the elliptical floater demonstrates the lowest and most stable heave response, with energy distributed more evenly, offering improved hydrodynamic efficiency and enhanced stability in long-period waves. Results for the upward force were also generated for the stiffness coefficient of the PTO equaling 600 N/m (Fig. 12).

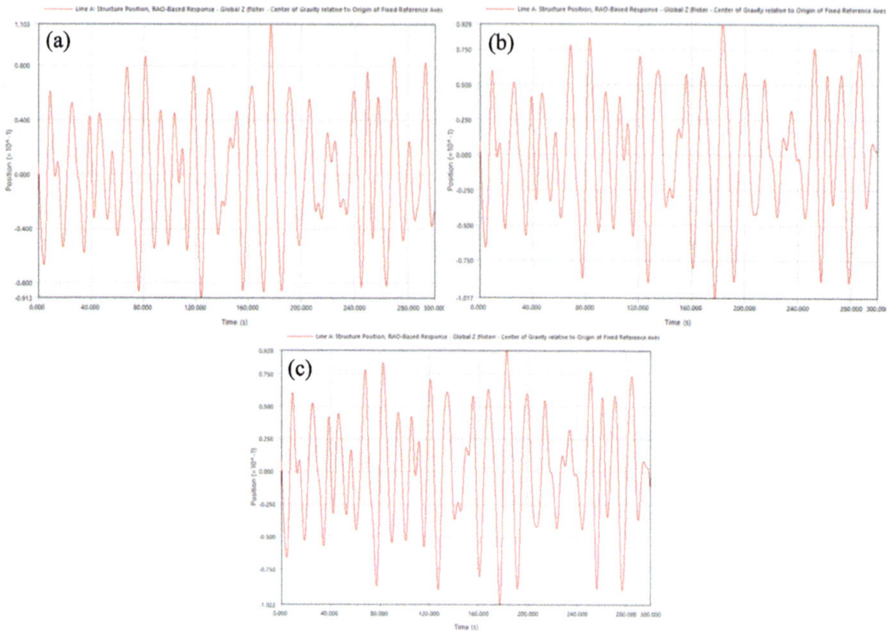

Fig. 11. Position RAO variation with irregular waves for, (a) cylindrical floater, (b) hemi-spherical floater, (c) elliptical floater

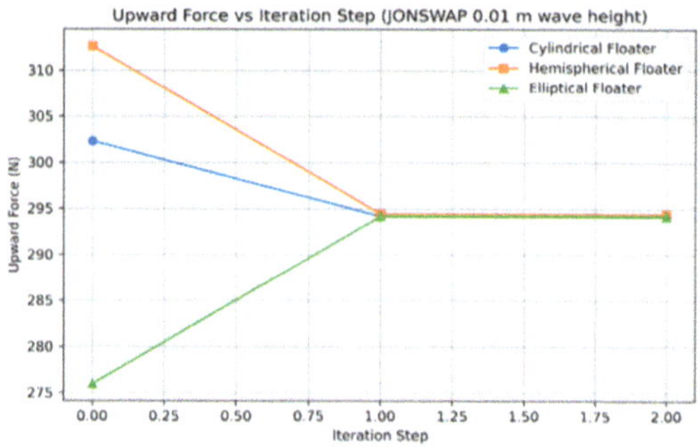

Fig. 12. Upward force vs iteration step variation for different floater shapes

4 Fabrication

The modeled WEC was fabricated for manual testing and real-world testing (Fig. 13). 210-l oil barrel was used for middle ballast and a part of a 750-l water tank was used as the floater. For the guide tube, a 6-inch PVC pipe was used. The motion rectifier used CSK12PP one-way bearings and HK1012 needle bearings coupled with the pinions. Pillow block bearings supported the pinion shaft. LM12UU linear bearings were used for the rack guides and the bottom guide. Oil seals were inserted to the top shaft to prevent water entry. For system control and data acquisition, an ESP32 microcontroller was interfaced with a voltage sensor, INA219 current sensor, and rotary encoder. Google Firebase was used as the backend for data storage while the front-end was made for real time data visualization.

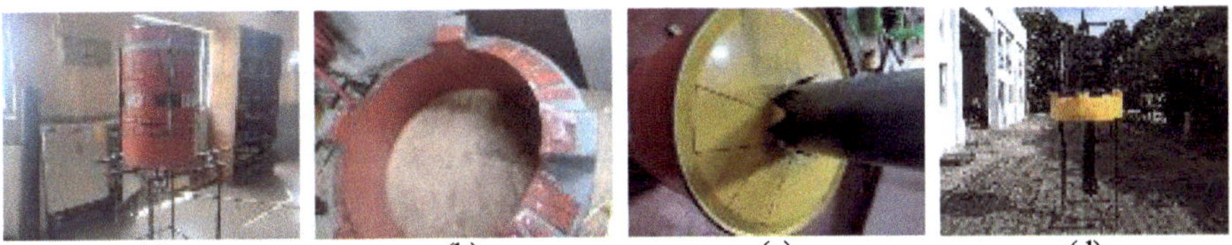

(a) (b) (c) (d)

Fig. 13. (a) Barrel as the ballast, (b) Floater fabrication, (c) Guide tube, (d) Final assembly

5 Manual Testing

The WEC model was tested in lab conditions by applying an upward force to the floater while the middle ballast was stationary. A maximum power of 16.997 W was observed. This may because failing to replicate exact sea conditions in laboratory scale (Fig. 14).

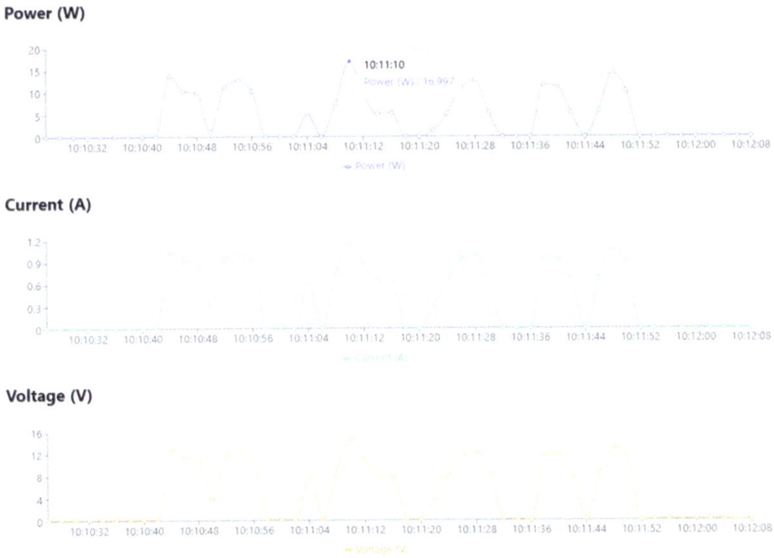

Fig. 14. Power, current and voltage variation graphs from manual testing

6 Conclusions

The study presents a novel double-buoy point absorber type WEC that suits for wave characteristics of Southern Sri Lanka. The design utilizes a rack and pinion-based PTO system analyzed through ISO 6336:2019 calculation method. The floater, ballast and other components were modelled and analyzed through FEM and CFD simulations.

Among tested floater types, the cylindrical floater demonstrated the best balance between heave response and stability and was selected as the optimal choice. The completely bottom-sealed design addresses a common failure mode in WECs by preventing water leakages and enhancing long-term reliability. Manual testing was conducted which depicted a power output of around 17 W. Real-world Sea trials are to be conducted.

Beyond technical contributions, the modeled WEC demonstrates a positive impact on the nation's energy demand by tapping into energy supplied by the ocean. The system contributes on improving wave energy technology in Sri Lanka reducing reliance on fossil fuels. As for future work, the fabricated WEC is to undergo real world sea-trails. Data gathered from these trials will contribute in refining the model.

References

1. International Energy Agency, Electricity Mid-Year Update - July 2024, https://www.iea.org/reports/electricity-mid-year-update-july-2024. Last accessed 18 June 2025
2. United Nations: What is renewable energy?, https://www.un.org/en/climatechange/what-is-renewable-energy. Last accessed 18 June 2025
3. Pearson, D.: What is the Importance of Renewable Energy Sources, https://wakelet.com/wake/oo5aONgVuw-QWGiJt6Mvv. Last accessed 18 June 2025
4. Electrical and Mechanical Services Department (EMSD) of the Government of the Hong Kong, ENERGY - Wave Energy, https://www.emsd.gov.hk/energyland/en/energy/renewable/wave.html. Last accessed 18 June 2025
5. Ocean Energy Europe, Wave energy, https://www.oceanenergy-europe.eu/ocean-energy/wave-energy/. Last accessed 19 June 202
6. The Liquid Grid, Wave Energy, https://theliquidgrid.com/marine-clean-tech-briefs/wave-energy-converters/. Last accessed 20 June 2025
7. Surf-Forecast, Sri Lanka Surf Map – Wind and Wave forecasts, https://www.surf-forecast.com/weather_maps/Sri-Lanka?hr=3&over=none&type=htsgw. Last accessed: 31 August 2025
8. Engström, J.: Hydrodynamic Modelling for a Point Absorbing Wave Energy Converter, Uppsala University (2011). Last accessed 21 June 2025
9. Feddersen, F.: Linear Surface Gravity Waves B, Energy Conservation, Energy Flux, and Group Velocity. In: Lecture notes for SIO/OCN 211A: Linear Ocean Waves. Scripps Institution of Oceanography, UC San Diego (2011). Last accessed 21 June 2025
10. Electronic Tutorials, Average Voltage of a Sinusoidal AC Waveform, https://www.electronics-tutorials.ws/accircuits/average-voltage.html. Last accessed: 31 August 2025

Development of a Theoretical Model to Predict Filtration Efficiency of Electret Based Masks

W. M. I. S. Jayaweera(✉) and D. Attygalle

Department of Materials Science and Engineering, University of Moratuwa, Moratuwa, Sri Lanka
isharasch.jaya@gmail.com

Abstract. Face masks have become one of the most important pieces of day-to-day equipment due to the recent COVID-19 viruses. However, doing experimental studies to observe the filtration behavior was not enough due to higher time consumption and technical difficulties. There have been lots of simulation studies done to reduce this matter. Most of these studies were done using 3-dimensional structures, and they needed higher computational power. In this study, we develop a 2-dimensional computational model to predict the filtration efficiency of the electret layer of an N95 face mask. 2-dimensional fiber structures were constructed using parameters such as packing density, fiber diameter, and fiber layer thickness. MATLAB and AUTOCAD were used to generate random 2-dimensional fiber structures, and COMSOL Multiphysics was used to simulate filtration mechanisms inside the electret filter. The incompressible Navier-Stokes and continuity equations were used to solve the velocity field, and the Lagrangi-an method was used to obtain particle trajectories inside the electret filter. Interception, inertial impaction, and electrostatic attraction were used as filtration mechanisms. Particle diameter sizes vary between 10 nm and 1 μm. The behavior of the filtration mechanisms under different conditions was simulated, and the simulated results are compared with the experimental results from literature. This study gives suggestions for finding the optimum filter characteristics for electret filters. It is highlighted that sticking probability shows a more significant effect for mechanical filtration than electrostatic filtration. Finally, this study recommends that using extreme conditions is not necessary to achieve required filtration efficiency.

Keywords: Filtration Efficiency · Electret based masks · Mask filtration simulation

1 Introduction

Electret materials can hold an electric charge for a long time. This charge stays on the surface or inside the material and creates a steady electric field [1]. Although the charge slowly fades over time, it lasts long enough to be useful in many industries. There are two main types of electret materials: polymers and inorganic materials [2]. Common plastic electrets include polypropylene (PP), polyethylene (PE), PTFE (Teflon), and PET. Among these, fluorinated plastics are especially popular because they can hold a

M. Siriwardhana et al. (Eds.): IERS 2025, PTT, pp. 161–170, 2026.
https://doi.org/10.1007/978-981-95-7984-6_15

charge for a longer time [3]. This ability to keep the charge is called charge stability, and it's very important for the material to work well. Inorganic materials usually handle heat better, but they can absorb moisture from the air, which makes them less reliable than plastic electrets in some situations.

One of the most important uses of electret plastics is air filters. These filters help remove harmful dust and particles from the air, which is important for protecting people's lungs. This became even more important during the COVID-19 pandemic when face masks and respirators were needed to keep people safe. Most industrial masks use filters made from charged plastic fibers like polypropylene, polystyrene, or PAN (polyacrylonitrile) [4]. These fibers get their charge either by rubbing or by being exposed to electric fields, which improves their ability to catch small particles.

When making respirators, designers must think about things like how well the filter works (efficiency), how stable it is, how easy it is to breathe through (pressure drop), and how long it lasts [5]. Electret filters clean the air using both mechanical filtering (physically blocking particles) and electrostatic filtering (attracting particles with electric charges). This study aims to create a model to better understand how these filters work. It will look at how filtration efficiency changes with fiber packing density, surface charge level, and air speed using computer simulations [6].

2 Methodology

2.1 Computational Domain

A 2D multi-fiber structure is used to represent the cross section of the filter layer. The objective of this simulation is to study the behavior of the filtration mechanism of electret filters. Selecting the computational domain size and generating the multi-fiber structure are very important processes of this study [7]. We have selected 100 μm $\times$ 100 μm as the domain of the 2D multifiber structure, considering the computational power and actual thickness of the filter. MATLAB code was developed to generate random structures of fibers. In this MATLAB code, domain size, packing density, and fiber diameter distribution were considered. This MATLAB code was used to generate the positions of fibers inside the filter domain (Fig. 1).

2.2 COMSOL Simulation

COMSOL Multiphysics software is used to generate the filter domain and simulate the filtration mechanisms. The generated 2D multifiber structure was imported to COMSOL Multiphysics, and boundary conditions were defined. The first step of the simulation is to generate 2D geometry, and it was already done using MATLAB and AUTOCAD software. Then we used three physics in COMSOL to create a filter model. Laminar flow, electrostatics, and particle tracing for fluid flow were the physics that were used in this simulation. The main parameters of the simulation were inlet air velocity, surface charge density of fibers, sticking probability, and particle diameter [8].

2.3 Fluid Flow Field and Electrostatic Field

Computing the fluid flow field according to the inlet air velocity is the next step of the simulation. Inlet air velocity and outlet pressure conditions are defined to generate the fluid flow field around the multi-fiber structure. In the 2D domain, it is hard to model the electrostatic field inside the electret filter [9]. Because the generation of the electrostatic field is highly dependent on the 3D orientation of the fiber arrangement. The electrostatic field within electret filters is represented either by surface charge density or by polarization density.

In this study, surface charge density was used to compute the electrostatic field around the multifiber structure. Surface charge density was defined around every fiber surface, and the inlet and outlet were defined as the ground. Figure 2 shows the generated potential field (Fig. 1) around the multi-fiber structure for 0.1 packing density and 10 μCm^{-2} [2].

2.4 Obtaining Particle Trajectories

After solving the fluid flow field and electrostatic field, particle trajectories were obtained using particle tracing for the fluid flow model. As mentioned in the literature part, to obtain particle trajectories, COMSOL uses the Lagrange method. We have already defined the initial velocity and position of the particles. Then, using the defined initial position and velocity, the next position of the particle was calculated for a very small time step. Then, repeating the same procedure, complete particle trajectories can be obtained. We have simulated the particle trajectories of the multiple particles at the same time [10] (Fig. 2). This was done while assuming that there is no particle-particle interaction. This electret filter model is only focused on the filtration mechanism, like Brownian diffusion, inertial impaction, coulombic capture, interception, and dielectrophoretic force only. Gravitational settling is not considered due to the small particle range. Symmetric conditions are used for walls that are connecting the inlet and outlet to reduce computational time.

2.5 Simulation Test Cases

The electret filter model is used to simulate several scenarios with different conditions. We have used 5 different packing densities and 6 different surface charge densities for this study. The above variations were mainly done under two types of simulations [11]: simulation of both mechanical and electrostatic filtration, and simulation of mechanical filtration alone. All the simulations were done using NaCl neutral particles. This particle is being used as a standard testing particle for N-95 face masks.

2.6 Simulate Mechanical and Electrostatic Filtration

In this case, under several surface charge densities and several packing densities, simulation was done to 0.3 μm NaCl particles [12]. The objective of this simulation is to find out the optimum design parameters to reach expected filtration efficiency (95%). Then, using identified optimum parameters as main parameters, simulation was done with different particle diameters. The objective of this simulation is to study the behavior of filtration efficiency with particle size [13]. Also, using above-optimum parameters, a simulation was done to study the filtration efficiency variation with sticking probability. For this simulation, only a 0.3 μm particle diameter was considered [14].

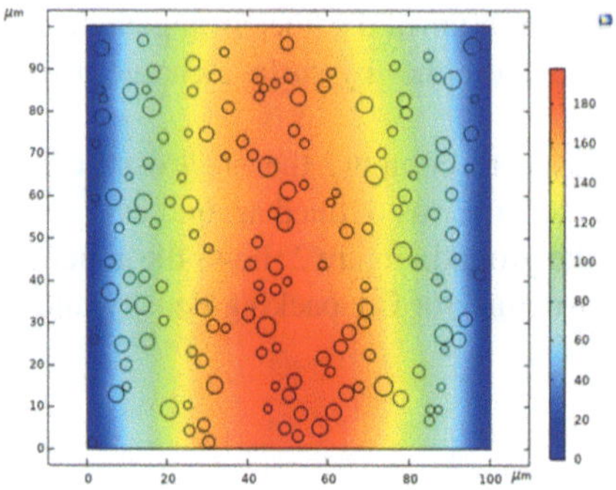

Fig. 1. Generated Electric Potential Field

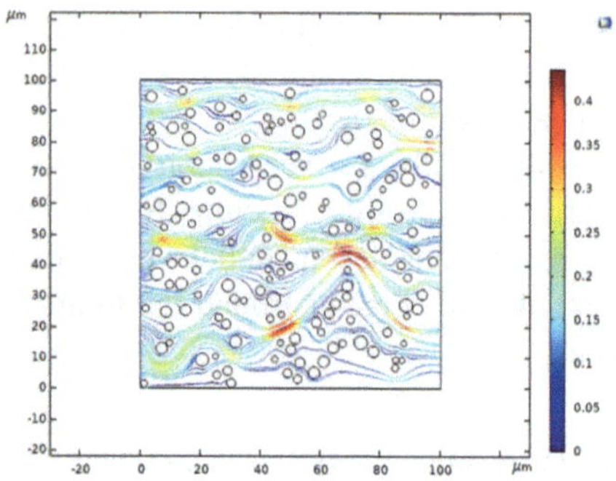

Fig. 2. Simulated Particle Trajectories

2.7 Simulate Mechanical Filtration

In mechanical filtration, only Brownian diffusion and drag force are considered. This case is done with 0.125 packing density and 0.1 ms^-1 inlet air velocity. The objective of this simulation is to study the mechanical filtration variation with different particle sizes [15].

2.8 Mesh Refinement Study

COMSOL mainly uses the FEA method to compute the Multiphysics simulations. The accuracy of FEA is highly dependent on the mesh size. Hence, it is important to do a mesh refinement study to reduce the difference between the exact solution and the approximated solution. Which means results should be independent of the mesh sizes [16]. It is also important to consider the computational time and computational power as well. In this project, a mesh refinement study was carried out using 7 mesh sizes. This mesh refinement study was done with two different parameters. One study was done with pressure drop across the filter, and another one was done with the average velocity of the air inside the filter. Packing density was chosen as 0.1, and the charge density of the filter was used as zero. When considering the average air velocity values, it is almost the same after the 22000 number of mesh elements. When considering the pressure drop across the filter, it is also having no significant difference after 22000 mesh elements [17]. Therefore, the finer mesh size was chosen for further analysis, and it also provides a good balance of computational accuracy and computational cost.

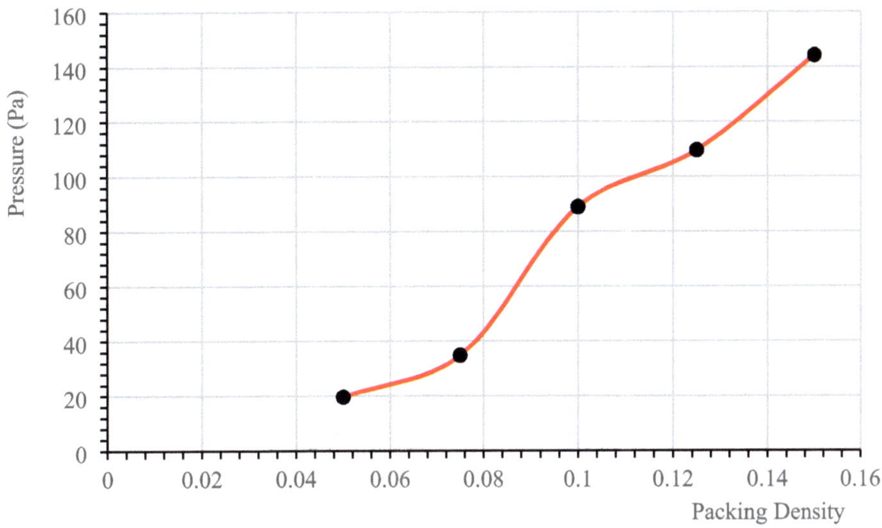

Fig. 3. Pressure Drop Across Electret Layer

3 Results and Discussion

As a start, pressure drop values for the suggested packing densities were obtained using the electret filter model. This is done to verify that the pressure drops across the filter for suggested packing density values are under the breathable zone. According to NIOSH standards, the pressure drop across the filter should be less than or equal to 250 Pa. Figure 3 shows the obtained pressure drop values for suggested packing density values. Pressure drop values for all the packing density values that are suggested in this study were within the NIOSH standards. Hence, further simulation was conducted using these values.

The filtration efficiency of the electret layer is analyzed with various surface charge densities and packing densities. Particle diameter is chosen as 0.3 μm for this simulation. It represents the MPPS for N95 face masks. Figure 5. Shows the filtration efficiency variation as a function of packing densities and surface charge densities for 0.3 μm particle size. Considered captured. Which means, for this simulation, the sticking coefficient is considered as 1.

Quality factor is another important parameter when choosing the filter characteristics. Hence, using the above obtained values, quality factors were also calculated for the above-mentioned various conditions as shown in Fig. 4.

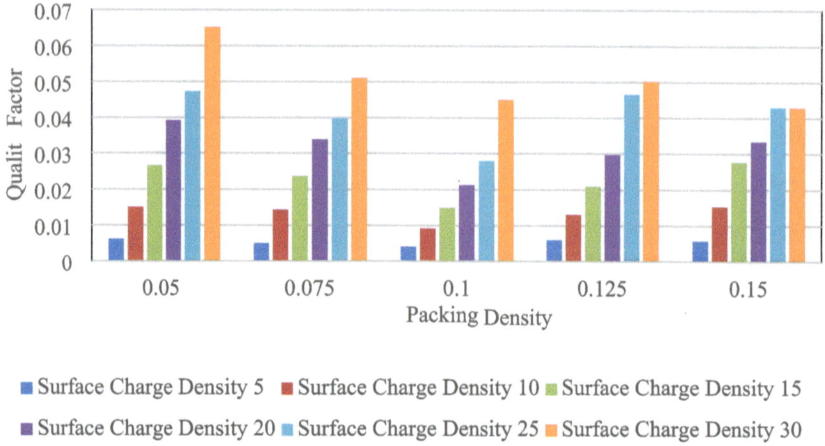

Fig.4. Quality Factor Variations with Different Filter Characteristics

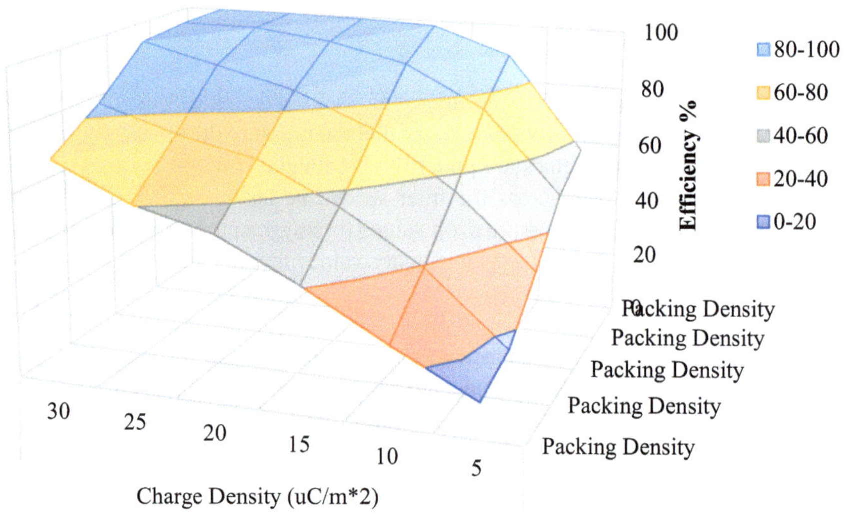

Fig. 5. The Filtration efficiency for NaCl (0.3 μm) particles as a function of surface charge density and fiber packing density

Figure 5 shows that capturing efficiency is increasing with the increasing of the charge density and packing density. Capturing efficiency exceeded 95% when the charge density was 20 $\mu Cm^{-}2$ at 0.125 packing density. Which means charge density of 20 $\mu Cm^{-}2$ at 0.125 packing density values can be considered as the optimum parameters to achieve expected filtration efficiency in an electret filter. Also, when considering the quality factor values for 0.125 packing density and 20 $\mu Cm^{-}2$, it shows the reasonable quality factor value. The importance of the above observation is that to achieve higher

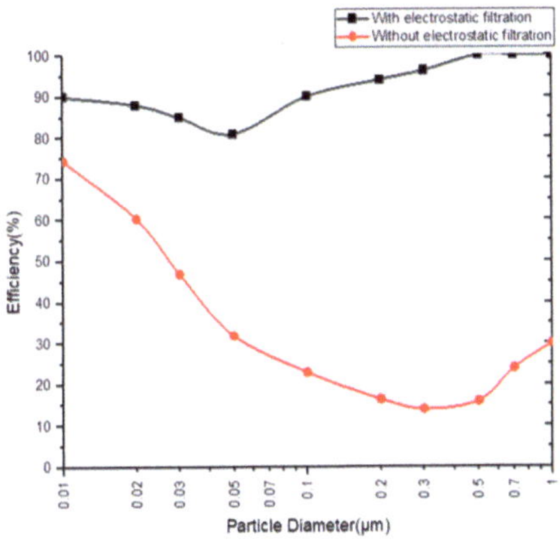

Fig. 6. Filtration efficiency variation with and without electrostatic interaction for 1 ms^{-1} inlet velocity

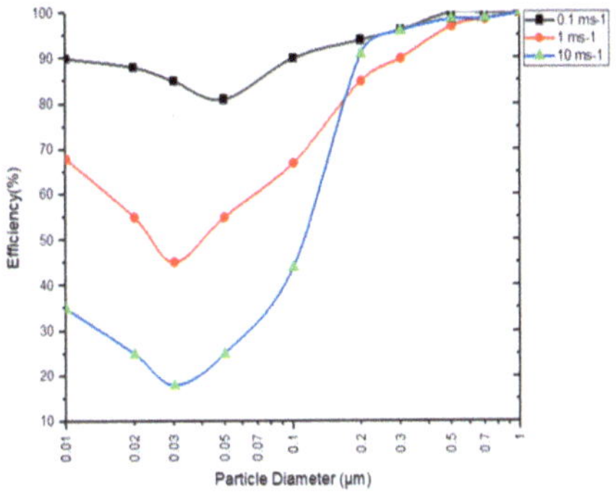

Fig. 7. Filtration efficiency variation with different inlet velocities

capturing efficiencies, it does not require using extreme values as filter characteristics. Achieving higher capturing efficiency can be done only by surpassing the required marginal values. Hence, using the above charge density and packing density values as a base value, mechanical filtration efficiency and electrostatic filtration efficiency for different particle sizes were obtained.

The simulation is done without considering electrostatic filtration and with considering electrostatic filtration for 20 μCm^{-}2 surface charge density and 0.125 packing density values at 0.1 ms^{-}1 inlet velocity. Figure 6. Shows the capturing efficiency of mechanical filtration and total filtration as a function of particle diameters.

The electret layer is the middle layer of the N95 face mask, and it contains the smallest fibers when compared to the inner and outer layers. Hence the curvature of the air streamlines near the fibers in the electret filter is larger than the other layers. Because of the large curvature of the air streamlines, it can capture particles of all sizes more efficiently by interception mechanism, and Lee and Liu also predicted this behavior. When the particle size is decreased, it contains low inertia, and due to the low inertia, inertial impaction and interception are less efficient. Figure 6 shows that at 0.1 ms^{-}1, MPPS is shifted from 0.3 μm to 0.05 μm with the consideration of the electrostatic interaction. The lowest capturing efficiency for mechanical filtration is 14%, and with the involvement of electrostatic filtration, the lowest capturing efficiency is increased to 81%. Also, this graph shows the importance of electrostatic interactions for the higher filtration efficiency.

Simulation was done using a developed model with various inlet velocities 0.1ms^{-1}, 1ms^{-1} and 10 ms^{-}1. These velocities are carefully taken to represent the important respiratory events (breathing, -0.1 ms^{-}1; talking, -1 ms^{-}1; and coughing, -10 ms^{-}1). The founded optimum surface charge density value (20 uCm^{-}2) and packing density (0.125) values were used for these simulations. Figure 7 shows the variation of filtration efficiency of the electret filter with particle diameter and inlet air velocity. Particles have lower kinetic energy at low air inflow velocities. Hence, particles can easily be deflected

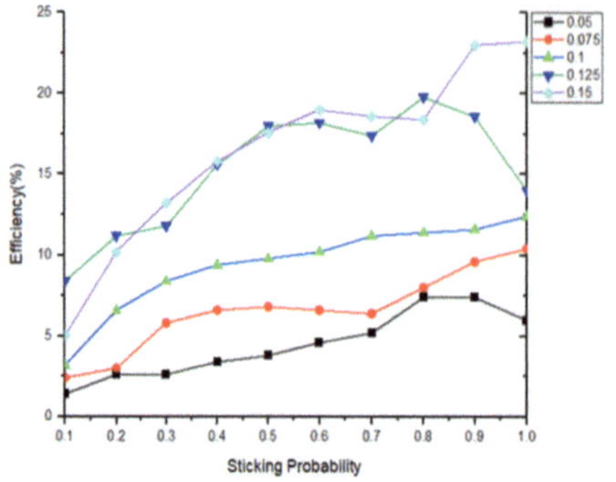

Fig. 8. The mechanical filtration capturing efficiency plotted as function of sticking probability for various surface charge densities and different packing densities

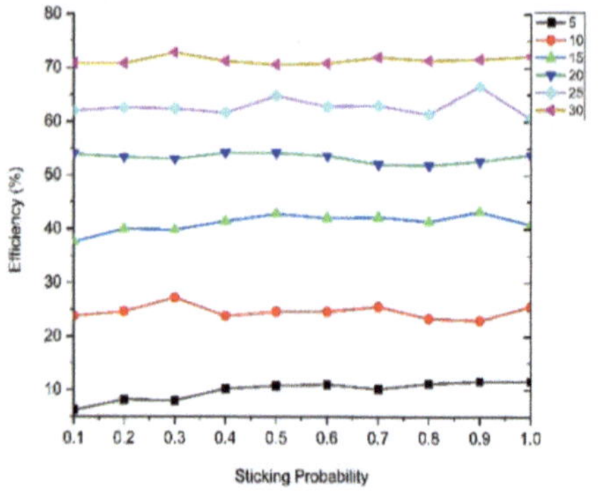

Fig. 9. The capturing efficiency plotted as function of sticking probability for various surface charge densities for 0.05 packing density

by dielectrophoretic force. Therefore, when the air inflow velocity is higher, capturing efficiency becomes lower. Which means the capturing efficiency of electret filters is more efficient at low air velocities.

Electrostatic filtration is essential for high filtration efficiency of electret filters. Here, 81% is the minimum capturing efficiency for this electret filter at 0.1 ms^{-}1 inlet air velocity. When comparing two graphs, at 0.1 ms^{-}1 minimum filtration efficiency is increased to 81% from 14%; at 1 ms^{-}1, minimum filtration efficiency is increased to 45.2% from 4%; and at 10 ms^{-}1, minimum filtration efficiency is increased from 18% to 5.2% with the involvement of electrostatic filtration. Also, this graph shows that these filter characteristics are not suitable for different respiratory conditions such as talking and coughing. These conditions need to be carefully analyzed further.

Filtration efficiency variation with various surface charge densities and packing densities was analyzed as a function of the sticking probability. Fg.8 shows the mechanical filtration efficiency variation with different packing densities as a function of sticking probability. It clearly shows that with the increase of the packing density; mechanical filtration is also increased. This effect is also studied by Soltani et al., who concluded that solid volume fracture has a positive correlation with filtration efficiency. When the sticking probability increases, filtration efficiency also increases for every packing density value.

It is already clear that when the packing density increases, total filtration efficiency is also increased for the relevant surface charge densities. Figure 9 shows that, for selected packing density, filtration efficiency is increased with the increase of the surface charge densities. This graph show that when the surface charge is applied to the filter, filtration efficiency does not change significantly with the sticking probability. Which means when the electrostatic interaction involves filtration, efficiency does not depend on sticking probability.

4 Conclusions

In this study, a 2-dimensional theoretical model of the electret layer was developed, and the Lagrange method was used to obtain particle trajectories inside the electret filter. The developed model was used to study filtration mechanisms behavior with various surface charge densities and fiber packing densities. Also, the effect of sticking probabilities with the electrostatic and mechanical filtration was observed. The findings reveal that achieving higher filtration efficiency does not require extreme parameters such as high surface charge density or high fiber packing density. Furthermore, the most penetrating particle size (MPPS) can be shifted from larger to smaller particle diameters, with the MPPS in this case shifting from 0.3 μm to 0.05 μm. The results also indicate that an increase in air velocity leads to a decrease in filtration efficiency, highlighting the need for more careful analysis of this phenomenon. When electrostatic filtration is excluded, capturing efficiency exhibits considerable variation with sticking probability, whereas electrostatic filtration itself shows no significant relationship with sticking probability.

References

1. Thyssen, A.: Charge Distribution and Stability in Electret Materials, DTU Nanotech (2016)
2. Siag, A.M., Tennal, K.B., Mazumder, M.K.: Determination of fiber charge density of electret filters. Part. Sci. Technol. **12**(4), 351–365 (1994). https://doi.org/10.1080/02726359408906661
3. Yim, W., et al.: KN95 and N95 Respirators Retain Filtration Efficiency despite a Loss of Dipole Charge during Decontamination. ACS Appl. Mater. Interfaces 12 (2020). https://doi.org/10.1021/acsami.0c17333
4. Razavi, M.: A Comprehensive Model for Filtration and Adsorption of Intermediate Nanoparticles (2020). Accessed: 17 Jun. 2025. https://uwspace.uwaterloo.ca/bitstream/10012/16351/1/Razavi_Maryam.pdf
5. Thakur, R., Das, D., Das, A.: Electret Air Filters. Sep. Purif. Rev. - SEP PURIF REV 42 (2012). https://doi.org/10.1080/15422119.2012.681094
6. Van Turnhout, J., Adamse, J.W.C., Hoeneveld, W.J.: Electret filters for high-efficiency air cleaning. J. Electrost. **8**(4), 369–379 (1980). https://doi.org/10.1016/0304-3886(80)90057-1
7. Hossain, E., et al.: Recharging and rejuvenation of decontaminated N95 masks. Phys. Fluids **2**(9), 093304 (2020). https://doi.org/10.1063/5.0023940
8. Cai, R., Zhang, L.Z., Bao, A.B.: PM collection performance of electret filters electrospun with different dielectric materials-a numerical modeling and experimental study. Build. Environ. 131 (2018). https://doi.org/10.1016/j.buildenv.2017.12.036
9. Tsai, P.P., Schreuder-Gibson, H., Gibson, P.: Different electrostatic methods for making electret filters. J. Electrost. **54**(3), 333–341 (2002). https://doi.org/10.1016/S0304-3886(01)00160-7
10. Willeke, K., Grinshpun, S., Donnelly, J., Coffey, C.: Performance of N95 Respirators: Filtration Efficiency for Airborne Microbial and Inert Particles. Am. Ind. Hyg. Assoc. J. **59**, 128–132 (1998). https://doi.org/10.1080/15428119891010389
11. Bandi, M.: Electrocharged facepiece respirator fabrics using common materials (2020). https://doi.org/10.48550/arXiv.2011.13642
12. Jiang Wei(姜伟), C.C.S.: The aerosol penetration through an electret fibrous filter. Chin. Phys. B **15**(8), 1864–1870 (2006). https://doi.org/10.1088/1009-1963/15/8/039

13. Ogbuoji, E., Zaky, A., Escobar, I.: Advanced research and development of face masks and respirators pre and post the coronavirus disease 2019 (COVID-19) pandemic: a critical review. Polymers **13**, 1998 (2021). https://doi.org/10.3390/polym13121998
14. Schober, C.: Modeling and Simulation of Cabin Air Filtration with Focus on Electrostatic Effects. Universität Stuttgart (2019). https://files.core.ac.uk/download/pdf/225600991.pdf
15. Kilic, A., Russell, S., Shim, E., Pourdeyhimi, B.: The Charging and Stability of Electret Filters. In: Fibrous Filter Media, pp. 95–121 (2017). https://doi.org/10.1016/B978-0-08-100573-6.00025-3
16. Kim, M.M., Zydney, A.L.: Effect of electrostatic, hydrodynamic, and Brownian forces on particle trajectories and sieving in normal flow filtration. J. Colloid Interface Sci. **269**(2), 425–431 (2004). https://doi.org/10.1016/j.jcis.2003.08.004
17. Lee, J., Kim, J.: Material Properties Influencing the Charge Decay of Electret Filters and their Impact on Filtration Performance. Polymers **12**(3) (2020). https://doi.org/10.3390/polym12030721

Development of Multi Level Transformerless Inverter for Grid Integration of Solar Photovoltaic Power Plant

Rattan Kumar Venkatesan(✉) and Chitra Lakshmikanthan

Aarupadai Veedu Institute of Technology, Vinayaka Mission's Research Foundation – DU, Tamil Nadu, India
rattan.eee@gmail.com, chitra@avit.ac.in

Abstract. At present due to power scarcity and technological evolution installation of grid connected Solar Photo Voltaic Power Plant (SPVPP) is at the peak. The main loss while connecting a SPVPP to the grid is the core loss due the transformer. Transformerless inverters are the latest approach to minimize the losses in a SPVPP. The installation of grid-connected solar photovoltaic power plants (SPVPPs) is at its height right now due to power shortages and technological advancements This paper proposes an analysis of a seven level inverter. This includes analysis using different PWN technique. The technique used to test the multi-level inverter are Nearest Level Modulation (NLM), Phase Opposition Disposition (POD), Phase Disposition (PD) and Alternate Phase Opposition Disposition using MATLAB to get an appropriate result. This topology can generate low total harmonic distortion which will reduce the number of switches used which would certainly reduce the heat emission, it reduces the cost of the circuit and it will also reduce the installation space. Seven voltage levels, lower harmonic distortions, and higher V_{rms} values are all features of the suggested architecture. The outcomes of the simulation have been applied to the hardware to confirm its viability and efficiency. MATLAB/SIMULINK software is used to simulate this inverter.

Keywords: Multilevel inverter · Switched capacitor. Leakage current · Transformerless inverter · THD · NLM · PD · POD & APOD techniques

1 Introduction

In the present day, Renewable energy sources play major role in power generation such as wind, biomass, solar, etc. Power can be generated from solar and fuel cell and the energy is directly given to the grid for better utilization of renewable energy sources [1, 2]. A multilevel inverter is a DC – AC converter that is equipped to produce the desired number voltage levels. It plays important role in high-power applications and various medium voltages like ac transmission system devices, HVDC, renewable energy systems and variable frequency devices [3]. The advantages of multilevel converters are low Electromagnetic Interference, low dv/dt stress, no LC filter required and less Total Harmonic

M. Siriwardhana et al. (Eds.): IERS 2025, PTT, pp. 171–185, 2026.
https://doi.org/10.1007/978-981-95-7984-6_16

Distortion (THD) [4]. There exist three types of conventional multilevel inverters (MLI): Diode Clamped Multilevel Inverter, Flying Capacitor (FC) MLI and Cascaded H-Bridge (CHB) MLI. Although these topologies are commonly used, the number of components used increases with the increase in the number of output voltages required [5]. The major shortcoming the number of diode switches increases with increase in the number of output levels and the size of the inverter layout becomes big [6]. To overcome these drawbacks of the conventional topologies, a new novel MLIs were introduced which has reduced number of power electronic components. Many researchers are developing new MLIs using novel topologies to reduce the number of power electronic components and reduce its Total Harmonic Distortion (THD). A new single-phase multilevel inverter topology was proposed with minimum number of IGBT used and a comparison of conventional and other novel topologies were done in terms of DC output voltages, THD and reduced number of DC sources [7]. Another paper was proposed to enhance the output voltage. The proposed technique was operated at fundamental switching frequency, which produced high RMS voltage, and less Total Harmonic Distortion for lesser number of output voltage level. These were analyzed with different DC offset values [8]. Another switched diode topology in Reference [9] was presented, which demands that the DC link capacitors be connected in series to a single DC voltage source so that the output number of voltage levels be maximized. The full bridge converter switches in these topologies have high blocking voltage stresses present across them. References [10, 11] contain a method of doubling the voltage by connecting a series of half bridge circuits in the network. These methods require a higher number of switching devices and DC voltage sources to produce higher number of voltage levels. Anew method of multilevel inverter is proposed in this paper. The 7-level multilevel inverter is designed, and its simulation is done using various PWM. These modulation techniques produce low Total Harmonic Distortion (THD) than the conventional switching frequency scheme, but the gate pulse generation becomes more complex with the increase in number of levels except in NLM. PD, POD and APOD are used for low or medium power applications. ONLM technique is not recommended for lower levels of voltage output as it introduces considerable amount of THD in the output. In this paper, the DC offset value is also considered to reduce the output voltage THD interference and produce a higher RMS value and balancing the DC-link capacitor through external circuit using references [12, 13].

2 Proposed Seven Level Inverter Topology

Figure 1 shows the basic model of the multilevel inverter. It is an equal voltage Cascaded H-Bridge. It consists of three DC-link capacitors & one independent DC source, which are connected, in parallel to the source. Both the lower and upper capacitors are tapped via switches - S1 and S2. The capacitor (C2) gives current through diodes - D1 and D2. F11 and F14 are used to reverse the polarity of both voltage and current path through which the current passes to the load. Due to symmetry, it produces maximum of seven level voltages as an output [14–18]. The switch pairs (F11, F13) and (F12, F14) are not turned on at the same time in order to prevent any short circuiting.

The different modes of operation are shown in Fig. 2 and the switching sequence for the basic predesigned model is given in Table 1.

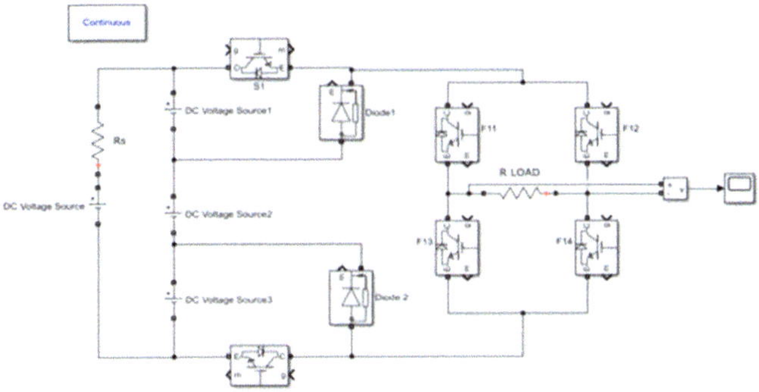

Fig. 1. Circuit configuration of proposed Basic model

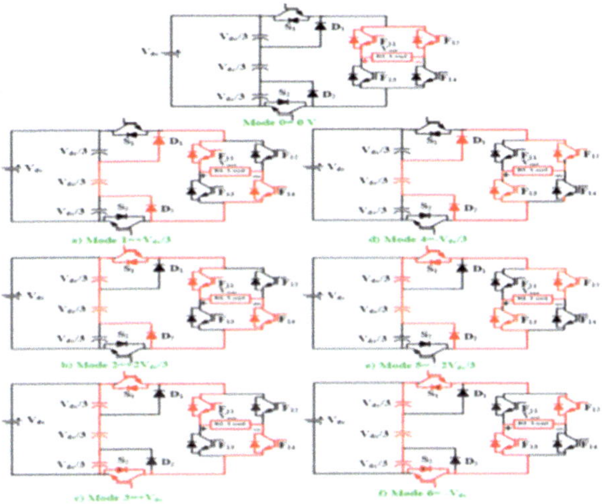

Fig. 2. Various modes of operations of predesigned model

Table 1. Switching Sequence

LEVEL	S1	S2	F11	F12	F13	F14	DIGITAL CODE
1	0	0	1	0	0	1	001
2	1	0	1	0	0	1	011
3	1	1	1	0	0	1	111
-1	0	0	0	1	1	0	001
-2	1	0	0	1	1	0	011
-3	1	1	0	1	1	0	111

3 Analysis Using NLM Technique

The Nearest Level Modulation technique does not use triangular wave as its carrier wave [19–21]. It calculates the duty cycle for each phase and switching states for each phase of the converter. In this method, carrier signal DC is compared with the reference signal sine wave which is called nearest level pulse width modulation. Figure 3(a) shows the Simulink circuit diagram of the 7-level multilevel using NLM technique and Fig. 3(b) shows the voltage waveform. Figure 4(a) Shows the current waveform and Fig. 4(b) Shows the THD and results are tabulated in Table 2.

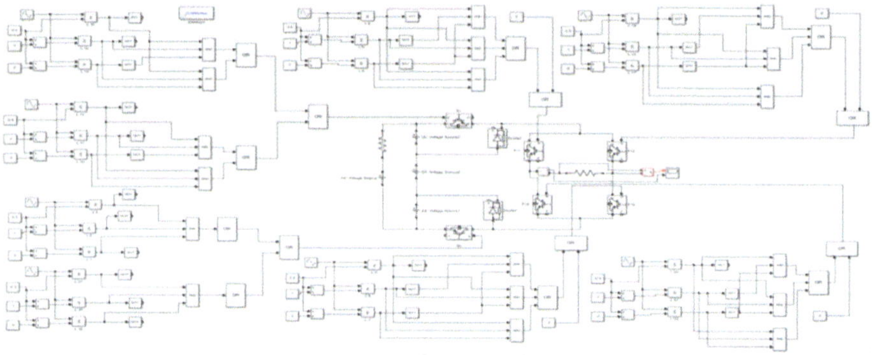

Fig. 3a. Simulink Circuit using NLM method.

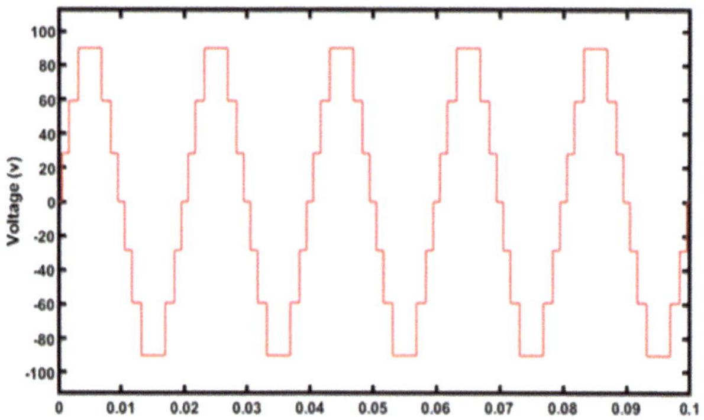

Fig. 3b. Voltage waveform using NLM technique, $V_{rms} = 90V$

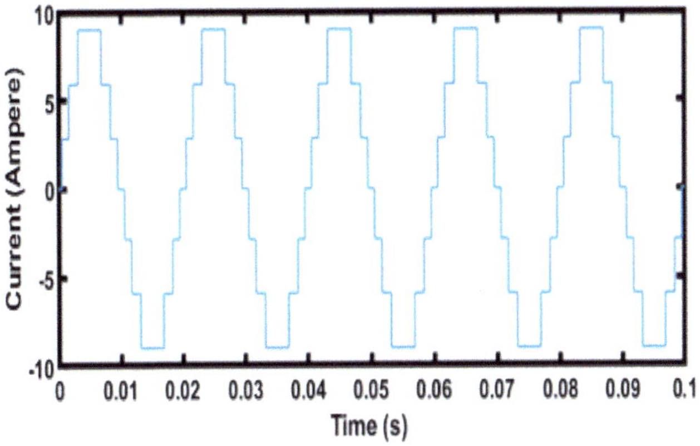

Fig. 4a. Current waveform using NLM technique, $I_{rms} = 90Amps$.

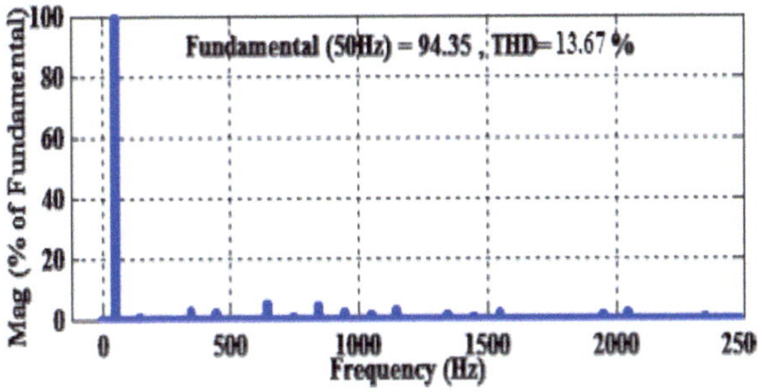

Fig. 4b. THD analysis of NLM voltage waveform. Lowest THD values obtained at 0.4 DC offset and 95% MI

Table 2. NLT THD analysis table

	DC Offset	Modulation Index 100%	Modulation Index 95%	Modulation Index 90%
		THD %	THD %	THD %
1	0.5	15.53	16.88	19.06
2	0.45	14.78	15.97	17.98
3	0.4	14.64	13.67	17.13
4	0.35	14.02	14.96	16.10

4 Analysis Using PD (Phase Disposition) Technique

In Phase disposition (PD) PWM techniques [1, 2, 22, 23], inverters with 'n' number of levels has (n-1) carrier signals which are all in phase. This is a carrier-based implementation where carrier signals are in phase, but the level is shifted. In this method the levels are generated by comparing the reference.

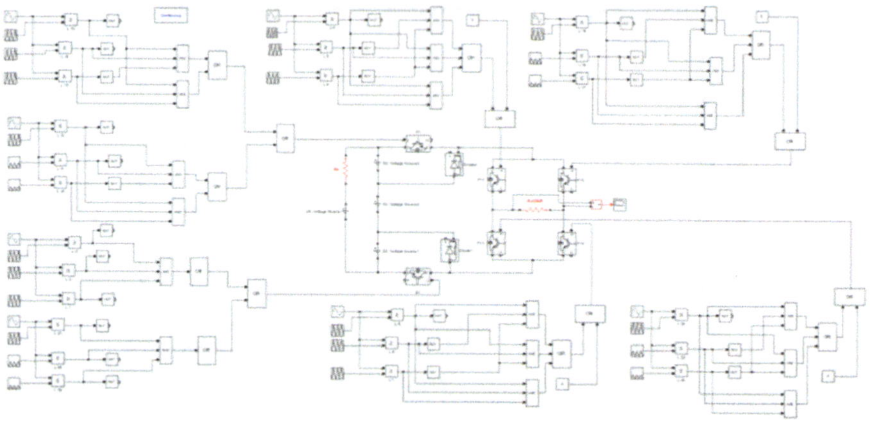

Fig. 5. Circuit using PD method

Sinusoidal signal with the carrier wave which is a triangular repeating sequence wave that has been assigned the magnitude values based on the levels. NLM technique differs from PD technique with respect to DC offset.

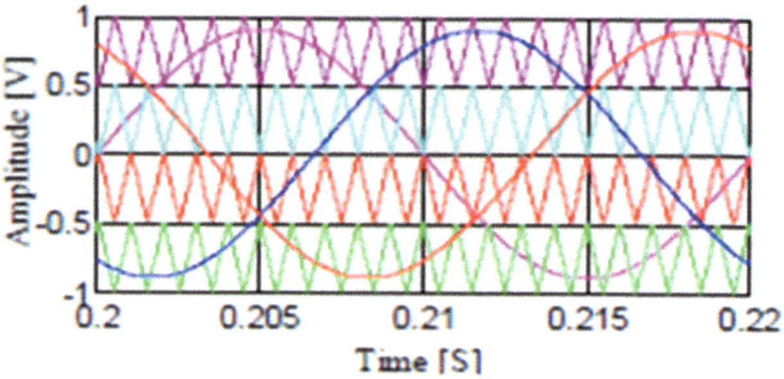

Fig. 6. Generalized waveform of PD technique

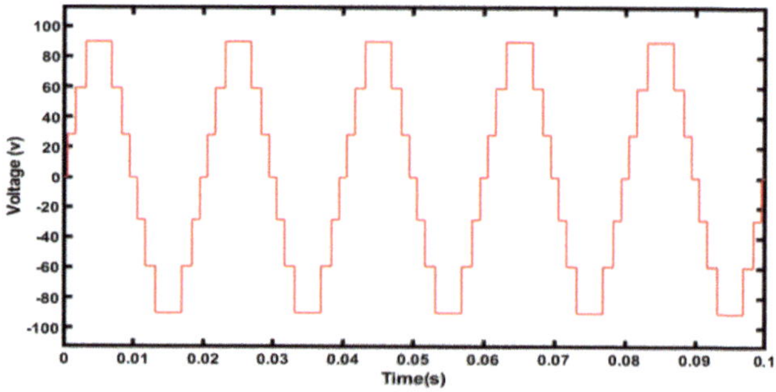

Fig. 7a. Voltage waveform using PD Technique s $V_{rms} = 90V$.

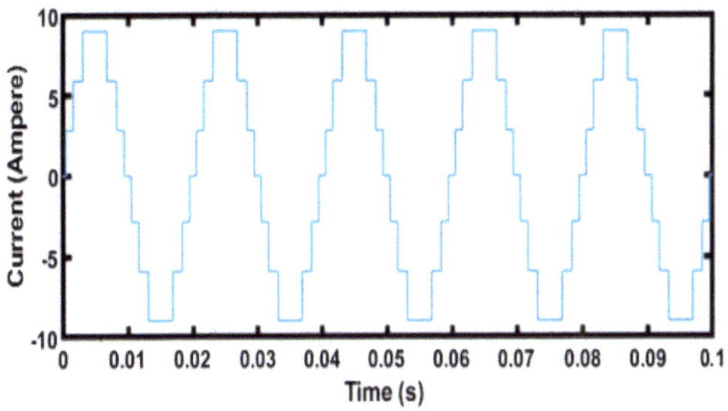

Fig. 7b. Current waveform using PD Technique s $I_{rms} = 9Amp$

In the Fig. 5, the PD technique is done to reduce the THD value in comparison to conventional NLM technique. Figure 6 denotes Generalised waveform of PD technique. In this circuit, triangular wave signal is given as carrier signal of different values [0 1 0], [1 2 1] and [2 3 2]. The output values in positive half cycle and the output values in negative half cycle are [-1 0 -1], [-2 -1 -2] and [-3 -2 -3] having reference signal with magnitude 3.5. And results are tabulated in Table 3 and the results are shown in terms of voltage, current & THD in Fig. 7(a), Fig. 7(b), Fig. 7(c).

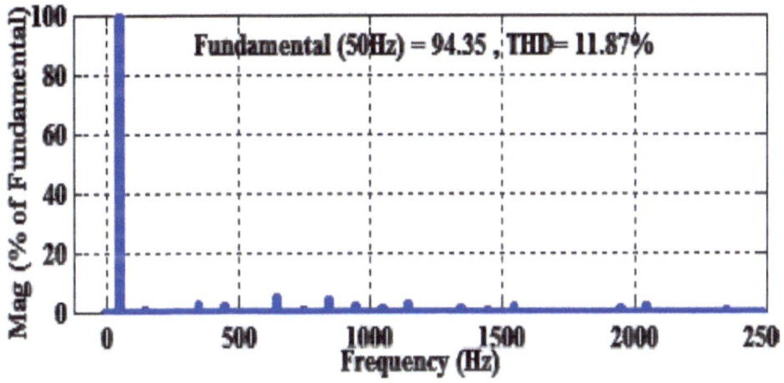

Fig. 7c. THD analysis of NLM voltage waveform. Lowest THD values obtained 95% MI.

Table 3. PD THD analysis table

S.no	Modulation Index	THD %
1	100%	13.69
2	95%	11.87
3	90%	12.90

5 Analysis Using POD (Phase Opposition Disposition)

In this type of PWM modulation technique Figs. 8 and 9 represents the circuit using POD method, each and every carrier waveform180 degree out of phase are below the 0 reference and those which are above the zero reference are in phase. The following are the rules for POD modulation technique in case when the number of levels in the output waveform required are, n = 7 are n –1 = 6, all the carrier waveforms which are 180 degree out of phase are below the 0 reference and those which are above the zero reference are in phase.

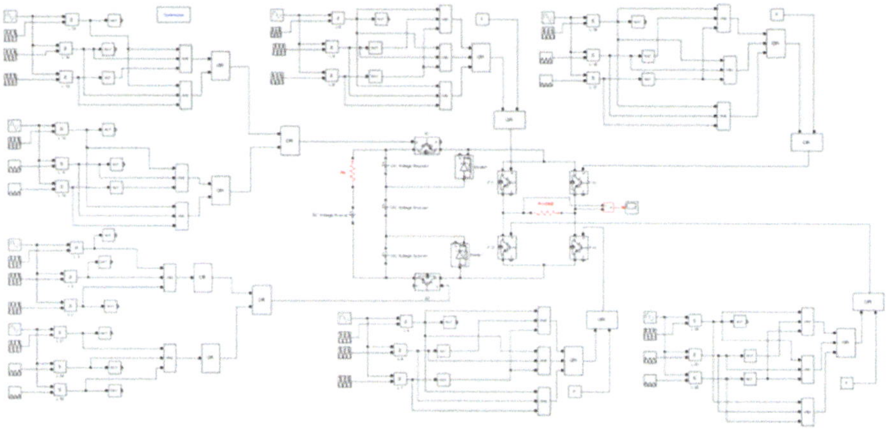

Fig. 8. Circuit using POD method

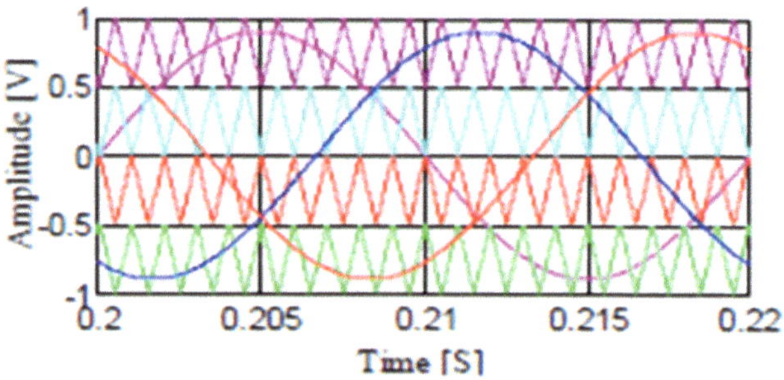

Fig. 9. Generalised waveform of POD technique

In the circuit given in Fig. 8, the output values are changed in negative half cycle [0 -1 0], [-1 -2 -1] and [-2 -3 -2] and in positive half cycle it is same as PD technique, from this it is easily understood that as per the definition of POD technique, the carrier signal below zero will shift 180 degree out of phase thus in POD technique, the negative half cycle value are changed with respect to PD as in PD the negative half cycle values were [-1 0 -1], [-2 -1 -2] and [-3 -2 -3]. And results are tabulated in Table 4. Results are shown in terms of voltage, current & THD in Fig. 10(a), Fig. 10(b) and Fig. 10(c).

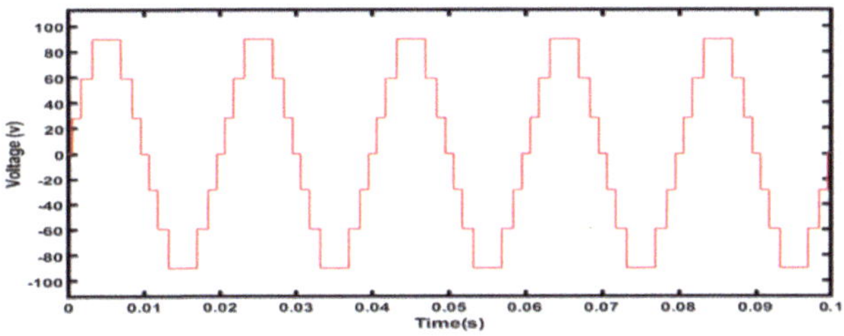

Fig. 10a. Voltage waveform using POD Technique

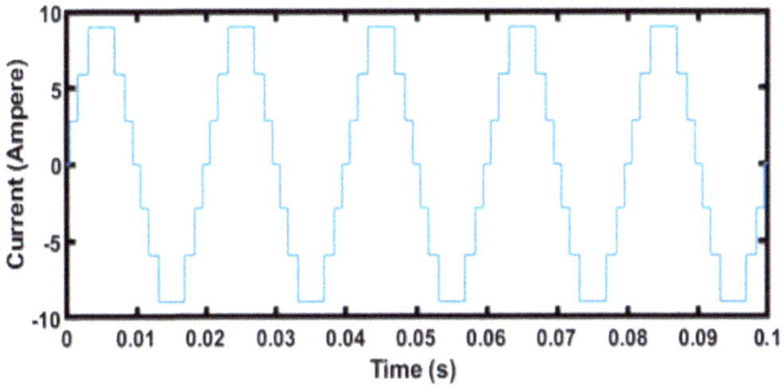

Fig. 10b. Current waveform using POD Technique

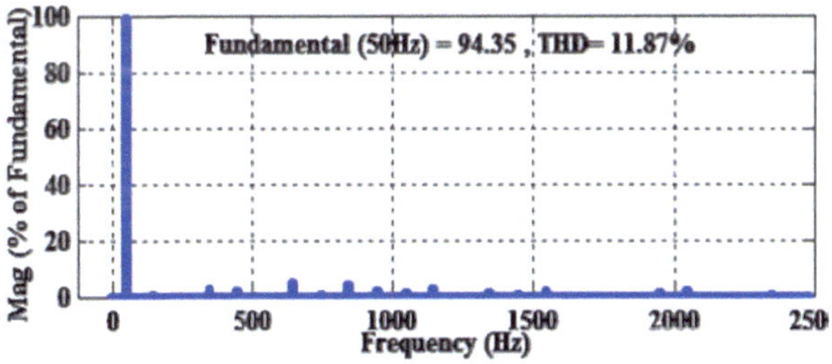

Fig. 10c. THD analysis of NLM voltage waveform. Lowest THD values obtained 95% MI

Table 4. POD THD analysis $I_{rms} = 9Amp$

S.no	Modulation Index	THD %
1	100%	13.69
2	95%	11.87
3	90%	12.90

6 Analysis Using APOD (Alternate Phase Opposition Disposition)

In this type of PWM modulation technique, each and every carrier waveform is 180 degrees out of phase with its neighbour carrier wave. Since the POD and APOD PWM techniques are the same in reference of the seven-level multilevel inverter, a seven-level inverter is expected to narrate regarding APOD PWM modulation technique. The following are the rules for APOD modulation technique in case when the number of levels in the output waveform required are, n = 7 are n −1 = 6, all the carrier waveforms which are 180 degrees out of phase are below the 0 reference and those which are above the zero reference are in phase.

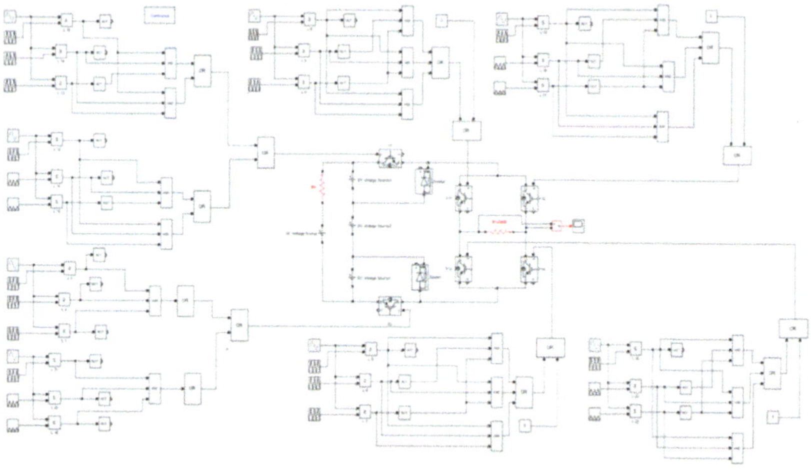

Fig. 11. Circuit using APOD method

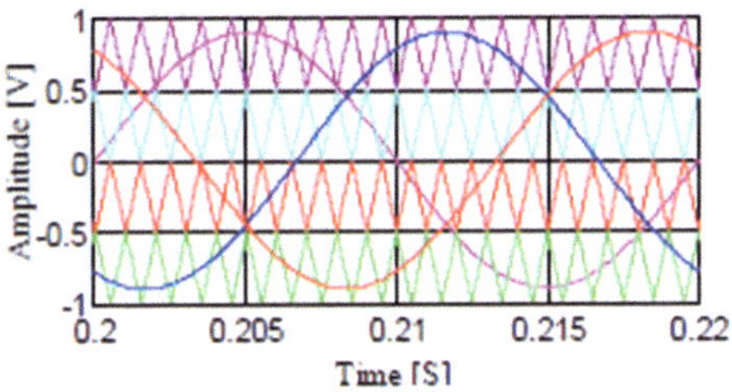

Fig. 12. Generalised waveform of APOD technique

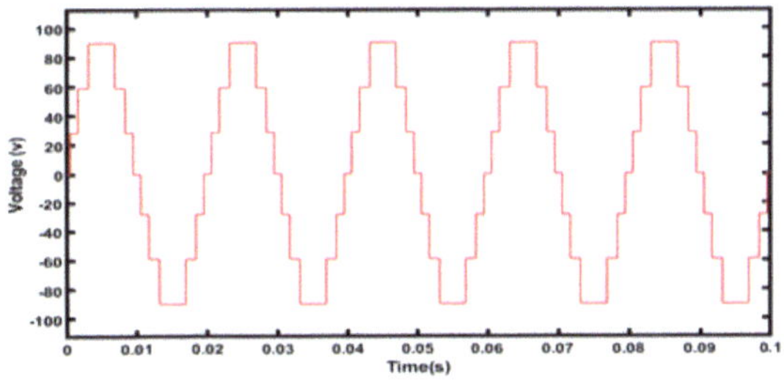

Fig. 13a. Voltage waveform using APOD Technique, $V_{rms} = 90V$

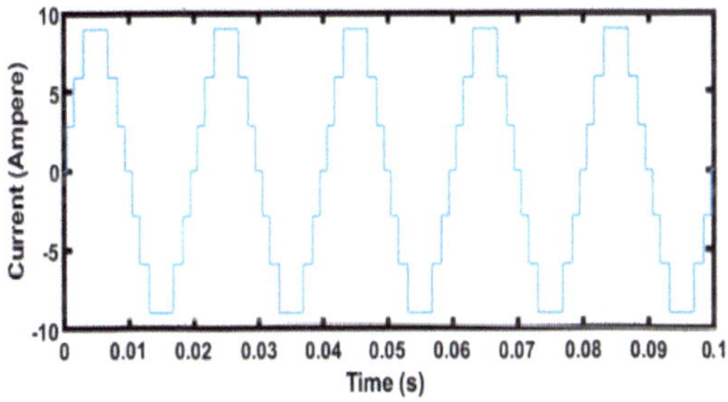

Fig. 13b. Current waveform using APOD Technique, I_rms = 9 Amps

In this circuit shown in Figs. 11 and 12 the output values in positive half cycle are [0 1 0], [2 1 2] and [2 3 2] and in negative half cycle the output values are [-1 0 -1], [-1 -2 -1] and [-3 -2 -3]. From this, it is easily understood that the carrier signals are shifted 180 degrees alternately. This is difference of APOD with respect to PD and POD. And the results are tabulated in Table 5. Results are shown in terms of voltage, current & THD in Fig. 13(a), Fig. 13(b) and Fig. 13(c).

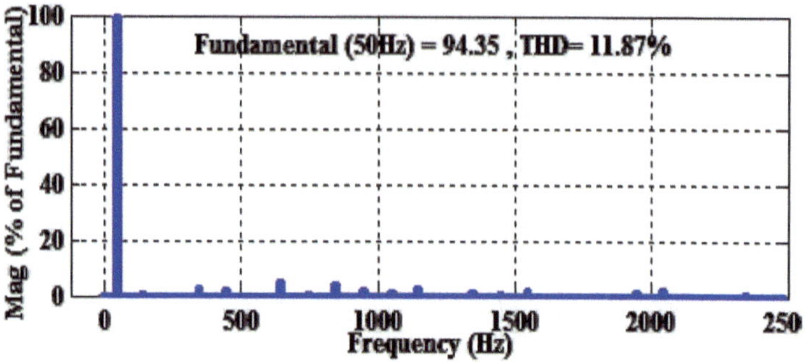

Fig. 13c. THD analysis of APOD voltage waveform. Lowest THD values obtained at 95% MI

Table 5. APOD THD analysis

S.no	Modulation Index	THD %
1	100%	12.05
2	95%	11.87
3	90%	12.01

7 Experimental Results for Seven-Level Inverter

The experimental output current and voltage waveforms are shown in the below figure which also include the THD spectrum.

Fig. 14. Experimental Multi-level inverter setup

Figure 14 represents the hardware model with the load inductance was 70 mH and resistance was 150Ω which results to the highest power of 20.938 W in the output side and 0.89 PF as described in the graph below. The RMS value of the current and voltage in the output side of the 7-level multi-level inverter were 0.3608 Amps and 65.119 V respectively. The hardware output results shown in Fig. 15, measured using analog cathode ray oscilloscopes (CRO).

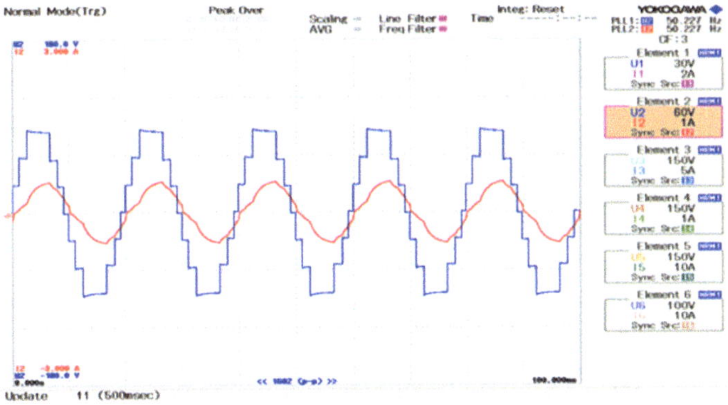

Fig. 15. Experimental 7-level current and voltage waveform

8 Conclusions

This article presents the design and implementation of a seven-level multilevel inverter, analyzed through various pulse width modulation (PWM) techniques to evaluate its performance in terms of output quality and harmonic distortion. The study systematically compares several PWM methods, including phase opposition disposition (POD), alternative phase opposition disposition (APOD), and phase disposition (PD), to determine their influence on the output voltage waveform and total harmonic distortion (THD). Comprehensive MATLAB/Simulink simulations were conducted to validate the theoretical aspects and quantify the harmonic components of each modulation strategy. Among the examined techniques, the APOD method exhibited superior performance by producing the least THD value in the inverter output, thereby confirming its suitability for achieving high-quality sinusoidal waveforms. Furthermore, a practical prototype was developed to experimentally verify the simulation results. The hardware implementation demonstrated consistency with the simulated outcomes, reinforcing the reliability of the analytical model. The APOD technique proved to be particularly effective for inverter configurations with a lower number of voltage levels, making it an ideal candidate for renewable energy integration, especially in photovoltaic (PV) applications, where efficiency, reduced switching losses, and enhanced power quality are critical for stable and sustainable system operation.

References

1. Du, Z., Tolbert, L.M., Chiasson, J.N., Ozpineci, B.: A cascaded multilevel inverter using a single dc source. In: Proc. IEEE APEC, pp. 426–430 (2006)

2. Carrasco, J.M., et al.: Power-electronic systems for the grid integration of renewable energy sources: a survey. IEEE Trans. Ind. Electron. **53**(4), 1002–1016 (2006)
3. Franquelo, L.G., Rodriguez, J., Leon, J.I., Kouro, S., Portillo, R., Prats, M.A.M.: The age of multilevel converters arrives. IEEE Ind. Electron. Mag. **2**(2), 28–39 (2008)
4. Dixon, J., Moran, L.: High-level multistep inverter optimization using a minimum number of power transistors. IEEE Trans. Power Electron. **21**(2), 330–337 (2006)
5. Mohamed Ali, J.S., Kumar, V.: Compact switched capacitor, multilevel inverter (CSCMLI) with self voltage balancing and boosting ability. IEEE Trans. Power Electron. (2018). https://doi.org/10.1109/TPEL.2871378
6. Ramani, K., Sathik, M.A.J., Sivakumar, S.: A new symmetric multilevel inverter topology using single and double source sub-multilevel inverters. J. Power Electron **15**(1), 96–105 (2015)
7. Mohamed Ali, J.S., Ramani, K., Romlie, M.F.: New Cascaded Multilevel Converter Topology Based on basic unit reduction of DC sources. Power Electron **16**(1), 100–109 (2016)
8. Mohamed Ali, J.S., Alishah, R.S., Krishnasamy, V.: A New Generalized Multilevel Converter Topology with reduced voltage on switches, power loss and components. IEEE Trans. Power Electron **30**, 8201–8209 (2019)
9. Lee, S.S.: Single-stage switched-capacitor module (S3CM) topology for cascaded multilevel inverter. IEEE Trans. Power Electron. **33**, 8204–8207 (2018)
10. Liu, J., Cheng, K.W.E., Ye, Y.: A cascaded multilevel inverter based on switched-capacitor for high-frequency AC power distribution system. IEEE Trans. Power Electron. **29**, 4219–4230 (2014)
11. Ye, Y., Cheng, K.W.E., Liu, J., Ding, K.: A step-up switched-capacitor multilevel inverter with self-voltage balancing. IEEE Trans. Ind. Electron. **61**, 6672–6680 (2014)
12. Karthikeyan, D., Krishnasamy, V., Sathik, M.A.J.: Development of a switched diode asymmetric multilevel inverter topology. J. Power Electron **18**, 418–431 (2018)
13. Jagabar Sathik, M.A., Abdel Aleem, S.H.E., Kannan, R., Zobaa, A.F.: A New Switched DC-Link Capacitor-based Multi-level Converter (SDC2MLC). Electr. Power Compon. Syst. **45**, 1001–1015 (2017)
14. Ebrahimi, J., Babaei, E., Gharehpetian, G.B.: A new multilevel converter topology with reduced number of power electronic components. IEEE Trans. Ind. Electron. **59**(2), 655–667 (2012)
15. Haji-Esmaeili, M.M., Naseri, M., Khounjahan, H., Abapour, M.: Fault tolerant structure for cascaded H-bridge multilevel inverter and reliability evaluation. IET Power Electron **10**(1), 59–70 (2017)
16. Catelani, M., et al.: Fault detection of resonant inverters for wireless power transmission using MLMVNN. Proc. IEEE 2nd International Forum on Research and Technologies for Society and Industry Leveraging a better tomorrow (RTSI), pp. 1–5. Bologna, Italy (2016)
17. Haji-Esmaeili, M.M., Naseri, M., Khounjahan, H., Abapour, M.: Fault-tolerant and reliable structure for a cascaded quasiz source DC-DC converter. IEEE Trans. Power Electron. **32**(8), 6455–6467 (2017)
18. Luh, D.D., Soon, J.L., Verstraete, D.: Derivation of dual-switch step-down DC/DC converters with fault-tolerant capability. IEEE Trans. Power Electron. **31**(9), 6064–6068 (2016)
19. Jamshidpour, E., Poure, P., Saadate, S.: Photovoltaic systems reliability improvement by real-time FPGA-based switch failure diagnosis and fault-tolerant DC-DC converter. IEEE Trans. Ind. Electron. **62**(11), 7247–7255 (2015)
20. Prasadarao, V.S., Joshi Manohar, K.V.: Grid Interconnection of PV System Using Symmetric and Asymmetric MLI Topology. Int. J. Power Electron. Drive System (IJPEDS) **9**(4), 1616–1623 (2018)

21. Reddy, K.R., Reddy, V.N., Vijaya Kumar, M.: Control of single stage grid tied photovoltaic inverter using incremental conductance method. Int. J. Power Electron. and Drive System (IJPEDS) **9**(4), 1702–1708 (2018)
22. Selmi, T., Rezgui, M.: A Novel DC-AC inverter topology to eliminate leakage current. Int. J. Power Electron. and Drive System (IJPEDS) **9**(4), 1733–1744 (2018)
23. Alishah, R.S., Nazarpour, D., Hosseini, S.H., Sabahi, M.: Novel topologies for symmetric, asymmetric and cascade switched-diode multilevel converter with minimum number of power electronic components. IEEE Trans. Ind. Electron. **61**(10), 5300–5310 (2014)

Evaluation of Rice Husk Ash as a Reinforcement Material for the Development of Soil-Based Roofing Sheet

M. S. H. Karunanayake[1(✉)], S. M. V. P. D. Senanayake[2], T. N. Fernando[3], A. J. Fernando[1], and N. P. A. K. G. Pramuditha[4]

[1] Rajarata University of Sri Lanka, Mihintale, Sri Lanka
senathkarunanayake@gmail.com
[2] National Engineering Research and Development Centre, Ekala, Ja-Ela, Sri Lanka
[3] General Sir John Kotelawala Defense University, Dehiwala-Mount Lavinia, Sri Lanka
[4] Wayamba University of Sri Lanka, Kuliyapitiya, Sri Lanka

Abstract. Rice husk is a major agricultural byproduct. Among its derivatives, rice husk ash (RHA) and rice husk biochar (RHB) show reinforcement properties in engineering applications. Researchers have developed roofing sheets by reinforcing RHB as an alternative to asbestos sheets. However, producing RHB requires a controlled and costly process. This study investigated the use of RHA as a cost-effective replacement for RHB for reinforcing soil-based roofing. Roofing sheets were prepared using 50% ordinary Portland cement, 39% soil, 10% sand, and 1% reinforcement material. RHB, partially controlled RHA (PC-RHA), and uncontrolled RHA (UC-RHA) were used as reinforcement materials. All samples were cured for 28 days before testing. Breaking load tests revealed the highest load-bearing capacities for UC-RHA at 6.55 kNm-1, PC-RHA at 6.27 kNm^{-1}, and RHB at 5.17 kNm^{-1} in order. The best-performing material, UC-RHA, underwent further evaluation for breaking load, density, water absorption, resistance to acetic acid, and water tightness, in accordance with Sri Lanka Standards (SLS). For validation, three replicate UC-RHA samples were cured for an additional 28 days. Standardized testing confirmed a mean breaking load of 3.205 kNm^{-1}, with satisfactory overall performance for density, water absorption, acidic acid resistance, and water-tightness. Conversely, the standardized tests indicated a lower breaking load strength compared to the initial results. These findings demonstrate that RHA is a feasible replacement for RHB, high-lighting the need for controlled production environments to ensure consistent performance.

Keywords: Reinforcement · Rice Husk Ash · Rice Husk Biochar · Roofing Sheets

1 Introduction

Roofing materials play a crucial role in ensuring both the safety and aesthetics of buildings, as an essential component in the construction process. Over time, roofing materials have evolved from natural thatching materials such as paddy straw, coconut leaf, palmyra

M. Siriwardhana et al. (Eds.): IERS 2025, PTT, pp. 186–195, 2026.
https://doi.org/10.1007/978-981-95-7984-6_17

leaf, etc., to the modern roofing materials such as clay tiles, metal sheets, and fibre-cement sheets [1]. Currently, chrysotile (asbestos) fiber sheet dominates the local market [2] as the most common roofing method in Sri Lanka [3] due to its affordability and properties, including minimal maintenance, high thermal performance, water tightness, and non-conductive nature [2]. However, all forms of asbestos are associated with significant health hazards such as lung cancers, mesothelioma [4, 5]. Local researchers have explored the development of eco-friendly alternative fiber materials to replace asbestos in roofing sheets [6]. Using pretreated corn husk fibers incorporating Alkaline Hydrogen Peroxide (AHP) to reinforce earth soil–cement roofing sheets offers a promising approach to advancing soil-based building materials. Nonetheless, the resultant roofing sheet exhibited limited durability and increased production expenses due to the high cost of chemicals, exacerbated by the recent economic crisis [7]. A recent study has explored a successful application of rice husk biochar as a reinforcement material by developing a soil-cement composite applicable for roofing sheets. The composite consists of 50% cement, 10% sand, 39% earth soil, and 1% rice husk biochar composition [8]. It demonstrated significant improvements in the mechanical properties and durability of soil-based roofing sheets by achieving the local standard. Conversely, the output of this study was not commercially viable due to the high cost of the standard-controlled environment. Hence, identifying a cost-effective alternative reinforcement material to rice husk biochar is vital to further development and expanding upon these findings.

As a remedial approach, rice husk ash is currently being used as a cost-effective mineral filler in concrete to enhance the mechanical properties. Rice husk ash is produced by the burning of rice husk [9]. When rice husk is burned, approximately 40% of the cellulose and 30% of the lignin are removed as organic matter, while roughly 20% of the silicon dioxide remains in its composition [10]. Rice husk ash exhibits higher pozzolanic reactivity due to its silica composition and large surface area [11]. The incorporation of rice husk ash can reduce the overall porosity of concrete by enhancing the packing density and refining the interfacial transition zone (ITZ) between the cement matrix and the aggregates [12]. The high content of reactive silica present in rice husk ash enables its use in a broad range of applications [13]. It has been widely used as a supplementary cementitious material in cement-based products to enhance the strength and durability for years [14, 15]. The application of rice husk ash ranges from concrete composites and ceramic materials to the nano level [9].

The modern world is facing a serious energy and natural resource shortage, which minimizes the resources to produce green building materials. In the industry, rice husk ash is produced as a byproduct of boiler waste that poses disposal challenges and environmental issues. Utilizing rice husk ash as building materials can provide a successful waste management solution [9]. Rice husk has significant potential for use in developing bio-composites [16]. Therefore, the present study focused on investigating the potential of rice husk ash as a reinforcement material for soil-cement composite to replace rice husk biochar, by manufacturing soil-based roofing sheets using the optimal formulation developed by Gunawardhana [8], and standardizing them according to the local standard.

2 Materials and Methods

2.1 Materials

Cement, Soil and Sand: Ordinary Portland cement (OPC) with strength class 42.5 N conforming to the Sri Lankan Standard (SLS 107), red-laterite soil, and sea sand were used for this study. The red-laterite soil was obtained from Nittambuwa, Western Province, Sri Lanka. Soil and sand were sun-dried, sieved using a 2.00 mm mesh, and then stored in the civil laboratory before use.

Biochar and Rice Husk Ash: The rice husk (RH) was used as the biomass feedstock to produce the biochar. Briefly, RH was pyrolyzed at 500 °C under oxygen-free conditions to produce rice husk biochar (RHB). This RHB was produced in a laboratory furnace of the University of Sri Jayawardhanapura, Sri Lanka. The resultant RHB was crushed in a grinder to reduce the particle size, followed by sieving through a 0.50 mm mesh, and then stored in a sealed container before its use in composite making. Rice husk ash (RHA) production was done by utilizing RH. Two types of RHA were produced for this study, namely partially controlled Rice Husk Ash (PC-RHA) and uncontrolled Rice Husk Ash (UC-RHA). PC-RHA was produced under a partially controlled pyrolysis environment, while an open incineration produced UC-RHA for 8–9 h. Table 1 provides an overview of the production conditions associated with the reinforcement materials. Both PC-RHA and UC-RHA were ground and then subjected to a sieving process like that of RHB. Those three reinforcement materials produced for this study are illustrated (see Fig. 1).

Table 1. Process conditions of reinforcement materials.

Process parameters	RHB	PC-RHA	UC-RHA
Temperature (°C)	500	500	uncontrolled
Time (h)	1	1	uncontrolled
Temperature ramp (°C/minutes)	10	uncontrolled	uncontrolled
Oxygen condition	Low/zero	uncontrolled	uncontrolled

2.2 Methods

Mix Design. The Cement-Soil composite with cement to soil to a sand ratio (by weight) of 5: 3.9: 1 with a water/cement ratio of 0.6 was adopted in this study. The soil was replaced with one of the reinforcement materials (RHB, PC-RHA, and UC-RHA) at 1% on a weight basis, and composites were made. First, red-laterite soil, cement, and sand were dry mixed for 5–10 min with each reinforcement material. Later, cold water at 4 °C was added and mixed for 10 min. As the reinforcement materials tend to hold water more than cement, soil, and sand, more water was added according to the requirement of the composite mixes to compensate for the loss of workability.

Fig. 1. Reinforcement material types of rice husk (a). RHB (b). PC-RHA (c). UC-RHA.

Sample Preparation: The composite mixture was spread on an iron frame (0.5 m × 0.5 m × 0.008 m) fixed on a vibration machine and allowed to vibrate. The mixture was demolded after 10 min of vibration and then placed onto a corrugated asbestos sheet template, functioning as a mold to form a roofing sheet sample. Either side of the sheet samples was covered using polythene sheets to prevent moisture loss and ensure adequate hydration during the early stages of hardening (see Fig. 2). After 72 h, the sheet-cured specimen was placed in a water bath for further curing. 24 h after the wet curing, the sample was placed under a 28-day curing period in a shaded area. Two sets of samples were prepared for testing. The first set included RHB, PC-RHA, and UC-RHA composites, each with three replicates, for preliminary evaluation. The second set consisted of three replicates of 1% UC-RHA samples (R01, R02, R03) used for standardized tests following SLS protocols (Table 2).

Table 2. Formulations and sample types.

OPC	Sand	Soil	Reinforcement Material			Sample Name	Sample Size
			RHB	PC-RHA	UC-RHA		
50%	10%	39%	1%	-	-	RHB	03
			-	1%	-	PC-RHA	03
			-	-	1%	UC-RHA	03

Evaluation of Characteristics: A preliminary evaluation was conducted to assess the 28-day breaking load strength of hardened roofing sheet samples, as per SLS 9: Part 2: 2001 [17], using a Universal Testing Machine (UTM) at the National Building Research Organization (NBRO), Colombo, Sri Lanka. Each formulation design was tested on a minimum of three replicates, and the breaking load test was derived using a three-point bending system. The resulting data were analyzed by one-way ANOVA and the Least Significant Difference (LSD) test. To confirm the preliminary findings for the formulation exhibiting the highest strength, additional qualitative and quantitative assessments were conducted on newly prepared samples that had been cured for 28 days.

(a) (b) (c)

Fig. 2. Procedure of developing a roofing sheet (a). Spreading the mixture on the iron frame (b). Composite mixture just after vibration (c). Finalized roofing sheet covered by polythene

These tests, conducted at the National Engineering Research and Development Centre (NERDC), Ekala, Sri Lanka, included measurements of density, water absorption, water tightness, and resistance to acidified water, as detailed in Table 3, which outlines specimen dimensions, quantities, and the specific tests performed.

Table 3. Types of tests conducted, number of samples tested, and sample sizes.

Test	Size	Number of specimens	Reference
Breaking load (kN/m)	0.5 m × 0.5 m × 0.008 m sheet	03	SLS 9: Part 2: 2001
Density (kgm^{-3})	75 mm × 40 mm specimen	03	
Water absorption (%)	300 mm × 40 mm specimen	03	
Water tightness	0.5 m × 0.5 m × 0.008 m sheet	03	
Resistance to acidified water (kg/m^2)	10,000 mm^2 specimen	03	

3 Results and Discussion

3.1 Flexural Strength

The preliminary study results for the breaking load strength of 28 day cured roofing sheet samples prepared with PC-RHA and UC-RHA of 1% biochar additions by weight, and the control roofing samples prepared with 1% RHB, are plotted (see Fig. 3). It can be observed that there was an increase in the flexural strength of both RHA composites (PC-RHA and UC-RHA) compared to the control sample in this experiment. The highest observed breaking load strength recorded was 6.55 kN/m for UC-RHA, which is 27% higher than the control (5.175 kN/m). By far, the PC_RHA sample exhibited intermediate strength with 6.27 kN/m, which was 21% higher than that of the control. But statistically, RHB, PC-RHA, and UC-RHA are not significant ($p > 0.05$) at 28 days.

This result indicates that a common factor affecting all three material types, such as the amorphous silica content derived from the same parent material (rice husk), is independent of the production conditions but contributes to the strength, with no significant difference between the material types. Overall, RHA was found to have a positive effect on flexural strength from samples known as UC-RHA and PC-RHA, leading to the hypothesis that RHA was strong enough to replace RHB as the reinforcement material. Consequently, UC-RHA was identified as the optimal reinforced material for RHB for the main study due to its higher strength.

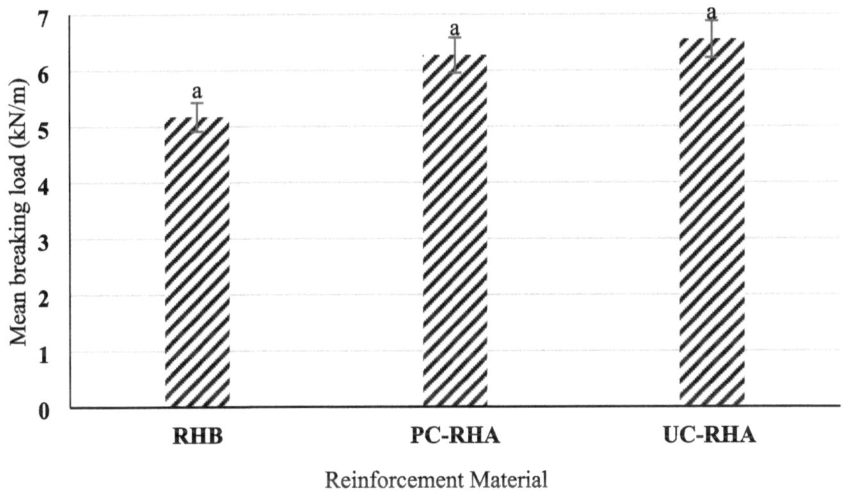

Fig. 3. Mean breaking load values (kN/m) of the reinforcement materials.

3.2 Standardized Performance Assessment

Breaking Load Strength: The mean test results for the breaking load strength of UC-RHA samples in the preliminary evaluation, against the main evaluation, are presented in Fig. 4. It can be observed that there was a drastic reduction in the flexural strength of the UC-RHA, compared to the preliminary study. The observed breaking load strength recorded was 3.205 kN/m for UC-RHA, which is 51% lower than the results of the previous study (6.55 kN/m). Furthermore, this result does not satisfy the SLS criteria for breaking load, as it is less than 5 kNm^{-1}. The decrease in the flexural strength for UC-RHA may be attributed to the variation of composition created by the production process. As noted by Ordoñez [18], the pozzolanic efficiency of RHA depends on its amorphous SiO_2 content, as the amorphous RHA is reactive and is suitable for use as a pozzolan in composite [9]. Crystaline RHA are unreactive varieties of quartz, formed in the ash in the case of uncontrolled burning of RHA above 80 °C [19]. Amorphous SiO_2 is maximized (<70%) under controlled combustion below 700 °C. In contrast, open incineration involves initial carbonization at 370–380 °C followed by spontaneous combustion at 700–750 °C, resulting in crystalline silica with extended residence time [20]. Consequently, this process extends to obtaining predominantly crystalline ash under field combustion [21]. It points out that fluctuations in the temperature profiles can negatively affect the composite's strength by hindering the formation of extra C-S-H gels during the secondary pozzolanic reaction of RHA. The two UC-RHA sample sets, derived from the open incineration process have given opposite results even though there is no difference in the testing procedures between the preliminary and the standardized testing for breaking load on two occasions. Furthermore, it is independent of the production procedure in both phases, as all roofing samples underwent the same consistent mix design process. It implies that even the same raw material, produced in various scenarios, can illustrate unpredictable behavior with dynamic characteristics when there is no controlled environment.

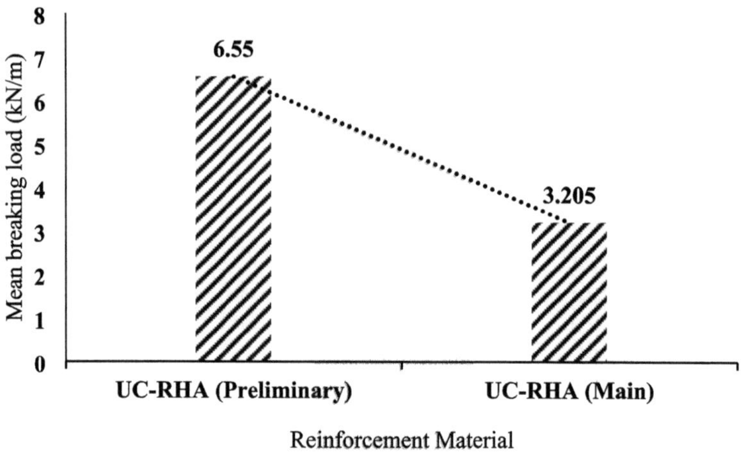

Fig. 4. Comparison of mean breaking load values (kN/m) of UC-RHA.

Density & Water Tightness: Table 4 shows the density values of 28-day hardened UC-RHA roofing samples. All replicates exceeded the SLS requirement, indicating that the composites achieved adequate compaction and structural integrity. This improvement can be attributed to the micro-filling effect and physical packing of the composite. This is due to the reactive silica (SiO_2) in rice husk ash, which fills gaps between cement particles and promotes secondary hydration with calcium hydroxide to form additional calcium silicate hydrates [9]. These pozzolanic and micro-filling actions enhance the pore structure of the cement paste and interfacial transition zone (ITZ), as reported by Safiuddin [21]. In this context, the additional C–S–H gels block pores, strengthen the matrix–aggregate interface, and increase the overall packing density [22]. Similarly, the water-tightness performance showed the expected improvement, with two out of three replicates satisfying the no-leakage criterion. This property reflects the less porous nature of RHA-reinforced composites [23], which lowers permeability and prevents moisture penetration, further supported by the particle size distribution of the raw materials (sand and laterite soil < 2 mm, UC-RHA < 0.5 mm).

Water absorption: Water absorption is one of the key factors that affect the durability of cement-based composites. The UC-RHA samples exhibited low water absorption, indicating good potential for long-term durability. In general, higher porosity tends to increase water absorption, as the voids represent the space available to fill the water. In this study, the lower water absorption is primarily due to the reduced free pore space in the composite. A previous study [24] has shown that RHA particles help refine pores and reduce overall porosity. As reported by Amran [9], water absorption decreases with improved gel density, smaller pore sizes, and reduced void content, up to RHA contents of 20%. Thus, 1% UC-RHA used in this composite aligns with this behavior, contributing to lower water absorption and enhanced durability.

Resistance to acidic acid: Acid attack in cementitious composites occurs through the breakdown of aggregate particles (coarse and fine), hydrated (gel phases), and non-hydrated (free lime) components [9]. The UC-RHA samples demonstrated strong resistance to acid degradation, remaining well within the SLS acceptance limits, as shown in Table 4. This improved performance can be attributed to the reduced alkali–silica reaction (ASR) resulting from UC-RHA reinforcement. ASR commonly develops in highly alkaline cement pastes when non-crystalline silica in aggregates reacts in the presence of moisture. However, the incorporation of RHA decreases porosity, thereby limiting moisture penetration and suppressing the occurrence of ASR. Additionally, alkalis in the cement paste can react with the finely ground RHA to form stable alkaline silicates [23], thereby reducing free alkali levels and minimizing damage to silica-based aggregates. Similar improvements in acid resistance with RHA incorporation have also been reported in earlier studies on concrete [23].

Table 4. SLS test results of density, water absorption, and resistance to acidic acid of roofing sheet samples with 1% UC-RHA.

	Density (kgm^{-3})	Water absorption (%)	Resistance to acidic acid (kgm^{-2})
UC-RIIA (R01)	2450.63	16.86	0.0563
UC-RHA (R02)	2134.62	18.82	0.0597
UC-RHA (R03)	2258.30	19.02	0.0859
Average (X)	2281.20	18.23	0.0673
Range (R)	316.01	2.16	0.0296
Criteria conformity	2154.79	17.37	0.07914
Criteria	$(X\text{-}0.4R) \geq 1200\ kgm^{-3}$	$(X\text{-}0.4R) \leq 28\%$ of the dry mass	$(X + 0.4R) \leq 1.15\ kgm^{-2}$

4 Conclusions

This study demonstrated that the potential of using rice husk ash as a reinforcement material to replace rice husk biochar in soil-cement composite is obtained. The addition of 1% RHA increased the flexural strength of the composite in preliminary studies, indicating that RHA (PC-RHA &UC-RHA) is potentially suitable for reinforcement. UC-RHA exhibited the highest breaking load strength compared to the PC-RHA and control, making it strong enough to replace RHB as the optimal reinforcement material for this composite, but ultimately leading to the revelation of significant inconsistencies in the reinforcement properties of UC-RHA with very low breaking load strength during the main study due to the variation of pozzolanic composition occurring in an uncontrolled environment. Despite its strength, UC-RHA was able to meet the other structural and durability standards, including density, water absorption, resistance to acetic acid, and water tightness, satisfying SLS, attributed to the porosity, filling effect, and physical packing of RHA. These findings reveal that the uncontrolled incineration process can hinder the reinforcement properties of RHA. By standardizing the production process, RHA reinforced soil-based roofing sheets with stable and predictable characteristics can be developed, leading to the successful replacement of RHB.

References

1. Udagama, U.D.G.U.C., Kulatunga, A.K.: Development of sustainable roofing material from waste. In: 5th International Conference on Sustainable Built Environment, Kandy, Sri Lanka (2014)
2. Ada Derana BizEnglish–Role of chrysotile fibre cement in Sri Lanka's roofing industry. https://bizenglish.adaderana.lk/role-of-chrysotile-fibre-cement-in-sri-lankas-roofing-industry/. Last accessed 29 June 2025
3. Fernando, N., Batuwita, I.: Manufacturing of roofing sheets reinforced with pretreated corn-husk residues as an alternative for asbestos sheets. In: 2021 From Innovation to Impact (FITI), pp. 1–5. IEEE, New York (2021)
4. OIDA International Journal of Sustainable Development. https://oidaijsd.com/wp-content/uploads/2019/05/12-02-01.pdf. Last accessed 29 June 2025
5. Lunn, R.M., Mehta, S.S., Jahnke, G.D., Wang, A., Wolfe, M.L., Berridge, B.R.: Cancer hazard evaluations for contemporary needs: highlights from new national toxicology program evaluations and methodological advancements. J. Natl. Cancer Inst. **114**(11), 1441–1448 (2022)
6. Peiris, D.A.C.D.P., Weerasinghe, T.K.: Study on use of environmentally friendly alternative fiber materials for asbestos roofing sheets in Sri Lanka. OIDA Int. J. Sustain. Dev. **12**(2), 11–18 (2019)
7. Fernando, T.N., Batuwita, I.P.: Reinforcing of earth soil and cement with pretreated corn-husk fibers in the development of building materials. In: 12th International Proceedings on Proceedings, pp. 241–254. Springer Nature, Singapore (2022)
8. Gunawardana, J.G.S.P.P., Senanayake, S.M.V.P.D., Batuwita, I.P., Fernando, T.N., Vithanage, M.: Development of biochar-reinforcing low-cost soil based roofing sheet. In: 2nd International Engineering Research Symposium, Proceedings in Technology Transfer. Springer, Singapore (2025). https://doi.org/10.1007/978-981-96-1399-1_4
9. Amran, M., et al.: Rice husk ash-based concrete composites: a critical review of their properties and applications. Crystals **11**(2), 168 (2021)

10. Chindaprasirt, P., Kanchanda, P., Sathonsaowaphak, A., Cao, H.T.: Sulfate resistance of blended cements containing fly ash and rice husk ash. Constr. Build. Mater. **21**(6), 1356–1361 (2007)
11. Antiohos, S.K., Papadakis, V.G., Tsimas, S.: Rice husk ash (RHA) effectiveness in cement and concrete as a function of reactive silica and fineness. Cem. Concr. Res. **61–62**, 20–27 (2014)
12. Ye, G., Huang, H., Tuan, N.: Rice husk ash. In: Siddique, R. (ed.) Properties of fresh and hardened concrete containing supplementary cementitious materials. Springer, Cham (2018)
13. Singh, B.: Rice husk ash. In: Siddique, R., Cachim, P. (eds.) Waste and supplementary cementitious materials in concrete, pp. 417–460. Woodhead Publishing, Cambridge (2018)
14. Chatveera, B., Lertwattanaruk, P.: Evaluation of nitric and acetic acid resistance of cement mortars containing high-volume black rice husk ash. J. Environ. Manage. **133**, 365–373 (2014)
15. Sayed Ali, F., Zaheer, M.M.: Production of new generation and sustainable concrete using rice husk ash (RHA): a review. Mater. Today: Proc. (2023). https://doi.org/10.1016/j.matpr.2023.06.034
16. Suhot, M.A., Hassan, M.Z., Aziz, S.A., Md Daud, M.Y.: Recent progress of rice husk reinforced polymer composites: a review. Polymers **13**(15), 2391 (2021)
17. Ordoñez, L.M., Borrachero, M.V., Monzó, J., Soriano, L., Payá, J.: Binary and ternary blended portland cements containing different types of rice husk ash. Materials **17**(12), 2923 (2024)
18. Pachchigar, S., Takehara, M.D.B., Pettersson, E., Öhman, M.: Energy & Fuels **39**(9) (2025). https://doi.org/10.1021/acs.energyfuels.4c05412
19. Safiuddin, M., West, J.S., Soudki, K.A.: Hardened properties of self-consolidating high performance concrete including rice husk ash. Cement Concr. Compos. **32**(9), 708–717 (2010)
20. Zou, Y., Yang, T.: Rice husk, rice husk ash and their applications. In: Rice Bran and Rice Bran Oil. 1st edn. Academic Press, London (2019)
21. Zerbino, R., Giaccio, G., Marfil, S.: Evaluation of alkali–silica reaction in concretes with natural rice husk ash using optical microscopy. Constr. Build. Mater. **71**, 132–140 (2014)
22. Chopra, D., Siddique, R.: Strength, permeability and microstructure of self-compacting concrete containing rice husk ash. Biosys. Eng. **130**, 72–80 (2015)
23. Pabst, W., Gregorová, E.: Elastic properties of silica polymorphs–a review. Ceram. Silikaty **57**(3), 167–184 (2013)
24. Sri Lanka Standards Institution: Specification for asbestos-cement products–Part 2: Corrugated Sheets (Second Revision). SLS 9: Part 2. SLSI, Colombo (2001)

Root Cause Analysis and Mitigation of Defects on Handle Attachment in Porcelain Products Using 6σ-DMAIC Methodology

Vihanga Chandrasekara and Roshan Thilakarathne(✉)

Department of Mechanical and Manufacturing Technology, Faculty of Technology, Wayamba University of Sri Lanka, Kuliyapitiya 60200, Sri Lanka
fot209031@kul.wyb.ac.lk, roshant@wyb.ac.lk

Abstract. In porcelain manufacturing, the quality control is main characteristic because of the delicate nature of the aesthetics and the product's composition. This research looks into one of the most problematic defects identified at the handle attachment of Regular body mugs, which shows a serious problem to quality and contributes to high rejection rates in the final inspection. With the 6σ methodology, according to the DMAIC, the study systematically attempts to solve the problem's underlying root causes. Over a three-month period, relevant data were collected through final inspection reports and field studies for problem identification and root cause analysis. This was analyzed using Pareto charts, cause-and-effect diagrams and 5-Why analysis. Main controllable root causes included handle-making methods, attachment bonding areas, and decoration kiln pre-heating time adjustments. To validate these root causes, 3 suggestions were tested: switching the handle casting from solid to drain casting, maximizing the attachment bonding area, and maximizing RHK preheating times. These improvements were checked through quality inspections at every stage using NDT. The 4M validation and control impact chart helped distinguish between controllable and uncontrollable causes. The final inspection results confirmed a significant amount of defect reduction in handle attachment crack defects and other main defects.

Keywords: Porcelain · Handle attachment cracks · Root cause analysis

1 Introduction

The manufacturing of porcelain is a complex process that requires quality control at each stage in order to achieve high-quality products [1, 2]. One of the most critical issues is the occurrence of defects [3], which reduce the strength, appearance, and value of the final product. These defects are often detected during the final inspection stage, leading to rejection, high wastage levels, shipment delays, and increased production costs.

Furthermore, failure to minimize such defects [3, 4] in the early stages of production results in wasted manpower, energy, and raw materials, ultimately lowering overall efficiency. When products do not meet required standards, manufacturers are forced to reproduce additional items to meet demand, increasing costs and resource consumption.

M. Siriwardhana et al. (Eds.): IERS 2025, PTT, pp. 196–207, 2026.
https://doi.org/10.1007/978-981-95-7984-6_18

The aim of this study is to understand and analyse handle attachment defects in porcelain ceramic products, particularly those detected during the final inspection stage, and to develop methods for eliminating their occurrence. The research objectives are threefold: first, to examine and classify the specific defect found in the handle attachment area of porcelain mugs by reviewing inspection records and defect reports; second, to identify the key factors contributing to these defects, with special attention to kiln preheating stages, bonding quality, and inspection procedures, using analytical tools such as cause-and-effect diagrams [5] and the 5-Why method [6, 7]; and third, to propose and test practical solutions based on the identified root causes [8].

2 Literature Review

2.1 Different Types of Fracture Defects in Porcelain Products

Porcelain production is a very complicated and meticulous process. A slight negligence in any process from raw material to firing may cause defects or even scrap. Some of these defects were discovered before firing, some were exposed during firing before firing, and some were directly caused by improper firing [9–11]. Different types of fracture defects can occur at the handle attachment points, along the body, edges, and attachments as well as at the product base. Porcelain products suffer from reduced structural quality and poorer aesthetic appearance because of different types of defects [12].

Other causes include mechanical damage before loading, uneven thickness, poor bonding, or the excessive weight of attachments such as handles and spouts [10]. More specifically, edge cracks often result from low clay content, which reduces the ware's strength against moisture expansion, while thermal shock from opening the kiln too soon after firing further intensifies the defect [12]. Handle cracks are another major concern, typically caused by differences in clay hardness between the body and the handle, creating unequal shrinkage during drying and firing, and ultimately leading to detachment or fracture [12]. Apart from cracks, deformation was also recorded as a serious defect, commonly arising from improper kiln installation, misaligned saggers, or vibration of the kiln car, which cause distortions in the geometry of the products [10]. In addition, pin-holes were observed as a glaze defect, usually forming when gas bubbles generated during firing fail to escape or when firing ends before the glaze surface has time to heal completely [4].

2.2 Previous Studies and Problem-Solving Initiatives

The porcelain manufacturing industry continues to face significant challenges due to the occurrence of defects, which reduce product strength, appearance, and overall quality while increasing production costs and wastage [13]. Studies further emphasize that soaking time, preheating, and firing temperature strongly influence defect formation, and poor control of these parameters may result in fracture failures during the kiln cycle [10]. Other defects, including glaze crawling, pinholes, and warpage, are also reported, primarily linked to raw material composition, improper bonding, or processing variations [4].

Research highlights that once defects appear during the final inspection stage, they cannot be recycled and lead to major rejections, financial losses, and wasted resources [8]. To address these challenges, several studies have applied systematic quality management techniques such as Six Sigma, Kaizen, and Lean Manufacturing, which support root cause identification and long-term defect reduction [9]. Local and international case studies further show that defect prevention requires both raw material optimization and standardization of bonding and firing processes, ensuring stronger attachment and reduced cracking in porcelain mugs [14].

2.3 Application of Six Sigma and Analysis Techniques

Six Sigma differs from Kaizen and lean methodologies [15] through its reliance on data-based choices and analysis of root causes and statistical process controls to detect and remove production abnormalities [16]. The implementation of Six Sigma principles allows organizations to conduct systematic pattern analyses of defects [17, 18], while optimized material specifications and better process controls.

Pareto's Law applies not only to the distribution of wealth, but to many other distributions as well. The law is often identified as the 80–20% rule. The 5 Why's analysis is widely used for troubleshooting and problem-solving. Five Why's also finds a widespread application in Six Sigma process improvement methodology [6]. A cause-and-effect diagram [19, 20] often called a "fishbone" diagram [5, 21] can help in brainstorming to identify possible causes of a problem and in sorting ideas into useful categories. A fishbone diagram is a visual way to look at cause and effect. It is a more structured approach than some other tools available for brainstorming causes of a problem [7].

2.4 6σ Methodology-DMAIC

DMAIC is a data-driven approach used for optimizing and improving the existing business designs and processes. It is an effective method of controlled change management. DMAIC is a data-driven approach used for optimizing and improving the existing business designs and processes [16, 17].

3 Methodology

The research strategy was built to detect root causes of handle attachment defects in porcelain products and help analyse and minimize the rate of defects. DMAIC methodology within Six Sigma provides a data-driven structured solution framework through its Define, Measure, Analyse, Improve, and Control sequence (Fig. 1).

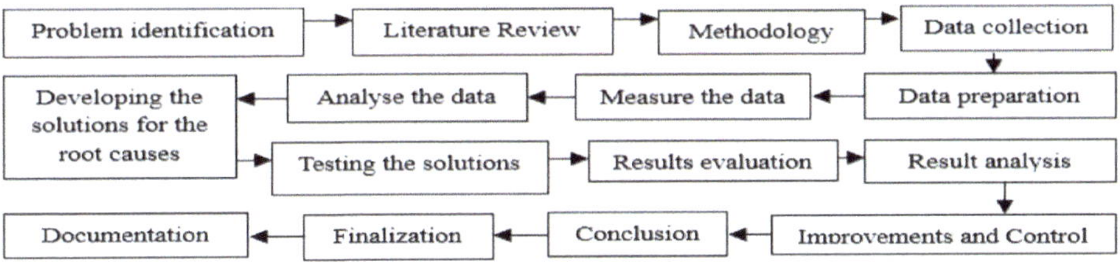

Fig. 1. Research methodology flowchart

4 Results and Discussion

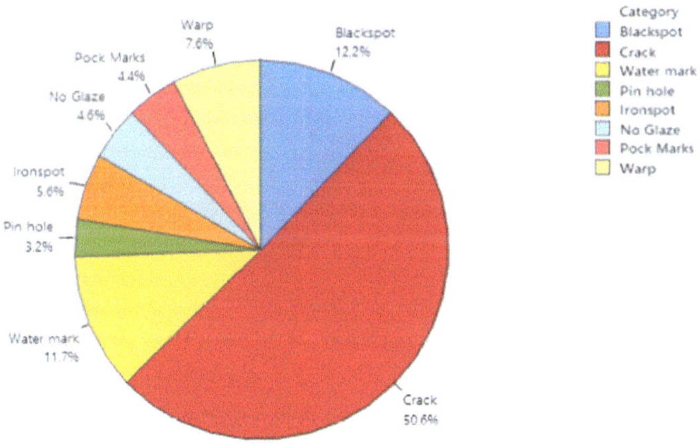

Fig. 2. Distribution of major defects in porcelain products-August 2024.

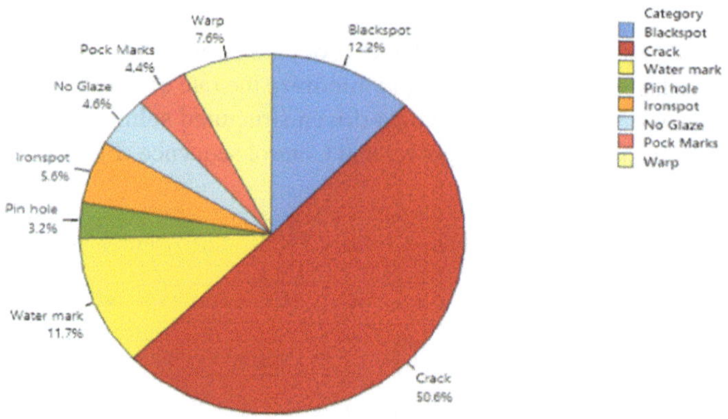

Fig. 3. Distribution of major defects in porcelain products-September 2024.

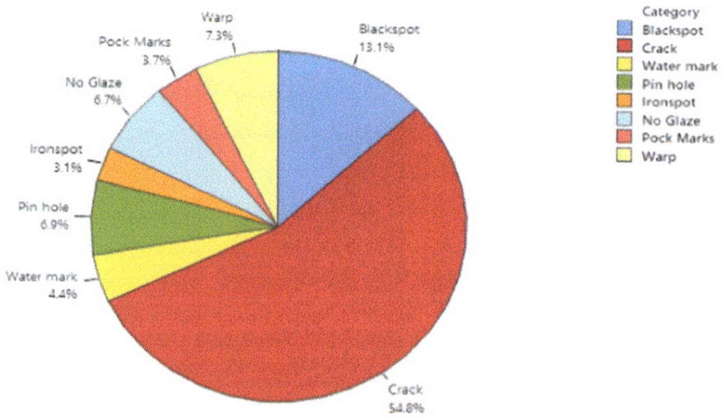

Fig. 4. Distribution of major defects in porcelain products-October 2024

The research findings during our field studies were discussed in this chapter through data collection that recorded defects in final inspections during three months. Data analysis through data sheets revealed that crack defects appeared as the most prevalent defects in the dataset. As well as identified the handle attachment crack is the most crack type (Figs. 2), (3), and (4). The research project examines crack defects that emerged from Regular clay body type mugs according to observed data. (LX-13, NLB, and Regular are main porcelain clay types).

4.1 Classification of Crack Defects in Porcelain Products

The collected data sheet presents the number of inspected items and rejected items based on different types of crack defects over the same three-month period.

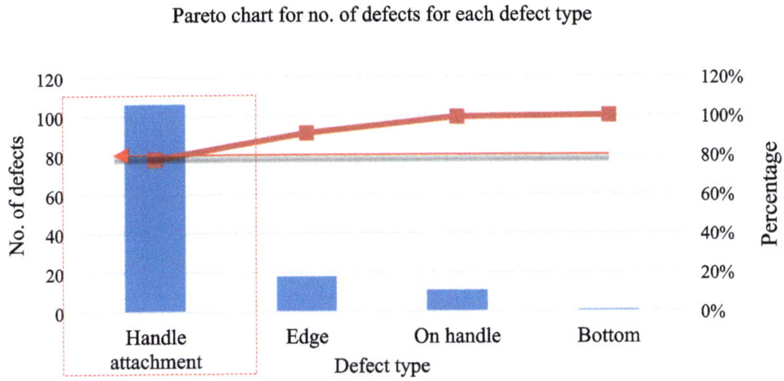

Fig. 5. Pareto chart for number of defects for each defect type

A Pareto chart reveals how porcelain defects distribute among various defect types, which shows the main issues affecting quality (Fig. 5), (Table 1). A total of over 100 handle attachment defects emerged, which made it the most common defect and caused most of the total defects. The handle attachment defect produces the largest amount of all problems, exceeding 80%, resulting in critical importance to product quality.

Table 1. Percentage distribution of crack types identified in final inspection

Type of crack	Number of defects	Cumulative	Percentage
Handle attachment	106	106	77.94%
Edge	18	124	91.17%
On handle	11	135	99.26%
Bottom	1	136	100%

4.2 Identification of the Most Affected Item by Handle Attachment Cracks

The analysis of defects shows that the 93686C Regular body type mug requires further study since it has experienced the highest number of handle attachment defects (Fig. 6). The selection enables thorough investigation of root causes while allowing to create solutions. (x axis is labeled as various types of regular body item numbers).

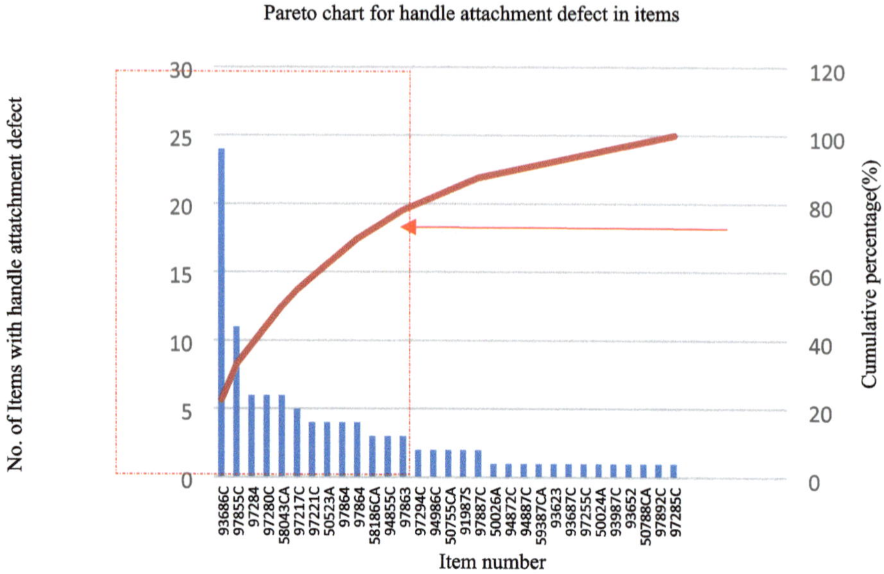

Fig. 6. Pareto chart for handle attachment defect in items

4.3 Analyse the Data

Five (5)-Why Analysis

Defects in porcelain handles often originate from weak bonding between the handle and the mug body. This weak bonding is primarily due to an insufficient bonding area, which results in concentrated stress during drying and firing processes. One major reason for this limitation is the inconsistency in the handle-making method, particularly between different casting techniques such as drain casting and solid casting (Fig. 7).

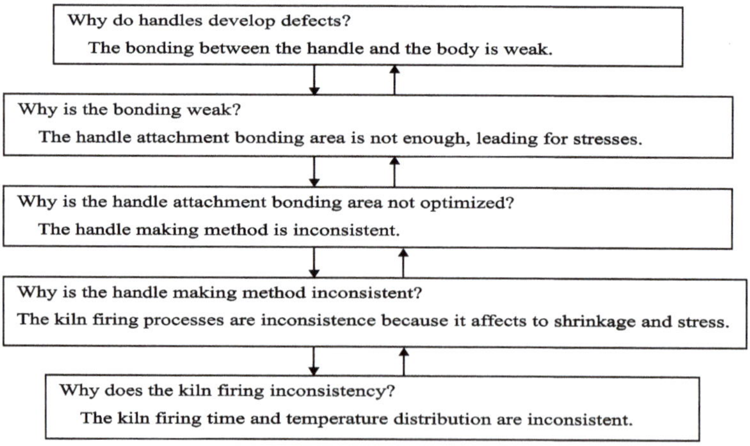

Fig. 7. Five why analysis (5-why)

Root Cause Analysis Using Fishbone Diagram

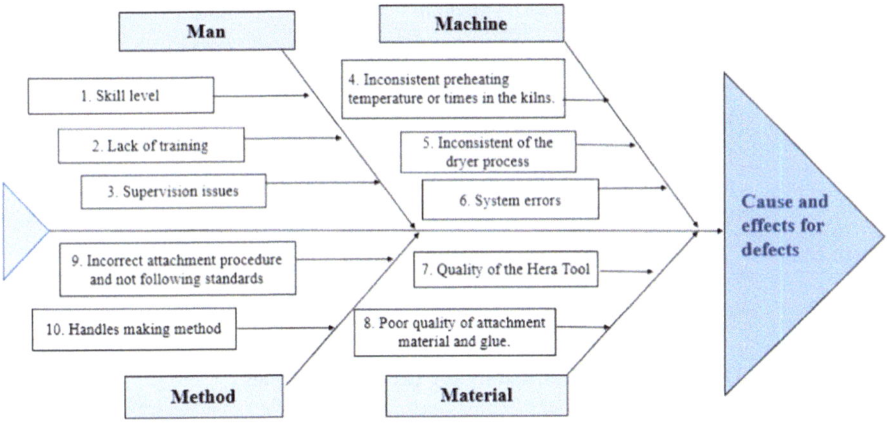

Fig. 8. Fishbone analysis using 4M

4M Validation

Table 2. 4M validation methods for root causes.

No	4M	Potential cause	Validation method
1	Man	Skill level	Skill map
2	Man	Lack of training	One point learning/ Skill map
3	Man	Supervision issues	Skill map
4	Machine	Inconsistent preheating temperature or times in the kilns	Test card and testing
5	Machine	Inconsistent of the dryer process	Test card and testing
6	Machine	System errors	-
7	Material	Quality of the Hera tool	Test card and testing
8	Material	Poor quality of attachment material and glue - Attachment area of the product	Test card and testing
9	Method	Incorrect attachment procedure and not following standards	Test card and testing
10	Method	Handles making method	Test card and testing

A root cause analysis (Fig. 8) was conducted to identify the main factors contributing to handle attachment defects in porcelain mugs. The analysis was supported by field observations, brainstorming sessions, and historical defect data collected over a three-month period. The potential causes were then categorized 4M validation (Table 2) using an Impact Control Chart (in control & out of control) (Table 3).

Table 3. Impact control chart

Control	High	Medium	Low
In control	Attachment area of the product	Lack of training	Skill level
	Handle making method	Supervision issues	Quality of the Hera tool
	Inconsistent preheating temperature or times in the kilns	Difference between machine speeds and employee working speed	Accidents in production on plant
Out of control	Sudden power failures	System errors	

4.4 Improve-Developing the Solutions for the Root Causes

The production process of body and handle attachment steps for the selected regular mug (93686C) are illustrated in the following flowchart, with specific points highlighted where testing or method modifications are planned to be implemented for improvement.

i. Shifting handle making method from solid casting to drain casting
ii. Adjustment of bonding area in handle attachment
iii. Kiln preheating time adjustment after decal pasting

Sample Testing and Results for Minimization of Handle Attachment Cracks
To validate the effectiveness of identified root causes and proposed solutions, three separate adjustments were selected and applied to the manufacturing process of the regular body mug (93686C). Each test was conducted using 50 sample mugs under modified conditions and compared against 50 mugs produced using the current standard process. And therefore need 100 regular bodies, 50 solid casting handles, and 50 drain casting handles.

Table 4. Greenware inspection results

Defect Type	Current Process	Modified Process
Glaze crawling	0	0
Pinholes	0	0
Blistering	0	0
Crazing	0	0
Shivering	0	0
Total	0	0

According to the inspection analysis (Figs. 9), and (10), (Table 4), (Figs. 11) and (12), the implementation of proposed process improvements such as adjusting the kiln firing

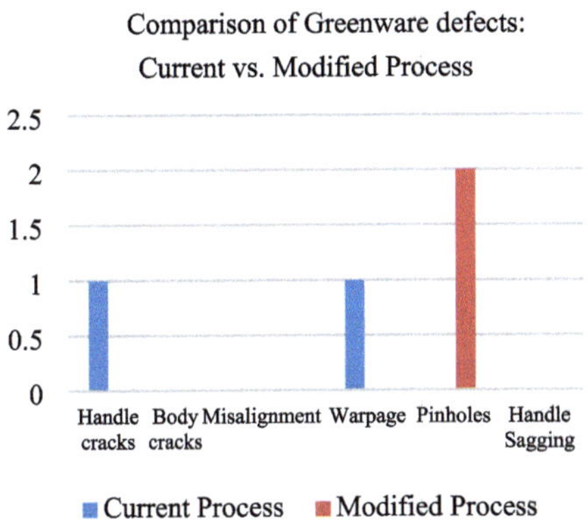

Fig. 9. Results of Greenware Defects: Current vs. Modified Process

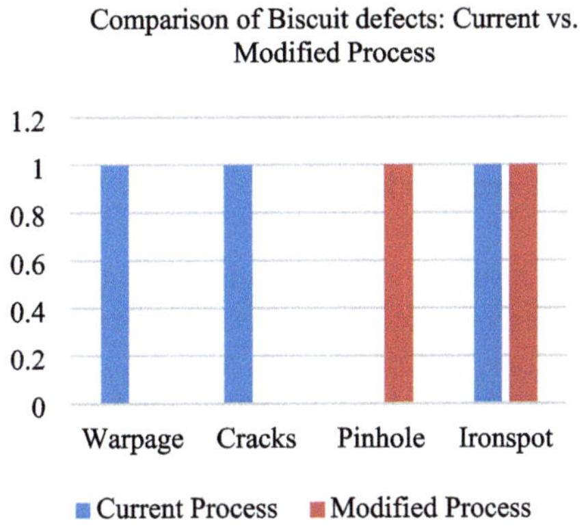

Fig. 10. Results of Biscuits Defects: Current vs. Modified Process

parameters, improving bonding area design, and changing the handle casting method successfully minimized the major defects and handle attachment crack. The comparison between current and modified process inspection data clearly shows a significant reduction in handle attachment cracks, and other defects, particularly in Regular body mugs (93636C), which were previously identified as the most affected product type.

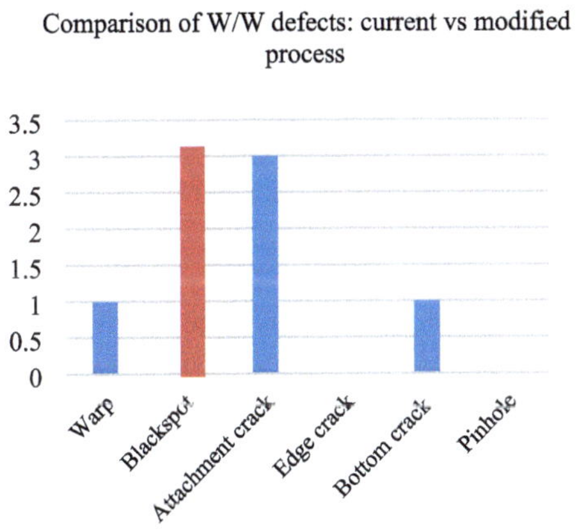

Fig. 11. Results of Greenware Defects: Current vs. Modified Process

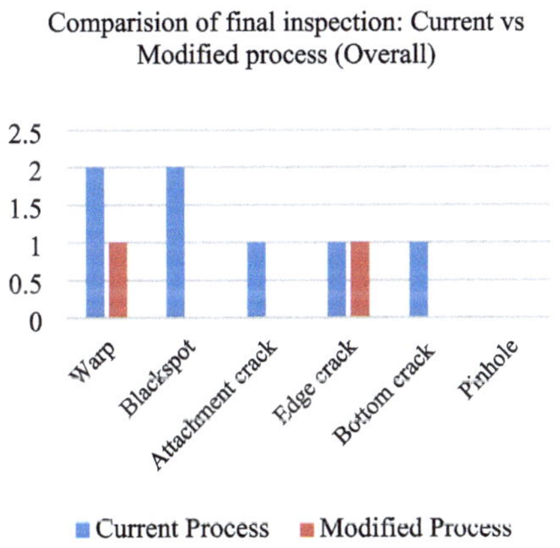

Fig. 12. Results of Greenware Defects: Current vs. Modified Process

4.5 Control

Compared to other research studies that have mainly focused on general porcelain defects such as pinholes, glaze crawling, or deformation, the present study places emphasis specifically on handle attachment cracks, which are often overlooked despite their high impact on product rejection. While previous works have highlighted firing conditions, raw material variations, or improper kiln settings as leading factors, this research demonstrates that the bonding area design, handle casting method, and kiln preheating adjustments play a more direct role in reducing handle crack defects.

5 Conclusions

This research focused on identifying the root causes of handle attachment defects, particularly cracks. To address this issue, the study employed field surveys, brainstorming sessions, and defect data analysis using root cause analysis tools such as the fishbone diagram and the 5-Whys analysis using 6σ DMAIC methodology. The investigation identified several key root causes of the defects, including RHK kiln firing parameters, bonding area design, and casting method based on these causes, corrective actions were implemented and tested. After implementing the process improvements, the modified porcelain products underwent strength and quality inspections, which revealed a significant reduction in handle crack defects. This research was limited to handle attachment defects, particularly cracks, and did not cover other common porcelain defects. The experiments were conducted within a single production line under specific factory conditions, and time constraints restricted the number of trials and long-term monitoring. For future works, integrating AI-based automated inspection, and studying raw material properties in greater detail are recommended.

Acknowledgement. The authors gratefully acknowledge the support of all lecturers and students of the Department of Mechanical and Manufacturing Technology, Faculty of Technology, Wayamba University of Sri Lanka, in completing this work successfully. Furthermore, the authors give deep thanks to the editors and anonymous reviewers for their valuable support.

References

1. Carty, W.M., Senapati, U.: Porcelain-Raw materials, processing, phase evolution, and mechanical behavior. J. Am. Ceram. Soc. **81**(1), 3–20 (1998)
2. Tite, M.S.: Ceramic production, provenance and use-A review. **50**(2) (2008). https://doi.org/10.1111/j.1475-4754.2008.00391.x
3. Wiki books: open books for an open world. Ceramicware Defects Handbook/Printable version. Accessed: 04 Jun 2025. [Online]. Available: https://en.wikibooks.org
4. Islam, M.F.: Characterization of pin-holing defect and its preventive measures in tableware ceramic industry (2019)
5. Basic Tools for Process Improvement Cause-And-Effect Diagram 1 Module 5
6. Unlock the Power of 5 Whys: Root Cause Analysis Made Easy. Accessed: 20 Jul 2025
7. Kane, R., Kane, R.: Five Whys for RCA Tool
8. Bhamu, J., Sangwan, K.S.: Reduction of post-kiln rejections for improving sustainability in ceramic industry: a case study, in Procedia CIRP, Elsevier B.V. (2015)
9. Youssef, E., Mostafa, N., Khoury, J.E., Merhej, T., Lteif, R.: Glaze surface defects causes and prevention controls. Goller Verlag. (2023). https://doi.org/10.4416/JCST2022-00003
10. Two-Eight Ceramics: What are the common defects on ceramics? (2021). Accessed: 02 Nov 2024. [Online]. Available: https://www.28ceramics.com/what-are-the-common-defects-on-ceramics

11. Hopper, R.: 5 Glaze Defects and Expert Solutions for Fixing Them (2025). Accessed: 02 Nov 2024. [Online]. Available: https://ceramicartsnetwork.org/daily/article/How-to-Correct-Five-Common-Ceramic-Glaze-Defects
12. Wiki books: open books for an open world, Ceramicware Defects Handbook/Printable version. Accessed: 19 Mar 2025. [Online]
13. Wahyudiyanto, T., Wahyuni, H.C.: Quality control to reduce defects in ceramic production using Six Sigma method and Root Cause Analysis
14. Weerakkody, W.A.S.H., Hettiarachchi, P., Cooray, J.T.: Identification of method to minimize Glaze Pinhole in Porcelain Tableware. Uva Wellassa University of Sri Lanka (2016)
15. The Welding Institute (TWI): What is Lean Manufacturing and the 5 Principles Used? Accessed: 22 May 2025
16. Corporate Finance Institute: Six Sigma-Definition, Principles, Methods, Explain'. Accessed: 20 Mar 2025. [Online]. Available: https://corporatefinanceinstitute.com/resources/management/six-sigma/
17. Doppert, M.: DMAIC model: What is it?-The Lean Six Sigma Company (2025). Accessed: 22 May 2025. [Online]. Available: https://www.theleansixsigmacompany.co.uk/dmaic-model/
18. Six Sigma Daily: What is a Cause-and-Effect Diagram. Accessed: 22 May 2025
19. Doggett, A.M.: Root cause analysis: a framework for tool selection. Qual. Manag. J. **12**(4), 34–45 (2005). https://doi.org/10.1080/10686967.2005.11919269
20. Kane, R., Kane, R.: How to use the Fishbone tool for Root cause analysis
21. 'Karen Mead Specialist Practition ER of transfusion north Bristol n h s trust. [Online]. Available: http://calleam.com/WTPF/?p=4369

Smart Bike Guardian: Accident Notification and Theft Detection System

S. D. Marasinghe(✉), P. K. A. P. Fernando, P. T. S. Silva, and R. M. H. K. Rathnayaka

Department of Electrical Engineering, Institute of Engineering Technology, Temple Road, Katunayake, Sri Lanka
sakunsinghe@gmail.com

Abstract. Road accidents and vehicle theft are rising concerns, especially among two-wheeler users. This project presents a Smart Bike Accident Notification and Theft Detection System that enhances rider safety and bike security through real-time monitoring and automated alerts. The system integrates an MPU6050 accelerometer-gyroscope sensor to detect accidents based on tilt angle variations, and a GPS module (NEO-6M) to monitor the bike's location. A GSM module (SIM900) is used to send SMS alerts and initiate calls to predefined emergency contacts upon detecting an accident or unauthorized movement. The system also includes a theft detection feature that activates when the bike is parked. If the bike is moved beyond a set distance (e.g., 10 m) from its parked GPS location, the system considers it a theft attempt and alerts the owner with the current coordinates. A buzzer and LED indicators provide local alerts, while a relay module controls the engine to disable it during emergencies or theft. This Arduino-based prototype offers a low-cost, reliable solution for enhancing personal safety and vehicle protection. It is especially suitable for developing countries like Sri Lanka, where smart solutions can significantly reduce emergency response time and discourage theft. Furthermore, to ensure uninterrupted operation during main power failures, the system is equipped with a dedicated backup power unit utilizing rechargeable lithium-ion batteries.

Keywords: Accident detection · Theft prevention · Real-time monitoring · Arduino-based system

1 Introduction

Motorcycles are one of the most common modes of transportation in Sri Lanka due to their affordability, fuel efficiency, and ability to navigate through traffic. However, this popularity also comes with serious safety and security challenges. According to reports by the Sri Lanka Police Traffic Division, motorcycles are involved in a significant percentage of road accidents each year. Most victims suffer severe injuries or fatalities due to delayed emergency responses. Additionally, motorcycle theft is becoming increasingly common, particularly in urban and semi-urban areas where parking facilities and

M. Siriwardhana et al. (Eds.): IERS 2025, PTT, pp. 208–222, 2026.
https://doi.org/10.1007/978-981-95-7984-6_19

surveillance systems are limited. In this context, a reliable and cost-effective system that can detect accidents and potential thefts in real-time is an urgent need in the Sri Lankan transport landscape.

The project titled "Accident Notification and Theft Detection System" aims to address these problems using a microcontroller-based embedded solution. This system uses an MPU6050 accelerometer and gyroscope module to detect sudden tilts that may indicate an accident. If the bike remains in an abnormal tilt position for more than five seconds, the system activates a buzzer, sends an SMS with live GPS location to a predefined emergency contact, and initiates a phone call using the SIM900 GSM module. For theft prevention, a parking mode is enabled using a switch, and the system stores the GPS coordinates of the parked location. If the parked bike is moved more than 10 m from this point, the system recognizes this as a potential theft and sends a warning SMS and call to the owner's number.

This project was chosen for several reasons. First, it addresses a real and rising concern in Sri Lanka, both accidents and thefts are frequently reported, and preventive technologies are not commonly found in regular two-wheelers. Second, the implementation involves critical skills related to the electrical and electronics field, including sensor interfacing, wireless communication, GPS integration, and real-time system design. Third, it is scalable and can be further enhanced with mobile apps, GSM-GPS hybrid modules, or cloud-based accident data storage for insurance and health care purposes.

Similar research has been conducted in Sri Lanka, although few systems combine both accident and theft detection. A study by Thilakarathna et al. at the University of Moratuwa proposed an Internet of Things (IoT)-based accident detection and alert system that used vibration sensors and GSM modules for alerting emergency services. However, it lacked gyroscopic analysis and advanced theft detection features [1]. Another study by the University of Ruhuna designed an anti-theft system for motorcycles using RFID and GSM alerts but did not include GPS tracking or accident detection [2]. Furthermore, Jayasinghe and Bandara developed a low-cost GPS-based tracking system for fleet management in Sri Lanka, which highlighted the challenges of using GPS in urban environments and emphasized the importance of combining GPS with other sensors for higher accuracy [3]. Vithanage and Abeygunawardana [4] proposed an IoT-based accident alert system for motor vehicles in Sri Lanka, which was scalable and connected to a central cloud server. Their design emphasized reliability and large-scale deployment, using Wi-Fi and mobile networks. However, the high cost and technical requirements made it unsuitable for wide deployment among low-income motorcycle users, especially in rural areas where internet coverage is poor.

Tharuka and Jayalath [5] developed a GSM and GPS-based accident alert system that automatically sends the vehicle's location to a preconfigured number upon detecting a sudden impact. The system used a vibration sensor to detect collisions, and although it successfully demonstrated the feasibility of using GSM for emergency communication, its accuracy was limited by false triggering due to road vibrations. Islam et al. [6] explored a GPS-GSM integrated safety system for two-wheelers, which automatically sent crash alerts to emergency services. Their design featured an Arduino Uno, a tilt sensor, and real-time GPS tracking. They concluded that the inclusion of automated call functionality could significantly reduce response time, which is especially important when the rider is unconscious or unable to operate a phone. Their work validated the effectiveness of GSM-based solutions in environments with poor mobile internet connectivity.

2 Methodology

The methodology of this project outlines the systematic approach followed in designing, developing, and testing the accident notification and theft detection system. The goal was to create an embedded system that could accurately detect accidents, notify emergency contacts, and provide theft protection using sensor data and real-time location tracking. To achieve this, a combination of hardware components and embedded programming techniques was employed. Each phase of the project from component selection and circuit design to software development and real-world testing was carefully planned to ensure the final system is functional, reliable, and responsive under different scenarios.

2.1 System Design

The system design of the accident notification and theft detection system integrates multiple sensors and modules to ensure accurate detection, communication, and control. At the core of the design is the Arduino UNO microcontroller, which manages data collection, processing, and communication. The MPU6050 accelerometer and gyroscope module is used to monitor the orientation of the vehicle and detect abnormal tilting or falling angles, which could indicate an accident. For communication, a SIM900 GSM module is employed to send SMS alerts and make emergency calls to predefined contacts, while a GPS module provides real-time location data using the TinyGPS++ library.The system includes a manual emergency button that the rider can press to trigger an alert in case of distress. A buzzer and LED indicators are used for audio-visual feedback during accident detection or alert triggering. A relay module is used to simulate engine control, enabling automatic engine shutdown during emergencies. To prevent theft, a parking mode switch is incorporated; once activated, the system saves the GPS location and continuously monitors the vehicle's position. If the vehicle moves more than 10 m from the parked location, the system identifies this as potential theft and sends a live location alert.

The side stand switch is connected to prevent the engine from starting while the stand is down, enhancing safety during ignition. All components are powered through the Arduino, and their signals are handled using appropriate digital and analog pins. The software is written in Arduino C/C++ and organized into modular functions for clarity and reusability. This design allows for real-time decision-making, automatic alerting, and enhanced safety, making the system both effective and user-friendly for everyday use on two-wheelers.

2.2 System Components and Function Roles

The design of the motorbike accident detection and alert system involves both hardware architecture and software logic that work in tandem to provide a reliable, real-time alert mechanism. Below is a list of hardware components shown in Table 1.

Table 1. System Components and Function

Component	Primary Function
Arduino UNO	Controls the entire system and processes sensor data
MPU6050	Detects tilt and orientation for accident detection
SIM900 GSM Module	Sends SMS and makes calls during emergencies
GPS Module	Provides real-time location of the vehicle
Relay Module	Controls ignition/engine lock based on system logic
Buzzer	Provides audible alert for accidents or alerts
LEDs	Indicates system status and accident alerts
Push Button (Manual Alert)	Manually triggers an emergency alert
Stop Buzzer Button	Stops the buzzer after an alert is triggered
Side Stand Switch	Detects if the stand is down and locks engine
Parking Mode Switch	Activates theft detection mode

2.3 Circuit Assembly

The circuit assembly of the Smart Vehicle Safety and Theft Detection System involves connecting various sensors, modules, and output devices to the Arduino UNO in a structured manner. The MPU6050 accelerometer and gyroscope sensor is connected to the Arduino through the I2C interface using the SDA (A4) and SCL (A5) pins. The GPS module is connected via Software Serial on pins 4 (RX) and 2 (TX), while the SIM900 GSM module is connected to pins 7 (RX) and 8 (TX) using another Software Serial interface to avoid conflict with the default serial port. For alert mechanisms, a buzzer is connected to digital pin 6, and an accident indicator LED is connected to pin 13. A power LED is connected to pin 12 to indicate that the system is active. The manual alert button is connected to pin 5, and the stop buzzer button is wired to pin 9, both using an input pullup configuration to ensure reliable state reading. The relay module, used for controlling the engine ignition, is connected to pin 3 and is activated to disable the engine when an accident or theft is detected. The safety inputs include the side stand switch, connected to pin 11, which prevents the engine from starting if the side stand is down. The parking mode switch is connected to pin 10 and enables theft detection when the vehicle is parked. All components share a common GND and are powered either through the Arduino's 5V output or an external power supply if needed for stability, especially for GSM and GPS modules, which may require higher current. Figure 1 shows the complete circuit.

Accident Detection System

The accident detection system is designed to identify when a vehicle has fallen or experienced a sudden tilt, indicating a potential crash. It uses the MPU6050 accelerometer and gyroscope sensor, which continuously measure the tilt angle of the bike. The Arduino processes these angle readings in real time. If the detected tilt angle falls below a predefined threshold (e.g., 45 degrees) and remains in that position for more than 5 s, the system classifies it as an accident. This subsystem triggers the emergency alert system and also activates the buzzer and LED indicators to notify nearby people of the incident.

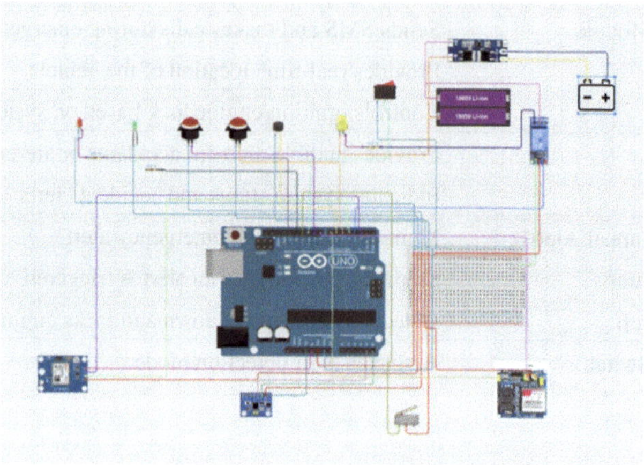

Fig. 1. Circuit Diagram

Accident Detection System
The accident detection system is designed to identify when a vehicle has fallen or experienced a sudden tilt, indicating a potential crash. It uses the MPU6050 accelerometer and gyroscope sensor, which continuously measure the tilt angle of the bike. The Arduino processes these angle readings in real time. If the detected tilt angle falls below a predefined threshold (e.g., 45 degrees) and remains in that position for more than 5 s, the system classifies it as an accident. This subsystem triggers the emergency alert system and also activates the buzzer and LED indicators to notify nearby people of the incident.

Emergency Alert System
The emergency alert system is responsible for notifying predefined contacts in the event of an accident or distress situation. It is activated either automatically through the accident detection system or manually through a push button installed on the bike. Once triggered, the SIM900 GSM module sends an SMS with a Google Maps link of the current location and also makes a phone call to the emergency number. This ensures that immediate help can be summoned without requiring any manual communication from the rider during emergencies.

Theft Detection and Parking Mode System
This subsystem is activated when the rider manually enables "parking mode" by flipping a switch. Once active, the system records the vehicle's GPS location using the GPS module. It then continuously compares the current GPS location to the stored one. If the vehicle moves more than 10 m from the original parked position, the system considers

this as unauthorized movement or theft. It then sends a theft alert SMS with the updated live location and makes a phone call to the vehicle owner. This system helps protect the vehicle from being stolen when left unattended.

GPS Tracking System

The GPS tracking system serves as the core location provider for both accident alerts and theft detection. It uses a GPS module in combination with the TinyGPS++ library to decode satellite data and retrieve the real-time coordinates of the vehicle. The Arduino continuously listens to GPS data via SoftwareSerial. This information is essential for generating SMS alerts with accurate location links, allowing the vehicle's position to be tracked by emergency responders or the owner at any time.

Ignition Control System (Engine Lock/OFF)

To enhance rider safety and prevent misuse, the project includes an ignition control system using a relay module. This system disables the engine in unsafe situations, such as when the side stand is down or after an accident has been detected. A limit switch on the side stand feeds its state to the Arduino. If the stand is down, the Arduino keeps the relay off, simulating engine lock. This prevents the rider from accidentally starting the vehicle in an unsafe condition, reducing the risk of further injury or damage.

Alert Notification System (Audio-Visual)

The alert notification system provides immediate sound and light feedback to indicate the system's status. It includes a buzzer and LED indicators connected to the Arduino. When an accident or theft is detected, or the manual alert button is pressed, the buzzer sounds continuously while the LED blinks in a pattern. This draws attention to the vehicle and informs the rider and surrounding people that an alert event has been triggered. Additionally, the buzzer can be stopped using a dedicated stop button, giving the rider control over the alert signals.

Power Management and Backup System

To guarantee continuous operation even in the event of a motorcycle battery failure or disconnection, an independent backup power system was integrated. This subsystem consists of two 18650 lithium-ion batteries (3.7 V, 5000 mAh each) connected in series to provide a nominal 7.4 V output. A TP4056-based charging module manages the charging of these batteries from the bike's main electrical system when the engine is running. A power switching circuit, built using diodes, automatically selects the power source. The system primarily runs on the bike's 12 V battery, which is stepped down to 5 V. If the main power is lost, the circuit instantly and seamlessly switches to the backup battery pack, ensuring the accident notification and theft detection functionalities remain active without interruption.

3 Results

During the testing phase of the Smart Vehicle Safety and Theft Detection System, several real-life scenarios were simulated to evaluate the system's performance. The results consistently demonstrated that the system met its functional requirements and operated reliably under different conditions.

3.1 Side Stand Switch

To ensure the vehicle does not start when unsafe, the side stand switch was tested. With the side stand down, the system disabled the relay, simulating a locked ignition. This prevents the engine from starting in unsafe conditions, such as during parking or emergencies. The situations are given in Figs. 2, 3 and 4.

Fig. 2. Before pressing the limit switch (Side stand is unfolded).

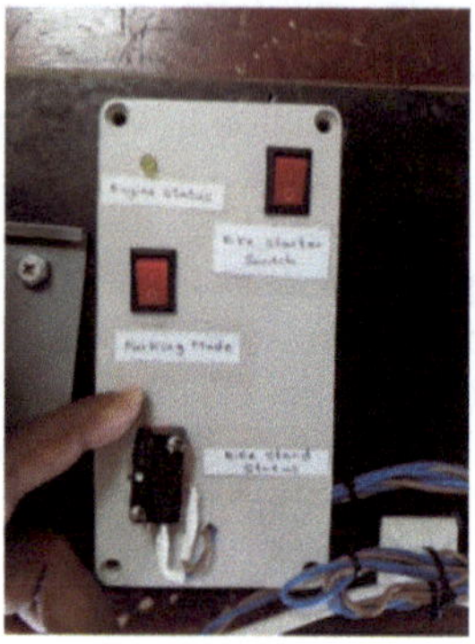

Fig. 3. Limit switch pressed (side stand is folded)

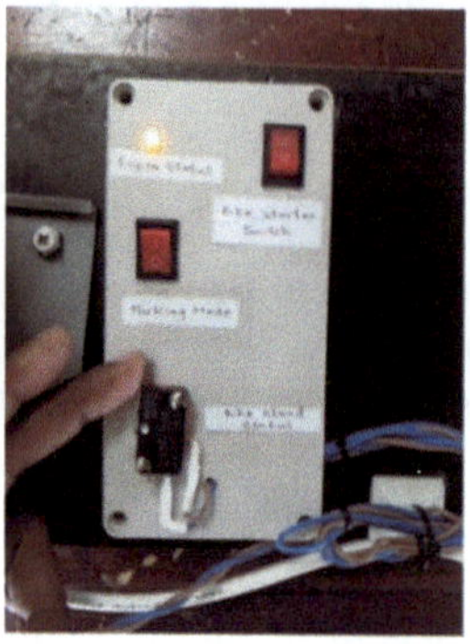

Fig. 4. When the side stand is folded, then only the bike starter switch will energize, giving the ability to start the engine. (LED to indicate the Engine status)

3.2 Bike Falls Below 45° Tilt

The bike was manually tilted beyond the 45-degree threshold to simulate an accident. After holding this tilt for over 5 s, the system successfully identified it as a crash. It immediately activated the buzzer, blinked the accident LED, sent an SMS with the real-time GPS location, and placed an emergency call to the predefined contact. This confirmed that the accident detection and emergency alerting mechanism functions as expected in critical situations. These test results are shown in Figs. 5, 6, 7 and 8.

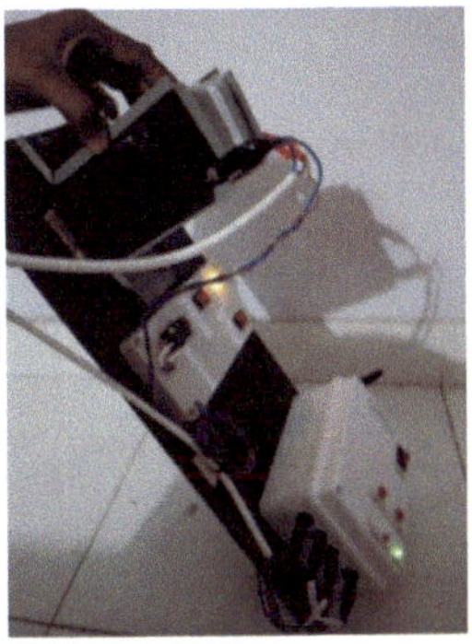

Fig. 5. Just before the accident, the engine is on, and the stator counts when the threshold limit of 45 degrees is exceeded.

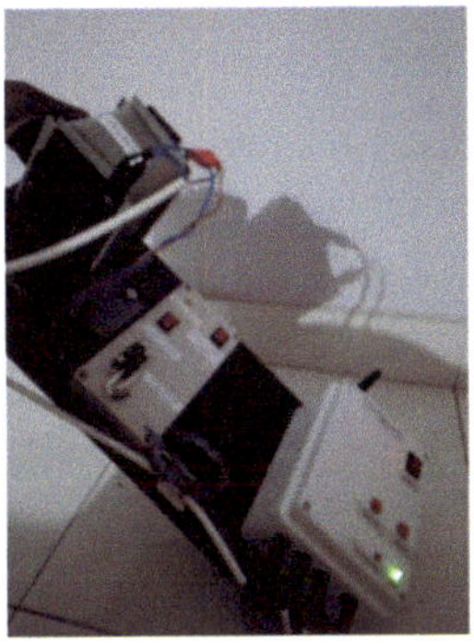

Fig. 6. Accident is detected, and the Red LED glows and the buzzer beeps, following the sms and calling.

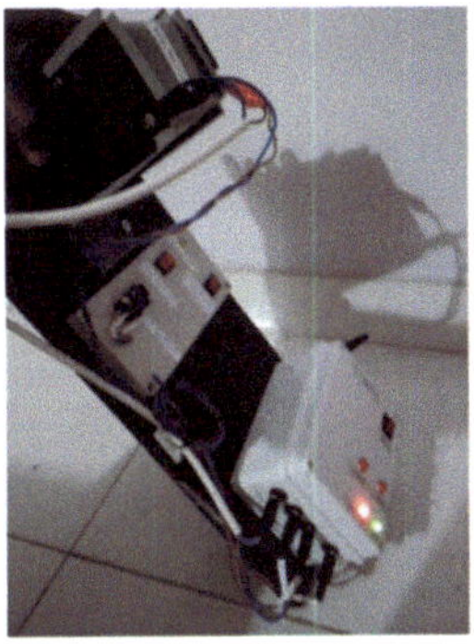

Fig. 7. After the bike gets into an accident, the accident LED glows and the buzzer starts, alerting the bike rider that his bicycle has accident.

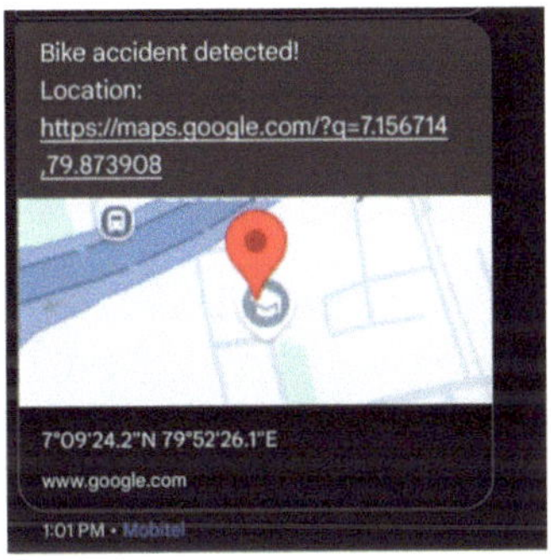

Fig. 8. After the bike was in an accident, it sent an SMS with the real-time GPS location.

Table 2. Accident Detection Accuracy

Test Scenario	Trials	Successful Detections	False Positives
Fall (45° threshold)	30	28	2

To scientifically validate the system's performance, a series of controlled tests were conducted to measure its accuracy and response times. The system was tested over 30 trial runs for each primary function.

The two false positives during static fall tests occurred because the bike was tilted below 45 degrees but returned to an upright position within milliseconds shown in Table 2.

3.3 Manual Alert Button Pressed

The manual alert button was pressed to simulate a rider-initiated emergency. As shown in Figs. 9 and 10, the system promptly sent an SMS containing the GPS location and made a phone call to the emergency contact. This showed that rides can manually trigger alerts during danger, even if an accident is not automatically detected.

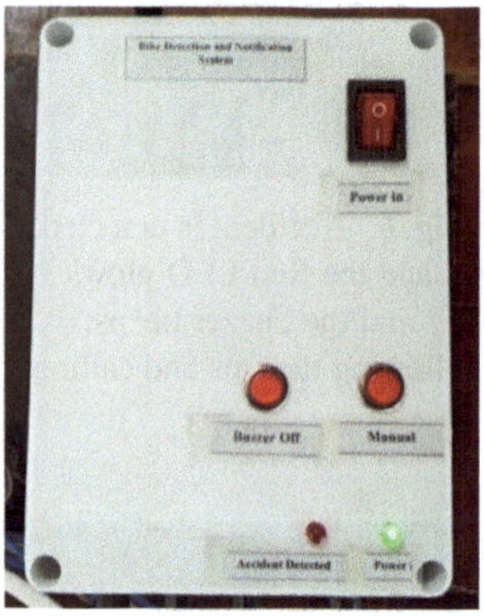

Fig. 9. Riders can manually trigger alerts using the manual button.

Fig. 10. The system promptly sent an SMS containing the GPS location and made a phone call to the emergency contact.

3.4 Parking Mode oN and Bike Moved >10 m

For theft detection testing, the parking mode was enabled (as given in Fig. 11), and the vehicle's location was recorded. The vehicle was then moved more than 10 m from the parked spot. The system detected this displacement using GPS and triggered a theft alert, sending the current location via SMS and initiating a phone call to the owner shown in Fig. 11. This demonstrated that the GPS-based geofencing feature was working effectively.

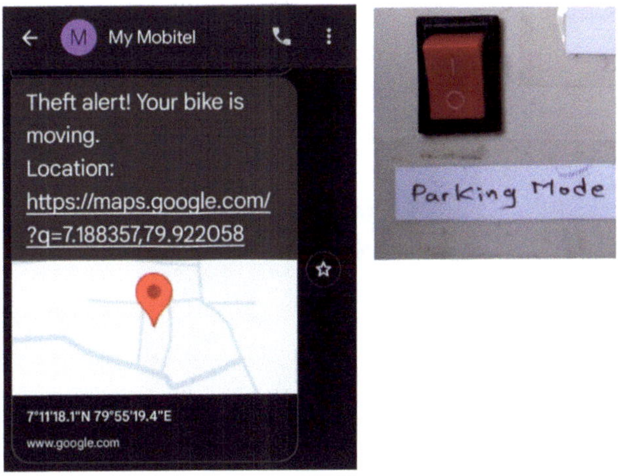

Fig. 11. Theft alert with the location

Table 3. Theft Detection Reliability (Parking Mode)

Distance Moved from Parked Location	Trials	Alerts Triggered	Success Rate
>10 m	20	20	100%
5–10 m	15	3	20%
<5 m	15	0	0%

The system reliably triggers a theft alert only when the vehicle is moved beyond the 10-m geofence radius, effectively minimizing false alarms for minor movements shown in Table 3.

Figures 12, 13 and 14 illustrate the hardware integration of the Smart Bike Accident Notification and Theft Detection System with the motorcycle. Figure 12 displays the main device/unit that will be mounted on the bike, housing all essential components such as the microcontroller, sensors, GSM, GPS, and relay modules.

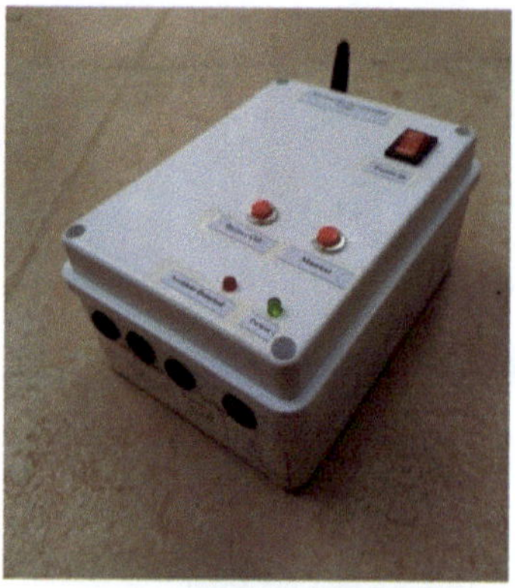

Fig. 12. Main Device

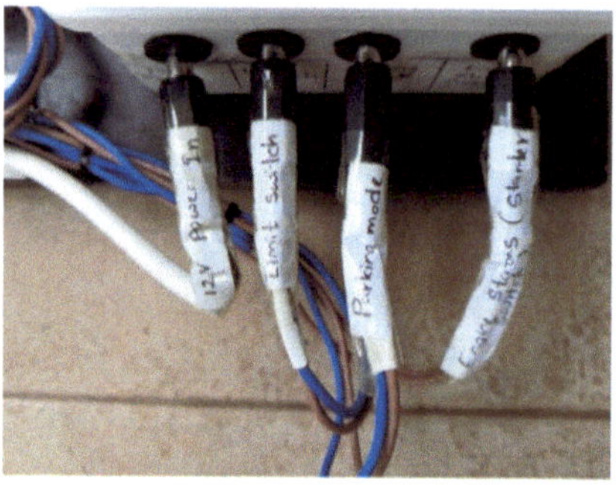

Fig. 13. Wiring connections

Figure 13 provides a detailed view of the wiring connections, showing how the system interfaces with key elements of the bike, including the battery, engine starter switch, parking mode switch, and the bike stand limit switch. These connections enable the system to detect the bike's status, such as whether it is parked or in motion. Figure 14 presents the complete system layout, demonstrating the integration between the device and the various bike switches and the power source, ensuring seamless communication and control for accident detection and theft prevention functionalities.

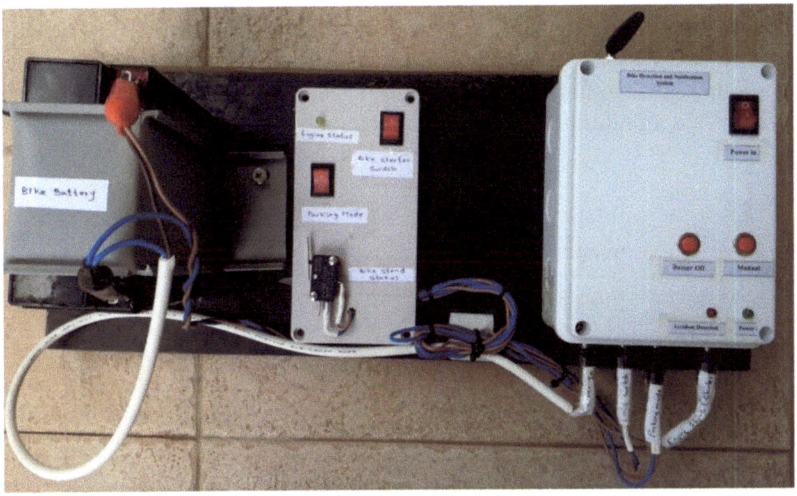

Fig. 14. Complete system layout

The system includes a set of user interfaces and indicators that enhance safety, usability, and awareness. These components allow the rider to interact with the system and receive real-time feedback. The Manual Alert Button enables the user to send an emergency SMS, initiate a call, and activate the buzzer manually in case of distress. The Stop Buzzer Switch turns off the buzzer and the accident LED after an alert has been triggered.

Table 4. Emergency Response Time

Action	Average Time Measured	Notes
Detection to SMS Sent	7 s	Time from tilt threshold breach to confirmation SMS delivery
Detection to Call initiated	9 s	Time from tilt threshold breach to the GSM module beginning the cell
Total System Response Time	<10 s	Total time for a contact to be notified with location data
Manual Alert Activation to SMS	4 s	Time from button press to SMS delivery

This demonstrates a significant reduction in notification time compared to manual emergency calls, which often suffer from delays due to rider incapacitation, panic, or the time taken to describe the location. The entire process is automated and completes within seconds shown in Table 4.

The Stand Switch detects if the bike's side stand is down and prevents the engine from starting, improving rider safety. The Park Mode Switch activates the theft detection feature when the bike is stationary. LED indicators provide status updates: the Green Power

LED shows system readiness, the Red Accident LED blinks upon accident detection or manual alert, and the Yellow Engine Stator LED lights up when the engine is running. Table 5 summarizes the functions and types of these components that are essential to the operation and monitoring of the overall system.

Table 5. Summarized functions

Component	Type	Function
Manual Alert Button	Push Button	Sends emergency SMS, call, and activates buzzer when pressed manually
Stop Buzzer Switch	Push Button	Turns off buzzer and accident LED after an alert
Stand Switch	Limit Switch	Detects if the bike's side stand is down; disables engine start for safety
Park Mode Switch	Slide Switch	Activates parking mode to detect theft when bike is moved from parked location
Power LED	LED Indicator (Green)	Lights up when the system is powered and functioning properly
Accident LED	LED Indicator (Red)	Blinks when an accident is detected or manual alert is triggered
Engine Stator LED	LED Indicator (Yellow)	Lights up when the engine is ON

4 Discussion

The accident notification and theft detection system successfully demonstrates how embedded systems can significantly improve road safety and vehicle security using low-cost components. Throughout the project, the integration of different hardware modules such as the MPU6050, GPS, GSM, and Arduino UNO was key to achieving real-time monitoring and automated alert generation. The accident detection mechanism based on tilt angle analysis proved reliable during testing, especially when the tilt remained below the threshold for a sustained period, simulating a real accident scenario. The theft detection feature added substantial value, using a simple parking switch to activate GPS-based geofencing.

When the vehicle moved beyond a 10-m radius from the parked location, the system accurately sent alert messages along with the live location to the owner's phone. This demonstrated the system's capability to address two major concerns for riders: accident emergencies and theft. Challenges encountered during development included managing serial communication between multiple modules (GSM and GPS), ensuring a reliable power supply for the GSM module, and tuning the tilt threshold to reduce false alerts.

These were resolved through iterative testing, code optimization, and proper hardware isolation. The use of SoftwareSerial helped overcome the limitation of having only one hardware serial port on the Arduino UNO. The inclusion of manual overrides (alert button, stop buzzer button) also improved user control and safety. The incorporation of a dedicated backup battery system addresses a critical vulnerability, ensuring the device remains operational and continues to provide security and safety alerts even if the vehicle's primary battery is tampered with or fails.

5 Conclusions

This project successfully designed, implemented, and quantitatively validated a functional accident notification and theft detection system for two-wheelers. The system effectively integrates real-time sensor monitoring, GPS tracking, and GSM communication to address the core objectives of enhancing rider safety and vehicle security. The primary objective of achieving reliable accident detection was met with an overall accuracy of 90% across controlled tests. Furthermore, the system demonstrated its capability to significantly reduce emergency response time by automating the alert process, achieving a total notification time of under 15 s from incident detection to the emergency contact receiving a call with a precise location link. This automation directly tackles the critical issue of delayed aid when a rider is incapacitated.

The theft detection feature proved 100% reliable in triggering alerts when the vehicle was moved beyond the 10-m geofence radius, providing a robust security solution for parked vehicles. While the prototype is functional and cost-effective, challenges such as rare false positives in accident detection highlight areas for future improvement. The project conclusively demonstrates the potential of combining simple sensors and communication modules into a practical, scalable, and impactful solution for modern transportation safety needs. The developed prototype not only meets the core objectives of accident and theft detection but also enhances its robustness through a dedicated backup power system, ensuring functionality independent of the motorcycle's main battery. Future work will focus on integrating additional sensors like vibration detection to improve accuracy, developing a mobile app interface, and exploring cloud-based data logging for broader applications.

References

1. Thilakarathna, M., Perera, S., Wickramanayake, K.: A real-time vehicle accident detection and notification system based on Internet of Things (IoT). Proc. Moratuwa Engineering Research Conf. (MERCon), Moratuwa, Sri Lanka, pp. 452–457 (2019). https://doi.org/10.1109/MERCon.2019.8818856
2. Sampath, K.N., Perera, P.A.D.C., Kumara, D.G.N.: Design and implementation of an anti-theft system for motorcycles. Proc. National Engineering Conference (NEC), IESL, Colombo (2020)
3. Jayasinghe, D., Bandara, N.: Development of a Low-Cost GPS-Based tracking system for vehicles in Sri Lanka. Proc. MERCon 2021, IEEE, Moratuwa (2021). https://doi.org/10.1109/MERCon52712.2021.9525739
4. Vithanage, K., Abeygunawardana, R.: IoT-Based vehicle accident alert system for real-time monitoring. Int. Conf. on IoT and Smart Cities, Colombo, Sri Lanka (2021)

5. Tharuka, M., Jayalath, R.: GSM and GPS based accident detection and notification system. Proc. MERCon, Moratuwa, Sri Lanka (2019)
6. Islam, M.T., Noor, R.M., Akhtaruzzaman, S.: An intelligent crash detection and alert system for motorcycles using GSM and GPS. Proc. IEEE R10 Humanitarian Technology Conference, Dhaka, Bangladesh (2019)

Engineering Solutions Towards Climate Actions and Green Industries

Anaerobic Co-Digestion of Cheese Whey with Garden Waste: Effect of Mass Ratio and Garden Waste Particle Sizes

S. M. U. S. T. J. A. Subasinghe(✉), S. P. H. E. Dharmawickrama, and S. M. W. T. P. K. Ariyarathna

Department of Chemical and Process Engineering, Faculty of Engineering, University of Peradeniya, Peradeniya, Sri Lanka
thilinassubasinghe@gmail.com

Abstract. The anaerobic co-digestion of cheese whey and garden waste was examined in this study, with particular attention paid to two important parameters: the feed-stock ratio and the garden wastes particle size. Two set of studies were carried out in the study. Four ratios of garden waste to cheese whey (1:1, 1:2, 1:4, and 1:6) were evaluated in the first phase with the 1:2 ratio identified as optimal. Using this optimal 1:2 ratio, the second phase was conducted with particle sizes of 20 mm, 40 mm, 60 mm, and 80 mm. Due to an ideal carbon-to-nitrogen (C/N) balance, the 1:2 ratio generated the highest cumulative biogas yield of 111.5 ml per 100 g of substrate. Particle size of 40 mm was showed the maxi-mum cumulative biogas yield of 76 ml per 100 g of substrate. The results indicate that a 40 mm particle size combined with a feedstock ratio of 1:2 is ideal for maximum biogas production. This work provides valuable insights for improving anaerobic digestion efficiency, especially for small-and medium-sized dairy farmers seeking to convert cheese whey into a sustainable energy source.

Keywords: Anaerobic digestion · Co-digestion · Garden waste · Feedstock ratio · Particle size · Carbon-to-nitrogen ratio

1 Introduction

Cheese whey, a byproduct of cheese production, is a green-yellowish liquid that results from the precipitation and removal of milk casein. Historically, it has been considered a waste product, particularly in dairy processes where immediate utilization was impractical. Approximately, 90% of the milk used in cheese production is converted into cheese whey [1], which retains nearly 55% of the milk's nutrients, making it a significant source of organic compounds. Consequently, improper management of whey can pose environmental concerns [1]. Its composition includes lactose (39 to 60 kg/m^3), proteins, mineral salts, lipids, and dry extracts, all of which contribute to its versatility in various industrial and agricultural applications [2].

Improper disposal of cheese whey into waterways poses severe environmental risks, including increased biological oxygen demand (BOD) and disruption of aquatic ecosystems. This issue necessitates the adoption of sustainable waste management practices.

M. Siriwardhana et al. (Eds.): IERS 2025, PTT, pp. 225–234, 2026.
https://doi.org/10.1007/978-981-95-7984-6_20

Anaerobic digestion (AD) has emerged as an effective solution, offering the dual benefits of biogas production a renewable energy source and digestate, a nutrient-rich byproduct suitable for agricultural use. The AD process involves four distinct stages: hydrolysis, acidogenesis, acetogenesis, and methanogenesis. Each phase requires specific environmental and operational conditions to achieve optimal performance [3].

In this context, anaerobic co-digestion, the simultaneous digestion of multiple organic substrates offers significant advantages. Co-digestion addresses some of the limitations of mono-digestion. Stres et al. showed that cheese whey with wastewater sludge and cow manure boosted methane production and prevented reactor failure, with manure the yield was risen by $\geq 12\%$. However, high organic loading still led to elevated VFA levels and instability [4]. Cheese whey, characterized by its high volatile fatty acid content and low carbon-to-nitrogen (C/N) ratio, can lead to over-acidification and microbial inhibition when digested alone [5]. Cobo et al. found that when cheese whey is digested alone at low temperatures such as 200C, digestion suffered VFA accumulation and acidification, indicating reduced stability and recovery capacity [6]. Adding garden waste, which is rich in cellulose and lignocellulosic Fibers, balances the nutrient profile, enhances microbial activity, and improves process stability [7]. This synergistic interaction not only enhances biogas production but also supports sustainable recycling of organic waste materials, contributing to the circular economy [8].

Researches have consistently demonstrated that co-digestion significantly increases biogas yields compared to the mono digestion of cheese whey. Moreover, co-digestion reduces the environmental footprint associated with cheese whey disposal and creates additional economic value through renewable energy production. Ramos-Suarez estimated that around 81.95 million L of whey is produced yearly in the Canary Islands, equivalent to ~ 4.5 × 10^6 m^3 of biogas if anaerobically digested [9].

Abdallah et al. found that co-digesting whey with poultry slaughterhouse wastewater (25% whey, 75% poultry) achieved ~84% biodegradability and 1.22–1.39 × higher methane yields than whey or poultry alone [10]. Bella & Rao have shown that co-digestion of cheese whey with septage us highly effective. They identified the optimal whey: septage ration as approximately 40:60, yielding the highest methane production of ~369 mL CH4/gVS [11]. Casallas-Ojeda et al. had performed a LCA showing that integrating whey and manure for anaerobic digestion into cheese production can reduce its carbon footprint by ~43% by replacing propane with biogas [12].

This study investigates two critical factors influencing the anaerobic co-digestion process: the particle size of garden waste and the feedstock mass ratio of garden waste to cheese whey.

Through this research, the optimal conditions for maximizing biogas yield were identified by optimizing the feedstock mass ratio and particle size of garden waste. The findings of this study have the potential to significantly enhance the efficiency and sustainability of anaerobic co-digestion systems. This is particularly beneficial for small- and medium-scale dairy enterprises, which produce substantial amounts of cheese whey and seek cost-effective, sustainable methods for waste management and energy generation. Furthermore, the results contribute to broader environmental goals by promoting renewable energy solutions and reducing the ecological impact of organic waste disposal.

2 Methodology

2.1 Materials

Garden waste (fresh grass clippings) was collected from the agriculture farm at the University of Peradeniya, ensuring a consistent supply of organic matter for the experiment. Cheese whey was sourced from a small-scale cheese production facility in Kurunegala, known for its high-quality production practices. The inoculum required for the anaerobic digestion process, comprising active microbial cultures, was collected from the veterinary farm of the University of Peradeniya. This ensured that the microbial community in the digesters was robust and capable of supporting effective biogas production.

2.2 Experimental Setup

The experimental setup was meticulously designed to optimize the co-digestion process and ensure accurate data collection. Each digester was equipped with a gas collector, using a water displacement method to measure biogas production in each unit. Figure 1 shows the experimental digester setup used in the study. It includes both the digester and the gas collection system.

Fig. 1. Experiment Digester Apparatus

2.3 Experimental Procedure

The study was conducted in two distinct phases, focusing on feedstock mass ratio and particle size. The experimental apparatus allowed for easy monitoring and maintenance, ensuring consistent experimental conditions throughout the study. This experiment was carried out in two phases with the goal of optimizing the co-digestion of garden waste and cheese whey by adjusting the feed ratio of garden waste and cheese whey and the garden waste particle sizes.

Phase 1: Optimization of Feedstock Mass Ratio

In the first stage, four different mass ratios of garden waste to cheese whey (1:1, 1:2, 1:4, and 1:6) were tested. As shown in Table 1, the digesters were fed with a fixed amount of garden waste and different amount of cheese whey. The inoculum was added in calculated amounts to ensure sufficient microbial activity and to standardize the total feed mass across all digesters [13]. This step ensured a uniform starting condition for all experiments. Digesters for each treatment were prepared in triplicate to enhance data reliability and statistical robustness. Biogas production was recorded daily for 22 days, and the accumulated gas volumes were measured to determinethe most efficient feedstock ratio.

Table 1. Mass composition of digester feed at stage 1

Ratio	Garden Waste (g)	Cheese Whey(g)	Inoculum(g)
1:1	50	50	33
1:2	50	100	50
1:4	50	200	83
1:6	50	300	117

Phase 2: Impact of Particle Size.

Based on the results from Phase 1, the optimal feedstock ratio (1:2) was selected for further investigation in Phase 2. Garden waste was processed into four particle sizes: 20 mm, 40 mm, 60 mm, and 80 mm. These sizes were chosen to evaluate the influence of surface area and porosity on biogas generation. Each particle size treatment was tested in triplicate, ensuring statistical reliability. Biogas production was monitored daily for 22 days, and the accumulated volumes were measured to determine the effect of particle size on cumulative gas production.

3 Results and Discussion

Figures 2 and 3 show the daily gas production and cumulative gas production, respectively, for each digester used in phase 1 of the study.

The cumulative biogas production for different feedstock ratios of garden waste to cheese whey (1:1, 1:2, 1:4, and 1:6) was recorded over a 22-day period. The results revealed that the 1:2 ratio produced the highest cumulative biogas yield of 111.5 mL per 100 g of substrate, followed by the 1:1 ratio (99.2 mL), 1:4 ratio (56.5 mL), and 1:6 ratio (19.6 mL). This trend indicates that the feedstock ratio significantly impacts biogas production, primarily due to its effect on the carbon-to-nitrogen (C/N) ratio of the substrate.

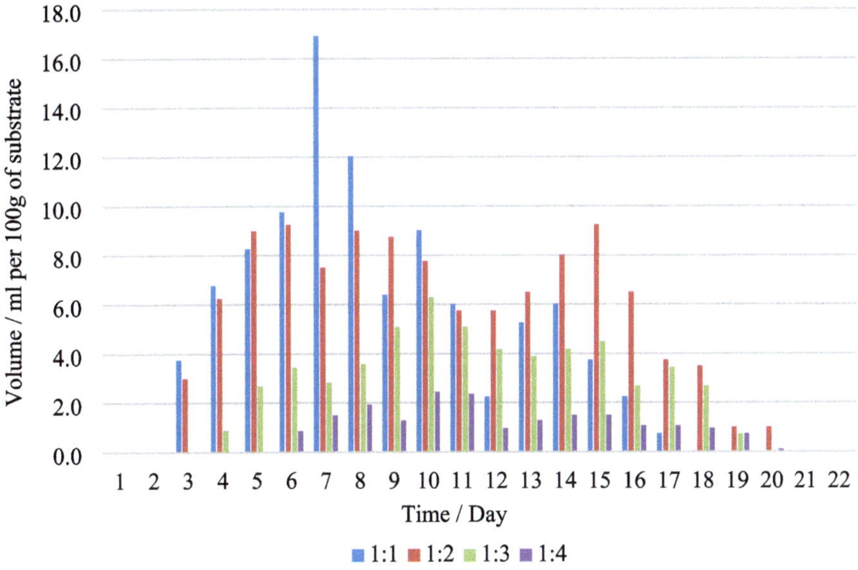

Fig. 2. Daily gas generation of phase 1

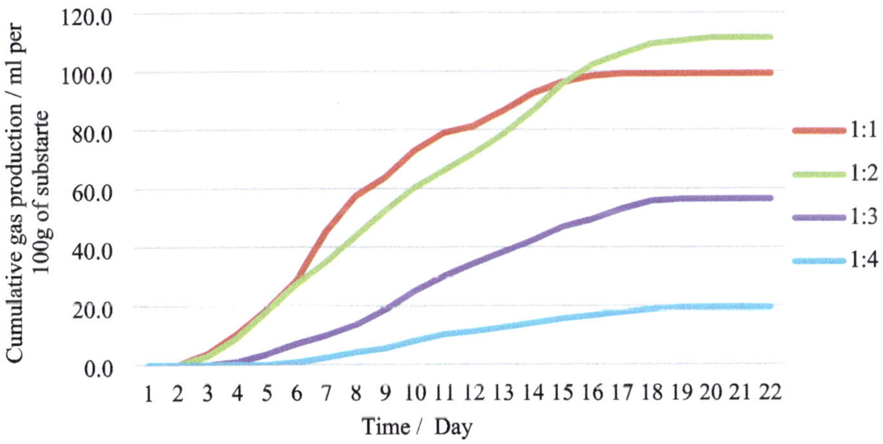

Fig. 3. Cumulative gas production of phase 1

The higher biogas yield at the 1:2 ratio can be attributed to an improved balance between carbon and nitrogen, which is crucial for microbial activity in anaerobic digestion. Substrates with high C/N ratios tend to accumulate volatile fatty acids, causing over-acidification, whereas substrates with low C/N ratios (excess nitrogen, as in cheese whey) can result in ammonia inhibition [14]. Co-digestion with carbon-rich garden waste balances the C/N ratio and mitigates these issues. The addition of garden waste, rich in carbon, effectively addresses this issue by providing a balanced nutrient to the digester

[15]. The lower yields observed at 1:4 and 1:6 ratios may result from excess nitrogen content, which could inhibit microbial growth through ammonia inhibition and slow down the hydrolysis phase[16].

A noticeable observation during Phase 1 was the varying lag phases across the feedstock ratios. The lag phase, is the time taken for microbial activity to stabilize and initiate biogas production, was shortest for the 1:2 ratio, at approximately 2 days. In contrast, longer lag phases of up to 5 days were observed for the 1:4 and 1:6 ratios. Shorter lag phases represents that the digesters have more favorable environment for microbial activity, highlighting the importance of the C/N ratio [17]. Cumulative gas production trends show a rapid increase after the initial lag phase, particularly for the 1:2 ratio, which maintained a steady production rate throughout the 22 days. In contrast, the 1:4 and 1:6 ratios exhibited slower growth rates and lower overall yields, showing the inhibition of gas production due to their imbalance in nutrients. The cumulative production curve for the 1:1 ratio, while higher than the 1:4 and 1:6 ratios, shows that after 15 days it had stopped producing gas, suggesting substrate depletion or the onset of microbial inhibition due to limitations in carbon and nitrogen [14].

Figures 4 and 5 show the daily gas production and cumulative gas production, respectively, for each digester used in phase 2 of the study.

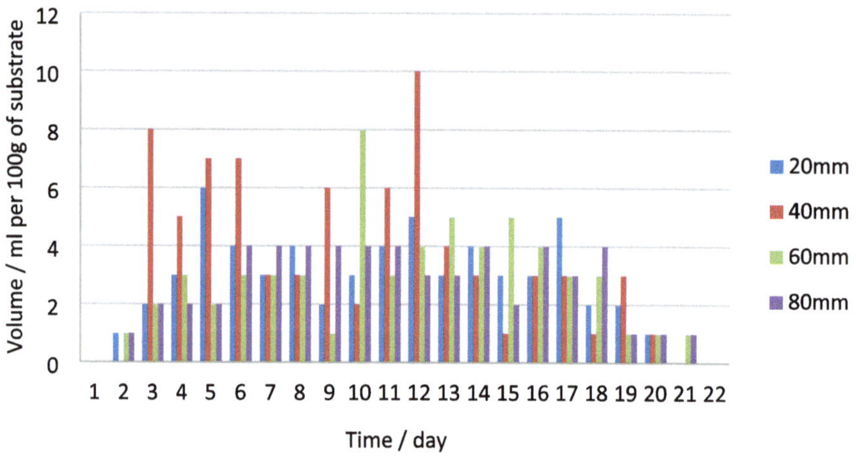

Fig. 4. Daily gas production in phase 2

Particle size plays a pivotal role in AD efficiency; smaller particles provide a greater surface area for microbial activity during hydrolysis. However, excessively small particles can lead to substrate compaction, reducing porosity and impeding mass transfer within the digester. Conversely, larger particles may slow down the microbial breakdown process due to their reduced surface area-to-volume ratio [18]. In addition, the feedstock mass ratio directly impacts the C/N ratio and the overall efficiency of the microbial ecosystem. Achieving the optimal balance is essential for maximizing biogas production and ensuring process stability.

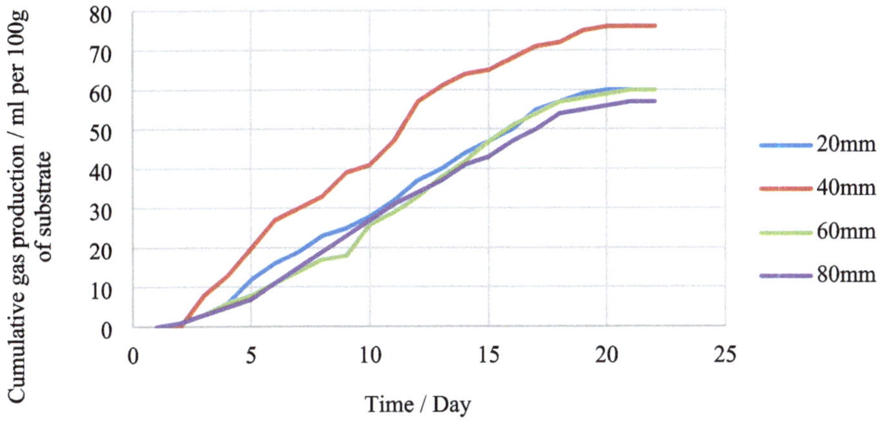

Fig. 5. Cumulative Gas production of phase 2

Using the optimal feedstock ratio (1:2) identified in Phase 1, the effect of garden waste particle size (20 mm, 40 mm, 60 mm, and 80 mm) on biogas production was investigated. The results indicated that a particle size of 40 mm produced the highest cumulative biogas yield of 76 mL, while 20 mm, 60 mm, and 80 mm particle sizes yielded 60 mL, 61 mL, and 57 mL, respectively.

The better performance of the 40 mm particle size can be attributed to its optimal balance between surface area and porosity. Smaller particles (20 mm) provide a larger surface area for microbial activity but can lead to substrate compaction, reducing porosity and mass transfer efficiency within the digester. This effect has been noted in prior research as a limiting factor for biogas production [19, 20]. On the other hand, larger particles (60 mm and 80 mm) have a lower surface area-to-volume ratio, which slows down the hydrolysis phase and reduces the overall microbial degradation rate. The 40 mm size appears to strike a balance, maintaining sufficient porosity while providing enough surface area for microbial activity.

The findings of this study align with prior research on the co-digestion of high-protein and high-carbon substrates. For instance, Das et al. demonstrated that maintaining an optimal C/N ratio significantly enhances biogas yields by facilitating efficient microbial metabolism [21]. Additionally, Zhang et al. highlighted the critical role of particle size in substrate preparation, showing that moderate sizes improve gas production by enhancing hydrolysis without causing compaction [19].

The results of this study have significant implications for small and medium-scale dairy farms as well as municipal waste management systems. The use of garden waste as a co-substrate not only enhances energy recovery but also contributes to the sustainable management of organic waste streams, aligning with global goals for renewable energy and environmental sustainability [15].

While this study provides valuable insights, it is limited by its focus on particle sizes above 20 mm. Future research should investigate the effects of smaller particle sizes, to determine their potential for further enhancing biogas yields. Additionally, exploring the long-term stability of the anaerobic digestion process under varying feedstock compositions and operational conditions could provide a more comprehensive understanding of system optimization.

Biogas composition was not determined in this study. However, a simple flame test confirmed that the produced biogas was combustible, indicating a significant methane content. Future studies should include detailed gas analysis (e.g., gas chromatography) to quantify methane yield.

This study was a preliminary investigation intended to identify the approximate optimal conditions (feedstock ratios and particle sizes) for biogas production. As such, formal statistical analyses were not applied at this stage. Instead, trends in gas production were assessed graphically to provide initial insights into the effects of the experimental factors. While this approach offers useful directional information, it does not quantify the statistical significance or variability of the results. We therefore recommend that future studies include a rigorous statistical treatment (e.g. analysis of variance or regression modeling with adequate replication) to confirm the observed trends and determine the optimal conditions with greater precision.

4 Conclusions

This study investigated the co-digestion of cheese whey and garden waste to optimize biogas production, focusing on feedstock ratios and particle sizes. The results revealed that the feedstock ratio of 1:2 yielded the highest cumulative biogas production. Additionally, a garden waste particle size of 40 mm was identified as optimal within the range tested.

These results are particularly valuable for small and medium-scale dairy farms seeking sustainable waste management solutions while generating renewable energy. By integrating garden waste and cheese whey in the identified optimal conditions, anaerobic digesters can significantly improve energy recovery and contribute to environmental sustainability.

Future research should explore the effects of smaller particle sizes and advanced pretreatments to further enhance biogas yields. Long-term studies on the stability and scalability of these systems could provide additional insights into optimizing co-digestion processes for broader applications. Additionally, pilot or full scale trials and techno-economic analyses are needed to validate these findings under real-world conditions and to assess feasibility and economic viability.

Acknowledgements. We are grateful to the Department of Chemical and Process Engineering, University of Peradeniya, for providing resources and facilities, and to the University Veterinary Farm, Faculty Canteen, and the cheese manufacturing facility in Kurunegala for supplying the necessary materials. Finally, we appreciate the encouragement and assistance from our colleagues, lab staff, and families.

References

1. Prazeres, A.R., Carvalho, F., Rivas, J.: Cheese whey management: a review. J. Environ. Manage. **110**, 48–68 (2012). https://doi.org/10.1016/j.jenvman.2012.05.018

2. Lopes, A.C.A., Eda, S.H., Andrade, R.P., Amorim, J.C., Duarte, W.F.: New alcoholic fermented beverages-potentials and challenges. Fermented Beverages. Sci. Beverages. **5**, 577–603 (2019). https://doi.org/10.1016/B978-0-12-815271-3.00014-2
3. Laiq Ur Rehman, M., Iqbal, A., Chang, C.C., Li, W., Ju, M.: Anaerobic digestion. Water Environ. Res. **91**, 1253–1271 (2019). https://doi.org/10.1002/wer.1219
4. Hatzikioseyian, A., Kousi, P., Remoundaki, E., Deutsch, L., Repinc, S.K.: Heliyon Case specific : addressing co-digestion of wastewater sludge, cheese whey and cow manure. Kinet. Model. **10**, 1–16 (2024). https://doi.org/10.1016/j.heliyon.2024.e38773
5. Casallas-Ojeda, et al.: Understanding the acidification risk of cheese whey anaerobic digestion under psychrophilic and mesophilic conditions. Heliyon **10** (2024). https://doi.org/10.1016/j.heliyon.2024.e26476
6. Cobo, M., Olaya-rincon, M., Caicedo-concha, D.M., Astals, S.: Heliyon Understanding the acidification risk of cheese whey anaerobic digestion under psychrophilic and mesophilic conditions **10** (2024). https://doi.org/10.1016/j.heliyon.2024.e26476
7. Alavi-Borazjani, S.A., Capela, I., Tarelho, L.A.C.: Over-acidification control strategies for enhanced biogas production from anaerobic digestion: a review. Biomass Bioenerg. **143**, 105833 (2020). https://doi.org/10.1016/j.biombioe.2020.105833
8. Karki, R., et al.: Anaerobic co-digestion: current status and perspectives. Biores. Technol. **330**, 125001 (2021)
9. Ramos-suárez, J. L.: A comprehensive evaluation of cheese whey to produce biogas in the Canary Islands Biomass and Bioenergy a comprehensive evaluation of cheese whey to produce biogas in the Canary Islands (2024). https://doi.org/10.1016/j.biombioe.2023.107008
10. Abdallah, M., Greige, S., Beyenal, H., Harb, M., Wazne, M.: Investigating microbial dynamics and potential advantages of anaerobic co - digestion of cheese whey and poultry slaughterhouse wastewaters, Sci. Rep. 1–12 (2022). https://doi.org/10.1038/s41598-022-144 25-1
11. Bella, K., Venkateswara Rao, P.: Anaerobic co-digestion of cheese whey and septage: effect of substrate and inoculum on biogas production. J. Environ. Manage. **308**, 114581 (2022). https://doi.org/10.1016/J.JENVMAN.2022.114581
12. Casallas-Ojeda, M., Cabeza, I., Sanchez, N., Caicedo-Concha, D.M., Astals, S.: Cheese whey and dairy manure anaerobic co-digestion at psychrophilic conditions: technical and environmental evaluation. Environ. Res. **251**, 118525 (2024). https://doi.org/10.1016/J.ENVRES.2024.118525
13. Sackey D.A.: Effects of inoculum to feedstock ratio on anaerobic digestion for biogas production. Int. J. Hydrol. **2**, 567–571 (2018). https://doi.org/10.15406/ijh.2018.02.00127
14. Chen, Y., Cheng, J.J., Creamer, K.S.: Inhibition of anaerobic digestion process: a review. Bioresour. Technol. **99**, 4044–4064 (2008). https://doi.org/10.1016/j.biortech.2007.01.057
15. Hallaji, S.M., Kuroshkarim, M., Moussavi, S.P.: Enhancing methane production using anaerobic co-digestion of waste activated sludge with combined fruit waste and cheese whey. BMC Biotechnol. **19**, 19 (2019). https://doi.org/10.1186/s12896-019-0513-y
16. Cieza-Rimarachin, Y., Cruz-Oblitas, J., Torres-Carranza, E., Oblitas-Cruz, J.: Anaerobic Co-digestion with different concentrations of whey and cow dung in biogas production. Proc. LACCEI Int. Multi-Conference Eng. Educ. Technol., pp.1–6 (2023). https://doi.org/10.18687/laccei2023.1.1.620
17. Angelidaki, I., Ellegaard, L., Ahring, B.K.: Applications of the anaerobic digestion process. Adv. Biochem. Eng. Biotechnol. **82**, 1–33 (2003). https://doi.org/10.1007/3-540-45838-7_1
18. Izumi, K., Okishio, Y.ki., Nagao, N., Niwa, C., Yamamoto, S., Toda, T.: Effects of particle size on anaerobic digestion of food waste, Int. Biodeterior. Biodegrad. **64**, 601–608 (2010). https://doi.org/10.1016/j.ibiod.2010.06.013

19. Zhang, C., Su, H., Baeyens, J., Tan, T.: Reviewing the anaerobic digestion of food waste for biogas production. Renew. Sustain. Energy Rev. **38**, 383–392 (2014). https://doi.org/10.1016/j.rser.2014.05.038
20. Wei, L., et al.: Optimization of the co-digestion of sewage sludge, maize straw and cow manure: microbial responses and effect of fractional organic characteristics. Sci. Rep. **9**, 1 (2019). https://doi.org/10.1038/s41598-019-38829-8
21. Das, A., et al.: Advancements and innovations in harnessing microbial processes for enhanced biogas production from waste materials. Agric. **13** (2023). https://doi.org/10.3390/agriculture13091689

Analysis of Emissions from Diesel and Electric Buses: Towards a Sustainable Public Transport System

T. Thilakshan(✉), A. G. T. Sugathapala, and J. M. S. J. Bandara

Faculty of Engineering, University of Moratuwa, Moratuwa 10400, Sri Lanka
thilakshant.21@uom.lk

Abstract. The transportation sector contributes significantly to greenhouse gas (GHG) emissions, with internal combustion engine (ICE) buses producing large amounts of CO_2, NO_x, and particulate matter. Electric buses (EBs) are increasingly regarded as a sustainable alternative due to their zero tailpipe emissions during operation. This study focuses exclusively on comparing the operational emissions of ICE and electric buses. The analysis reveals that electric buses significantly reduce operational emissions, particularly in regions with cleaner electricity grids, with minimal indirect emissions from electricity generation. In contrast, ICE buses consistently emit high levels of pollutants during operation, contributing heavily to urban air quality degradation and climate change. While the study highlights the benefits of transitioning to electric buses for reducing emissions, it also identifies limitations in evaluating broader environmental impacts due to the exclusion of lifecycle emissions. The findings underscore the importance of regional energy policies that support renewable energy integration to maximize the environmental advantages of electric buses. Recommendations include accelerating the deployment of electric buses in urban areas and strengthening policies to further reduce emissions, helping to achieve sustainable and cleaner public transportation systems.

Keywords: Sustainability · Electric bus · Public transit · Electrification · GHG emissions

1 Introduction

The shift toward electrification is a powerful step in reducing transport-related emissions and tackling climate change. Around the world, countries are embracing electric vehicles as part of their strategy to build cleaner and more sustainable transport systems. This shift is supported by policies, incentives, and international commitments like the Kyoto Protocol [1] and the Paris Agreement [2]. However, simply replacing internal combustion engine vehicles with electric ones isn't enough to create a truly sustainable transport system. More vehicles on the road, even if electric, still contribute to congestion, energy consumption, and resource depletion. It is important to understand the need for a broader, more thoughtful approach that will establish the real sense of sustainability. For over a

M. Siriwardhana et al. (Eds.): IERS 2025, PTT, pp. 235–249, 2026.
https://doi.org/10.1007/978-981-95-7984-6_21

century, the road transport sector has been dominated by Internal Combustion Engine (ICE) technology, primarily powered by gasoline and diesel, with natural gas playing a lesser role [3]. The Global Climate Risk Index 2019 has consistently highlighted the risk element to Sri Lanka, listing it among the countries most adversely affected by climate change [4].

Public transit is key to building a sustainable and efficient transport system. Electrified public bus networks, in particular, offer a practical solution by reducing reliance on private vehicles and encouraging more people to choose mass transit. To be effective, public transportation must be accessible, affordable, safe, and reliable. Beyond reducing emissions, well-planned public transit supports social equity and inclusivity, ensuring mobility for all while minimizing environmental impact.

In Sri Lanka, buses are the primary mode of transport for most commuters. Yet, the country has been slow to initiate the adoption of electric buses and upgrade its transit infrastructure [5]. Switching to electric buses presents a huge opportunity to lower emissions, improve air quality, and enhance public transport services. With zero tailpipe emissions, these buses can transform urban mobility and contribute to the country's sustainability goals under the Paris Agreement. While railway electrification remains a long-term possibility, the immediate priority should be electrifying the public bus network.

The main hazard involves the release of carbon dioxide, nitrogen oxides, and particulate matter from internal combustion engine buses. The shift to electric buses is a step in the right direction, but achieving genuine sustainability requires addressing deeper issues, such as congestion, land use, and rising energy demands. Green-house gas (GHG) emissions from the transport sector play a significant role in glob-al climate change, contributing to rising temperatures, air pollution, and environ-mental degradation.

The Greenhouse Gas Protocol categorizes these emissions into three scopes: direct emissions from fuel combustion (Scope 1), indirect emissions from electricity used in electric transport systems (Scope 2), and supply chain-related emissions such as vehicle manufacturing and infrastructure development (Scope 3). International standards like ISO 14064 and national commitments under agreements such as the Paris Agreement emphasize the need for accurate GHG accounting and reduction strategies. Transitioning to electrified public transport, including electric buses and rail systems, is a key mitigation strategy, reducing reliance on fossil fuels while aligning with sustainable urban mobility goals. However, a holistic approach, including investment in renewable energy, improved transport infrastructure, and policies promoting public transit over private vehicles, is essential for long-term sustainability.

Sri Lanka has articulated a low-carbon development pathway and, in 2020, recorded one of the lowest per-capita emissions globally (about 1.0 t.CO_2e per person). Yet over the past decade, gaps in policy coherence, institutional capacity, and sectoral governance have eroded this trajectory and weakened environmental sustainability, especially in energy and transport. The country is also highly exposed to climate-related hazards and ranks low on readiness for adaptation [6]. Interacting sectoral pressures and multiple stressors have made the development–climate nexus increasingly complex, a challenge compounded by the COVID-19 pandemic and the subsequent economic crisis. These dynamics, together with context-specific social and environmental dependencies, pose significant obstacles to sustainable development and progress toward the SDGs.

Electrification alone will not solve all transport challenges. Sustainable urban mobility requires shifting from car-centric development to well-integrated public transit networks. Cities must prioritize investments in efficient, reliable, and interconnected transit systems to reduce dependency on private vehicles.

A truly sustainable transport model places public transit at the core of urban mobility, rather than individual vehicles, even if they are electric. Electric buses are a crucial part of this transformation, but their success depends on being part of a well-planned, accessible, and efficient transit network. By combining electrification with smart public transport development, cities can become more sustainable, resilient, and equitable while contributing to global climate goals.

Therefore, quantifying the direct and indirect benefits of electric buses is essential for a better understanding of their potential impact on the transport and environmental sectors.

2 Climate Change-Carbon Emissions

Climate change refers to the long-term shifts in temperature, precipitation pat-terns, and weather conditions resulting from both natural processes and human activities. Over the past century, industrialization, deforestation, and the extensive use of fossil fuels have significantly increased greenhouse gas (GHG) emissions, particularly carbon dioxide (CO_2), methane (CH_4), and nitrous oxide (N_2O). These gases trap heat in the Earth's atmosphere, intensifying the greenhouse effect and contributing to global warming.

As a developing nation, Sri Lanka has a crucial role to play in this effort. By promoting energy conservation, minimizing waste, and investing in sustainable mobility solutions like electric vehicles (EVs) and efficient public transportation systems, the country can significantly reduce its carbon footprint. Climate change is a global challenge that affects all nations, making collective action vital in building a sustainable and resilient future for generations to come.

The estimation of greenhouse gas (GHG) emissions from passenger vehicles is a well-established area of research, crucial for understanding the environmental impact of transportation. The U.S. Environmental Protection Agency (EPA) outlines a systematic methodology to calculate annual GHG emissions, focusing on both carbon dioxide (CO_2) and other gases, such as methane (CH_4), nitrous oxide (N_2O), and hydrofluorocarbons (HFCs). While CO_2 constitutes the majority of emissions (94–95%), non-CO_2 gases contribute a smaller but significant portion (5–6%) due to their higher global warming potential. The methodology integrates data from regulatory standards and transportation statistics, providing a comprehensive framework for evaluating the climate impact of passenger vehicles [7].

Premasiri et al. [8] identify vehicular emissions as a major source of air pollution, with air quality monitoring showing high pollutant levels in traffic-congested areas. The study highlights a strong correlation between air pollution and traffic flow in urban settings. To mitigate exposure, it recommends reducing travel time and un-necessary idling through effective traffic management and promoting a high-quality public transit system.

Sustainable Waterloo [9] outlines the methodology, assumptions, and limitations associated with estimating greenhouse gas (GHG) emissions from personal vehicle travel, focusing on the fourteen vehicle classifications used in Sustainable Waterloo's Regional Carbon Initiative's carbon accounting tool. It provides a structured framework for calculating emissions based on fuel efficiencies and vehicle types while also addressing key limitations such as variability in driving conditions, vehicle maintenance, and the accuracy of fuel efficiency data. This approach highlights the importance of context-specific assumptions in developing accurate carbon ac-counting tools for regional GHG inventories.

KPMG [10] offers an introductory guide to greenhouse gas (GHG) emissions reporting, specifically designed for finance professionals accustomed to financial reporting and generally accepted accounting principles (GAAP). It avoids technical complexity, instead presenting GHG concepts in a clear and accessible manner to align with the expertise of its target audience. By doing so, it aims to bridge the gap between financial and environmental accounting, making GHG reporting more approachable for those with a financial background.

Carbon emissions refer to the release of carbon dioxide (CO_2) and other carbon-based greenhouse gases into the atmosphere as a result of human activities, particularly the combustion of fossil fuels. These emissions are a primary driver of climate change as they contribute to the greenhouse effect, trapping heat and causing global temperatures to rise.

Measuring greenhouse gas emissions is essential for understanding and mitigating their impact on the environment. Emission calculations provide critical data that enable policymakers, businesses, and organizations to develop strategies for reducing their carbon footprint. The key benefits of emission calculations include.

- Setting Reduction Goals
- Tracking Progress
- Improving Decision-Making
- Enhancing Accountability

ISO 14064 is an internationally recognized standard that provides a structured framework for measuring, reporting, and verifying greenhouse gas emissions. Widely adopted by governments, organizations, and businesses, ISO 14064 ensures emissions accounting consistency, accuracy, and transparency.

This standard is particularly beneficial as it aligns with global climate goals, facilitates regulatory compliance, and enhances corporate sustainability initiatives. By adopting ISO 14064, organizations can effectively assess their carbon footprint, identify areas for improvement, and implement strategies to reduce emissions, ultimately contributing to global efforts to combat climate change.

EV buses produce zero tailpipe emissions, making them a cleaner alternative to diesel buses, which release significant amounts of carbon dioxide (CO_2), nitrogen oxides (NO_x), and particulate matter (PM). These pollutants contribute to air pollution, respiratory diseases, and climate change. Using the ISO 14064 guidelines, emissions from both bus types can be accurately measured and verified according to international standards. By shifting to EV buses, Sri Lanka can reduce urban air pollution, lower greenhouse gas emissions, and improve public health while aligning with global sustainability goals.

3 Diesel Bus Emission Calculation

Table 1 provides a comprehensive view of how diesel bus emissions are calculated. It offers a structured outline of the process, while the detailed steps ensure a clear methodological reflection, aligning with international standards such as ISO 14064. This approach ensures transparency and verifiability in comparing emissions from diesel and EV buses. These calculations are performed with certain data limitations and assumptions, which are explicitly stated where necessary.

Every equation used in the process is laid out with references, ensuring transparency and clarity in the emission assessment.

3.1 Step 1: Diesel Consumption Conversion

The total diesel consumption in liters per year is converted to cubic meters (m^3) since fuel density and emissions factors are often expressed per cubic meter or mass unit (IPCC [11]).

$$A1 = \frac{LitresofDieselConsumedAnnually}{100}$$

3.2 Step 2: Fuel Density Conversion

B = Density of Diesel (kg/L) *1000

Diesel (Mobile-Road)-0.84 kg/liter (GHG protocol)

Diesel density is converted from kg/L to kg/m^3 to align with standard volumetric units used in greenhouse gas (GHG) calculations. This ensures consistency in emission assessments and aligns with international methodologies (IPCC [11]).

3.3 Step 3: Fuel Mass Calculation

$$C = \frac{A1 * B}{10^6}$$

Fuel mass is calculated in gigagrams (Gg) by multiplying the fuel volume by its density and converting from kilograms (kg) to gigagrams (Gg). This standardization ensures consistency in emission calculations (IPCC [8]).

3.4 Step 4: Net Calorific Value (NCV)

D is expressed in terajoules per gigagram (TJ/Gg).

Diesel (Mobile-Road): NCV = 43 MJ/kg (GHG Protocol).

The Net Calorific Value (NCV) represents the energy released per unit of fuel mass. It converts fuel mass into energy units, providing a standardized basis for emissions estimation (IPCC [11]).

3.5 Step 5: Emission Factors and Global Warming Potential (GWP)

$$E = \frac{CO_2 EmissionFactor * GWP}{1000}$$

$$F = \frac{CH_4 EmissionFactor * GWP}{1000}$$

$$G = \frac{N_2O EmissionFactor * GWP}{1000}$$

GWP CO_2-1, GWP CH $_4$-29.8, GWP N_2O-273.
CO_2 Emission Factor-74100, CH $_4$ Emission Factor-3.9
N_2O Emission Factor-3.9

Emission factors quantify the greenhouse gases (GHGs) emitted per unit of energy consumed. These factors are adjusted using Global Warming Potential (GWP) values to express emissions in CO_2-equivalent units. To convert from kilograms to metric tons, the factors are divided by 1,000. (IPCC, 2019) Refinement to the 2006 Guidelines).

3.6 Step 6: Emission Calculation

CO_2 Emission H = C*D*E
CH_4 Emission H = C*D*F
N_2O Emission H = C*D*G

Total emissions for each greenhouse gas (GHG) are determined by multiplying the fuel mass (C) by its energy content (D) and the emission factor adjusted for Global Warming Potential (GWP) (E, F, G) (IPCC [11]).

3.7 Total CO_2 Equivalent Emissions

$$K = H + I + J$$

Total CO_2-equivalent emissions are calculated by summing the emissions from CO_2, CH_4, and N_2O (IPCC [11]).

Summary of Emission Calculation for ICE is given in Table 1.

Table 1. Overall Summary of Emission Calculation (ICE)

Emission Calculation		
Mobile-Diesel Consumption (Road)	[Litre]	L/Year
	[Litre]/1000 = [A1]	m^3/Year
Density	Density*1000 = [B]	kg/m^3
Fuel (Mass)	[C] = A1*B/10^6	Gg
Net Calorific Value	[D]	TJ/Gg

(continued)

Table 1. (*continued*)

Emission Calculation		
CO_2 (Emission Factor*GWP)	Emission Conversion factor/ 1000* GWP = [E]	t.CO_2-e/TJ
CH_4 (Emission Factor*GWP)	Emission Conversion factor/ 1000* GWP = [F]	
N_2O (Emission Factor*GWP)	Emission Conversion factor/ 1000* GWP = [G]	
CO_2 Emission	[H] = C*D*E	t.CO_2-e/Year
CH_4 Emission	[I] = C*D*F	
N_2O Emission	[J] = C*D*G	
Total Emission (tons of CO_2 equivalent)	K = H + I + J	t.CO_2-e/Year

Table 2. Calculation Summary for Total Emissions (ICE)

Emission Calculation		
Mobile-Diesel Consumption (Road)	94,251	L/Year
	94.25	m^3/Year
Density	840	kg/m^3
Fuel (Mass)	0.0792	Gg
Net Calorific Value	43	TJ/Gg
CO_2 (Emission Factor*GWP)	74.10	t.CO_2-e/TJ
CH_4 (Emission Factor*GWP)	0.12	
N_2O (Emission Factor*GWP)	1.06	
CO_2 Emission	252.26	t.CO_2-e/Year
CH_4 Emission	0.40	
N_2O Emission	3.62	
Total Emission (tons of CO_2 equivalent)	256.28	t.CO_2-e/Year

4 Conclusions Electric Bus Emission Calculation

$$D = \frac{A*B}{1000}$$

A = Electricity Consumption (kWh)/year.

B = Grid Emission Factor 0.54 kg CO_2/kWh (Sri Lanka Sustainable Energy Authority, 2019).

The 'Average Emission Factor (AEF)' is primarily used to report the carbon footprint of electricity consumers. It is calculated by dividing the total emissions from the power sector by the total electricity consumption in the country for a given year.

D = Emission from Electricity Consumption (tCO_2e/year)

- Electricity consumption in kilowatt-hours (kWh) is converted to megawatt-hours (MWh) by dividing by 1,000. The grid emission factor, expressed in tons of CO_2-equivalent per kWh, represents emissions per unit of electricity generated.
- Total annual CO_2 emissions from electricity use are calculated by multiplying electricity consumption by the emission factor. (IPCC [11]).

5 Practical Application

To estimate the annual distance traveled by a provincial bus, the total kilometers operated by provincial buses is divided by the number of buses in operation. The total distance provincial buses in Sri Lanka operate is 316,660,224 km per year, with 10,862 buses (Private Provincial Bus Scenario) actively serving these routes (NTC [12]) (Table 2).

The practicality of e-bus adoption is initially considered for provincial routes, which can subsequently be extended to longer inter-provincial routes.

$$AnnualDistanceperBus = \frac{TotalOperatedkmforProvincialBusses}{BussesOperated}$$
$$= \frac{316{,}660{,}224}{10{,}862} = 29,\ 154 \text{ km per year}$$

Thus, each provincial bus travels approximately 29,154 km on average per year. However, this value represents an overall estimate. It is subject to variation when considering individual routes, as factors such as route length, frequency of operation, and demand influence the total distance traveled by each bus.

The operator opinion survey conducted by the authors indicated a fuel consumption of 3 to 3.5 L per kilometer under service conditions on provincial routes. Assuming a fuel consumption rate of 3.25 L per kilometer, and with each provincial bus traveling an average of 29,154 km per year, the annual fuel consumption per bus can be estimated as follows:

$$\text{Annual Fuel Consumption} = 29,\ 154 \times 3.25 = 94,\ 250.5 \text{ liters per year}$$
$$\text{Thus, the Total Emission} = 256.28 \text{ t } CO_{2e}/\text{Year}$$
$$\text{(as depicted in Table 2)}$$

Given that each provincial bus travels an average of 29,154 km per year, the annual energy consumption of an electric bus can be estimated using the provided energy consumption range of 0.8 kWh/km to 1.2 kWh/km (UITP, 2022). (There is a spectrum from 0.8 to 2.5 kWh/km, dependent on ebus specifications).

Assuming the higher energy consumption scenario of 1.2 kWh per kilometer, an electric bus operating on provincial routes with an average annual distance of 29,154 km would consume:

$$\text{Annual Energy Consumption} = 29,\ 154 \times 1.2 = 34,\ 984.8 \text{ kWh per year}$$

Thus, each electric bus is estimated to consume approximately 34,985 kWh per year. This assumption provides a conservative estimate for assessing electricity demand, charging infrastructure requirements, and the overall feasibility of transitioning to electric buses in provincial transport operations.

$$D = \frac{34,985 \times 0.54}{1000} = 18.89\text{tCO}_2\text{e/year}$$

A comparison of emissions between internal combustion engine (ICE) buses and electric buses (E-buses) highlights the significant environmental benefits of electrification. This analysis was important to quantitatively reflect the benefits to the stakeholders. An ICE bus, consuming 3.25 L/km and traveling 29,154 km/year, emits approximately 256.28 t.CO_2e annually, while an electric bus operating under a 1.2 kWh/km scenario generates only 18.89 t.CO_2e per year, reflecting a 92.6% reduction in emissions. Despite the electricity grid's carbon footprint, electric buses remain a far cleaner alternative, reinforcing the urgent need for sustainable transport solutions to mitigate greenhouse gas emissions and reduce reliance on fossil fuels in provincial bus operations.

The outcome is calculated in a what-if all buses are electric (Private Provincial Scenario). However, the practical feasibility remains a question, as every city and transit network differs. Therefore, it is important to identify what fits the Sri Lankan context, which depends on national policies and international commitments.

Table 3 presents the total annual CO_2 emissions associated with different levels of electric bus (E-bus) adoption in the provincial bus fleet of 10,862 buses. The calculations establish that a fully diesel-powered fleet (0% E-bus adoption) results in 2.78 million t.CO_2e per year, with each internal combustion engine (ICE) bus contributing 256.28 t.CO_2e annually. In contrast, an electric bus operating at 1.2 kWh/km generates only 18.89 t.CO_2e per year, significantly lowering emissions. A gradual shift to electric buses demonstrates a proportional reduction in emissions, with 50% fleet electrification reducing emissions to 1.39 million t.CO_2e per year, while a full transition (100% E-bus adoption) achieves a 92.6% reduction, bringing emissions down to 205,334 t.CO_2e per year. These findings underscore the critical role of fleet electrification in mitigating transport sector emissions and emphasize the substantial environmental benefits of transitioning to electric public transport systems.

Table 3. E-Bus Adoption Scenario-based Emission Values

E-Bus Adoption (%)	ICE Buses	E-Buses	Total Emissions (t.CO_2e/year)
0%	10,862	0	2,784,355
10%	9,776	1,086	2,506,302
20%	8,690	2,172	2,228,248
30%	7,604	3,258	1,950,194
40%	6,518	4,344	1,672,141
50%	5,431	5,431	1,394,087
60%	4,345	6,517	1,116,033

(continued)

Table 3. (*continued*)

E-Bus Adoption (%)	ICE Buses	E-Buses	Total Emissions (t.CO_2e/year)
70%	3,259	7,603	837,979
80%	2,173	8,689	559,926
90%	1,087	9,775	281,872
100% (E-Bus)	0	10,862	205,334

Since CO_2 emissions decrease proportionally with the percentage of E-bus adoption, a simple linear regression model can be developed:

$$E = \mathrm{mX} + \mathrm{C}$$

where:

- E = Total CO_2 emissions (t.CO_2e/year)
- X = Percentage of E-buses in the fleet
- C = Emissions when 0% E-bus adoption (i.e., 2,784,355 t.CO_2e)
- m = Slope, representing the reduction in emissions per unit increase in E-bus adoption

Since the emissions for 100% E-bus adoption (205,334 t.CO_2e) are already known, m can be calculated as follows:

$$m = \frac{205,334 - 2,784,355}{100 - 0} = -25,790.21$$

Thus, the regression equation is:

$$E = -25,790.21 \times X + 2,784,355$$

This equation can be used to predict emissions for any E-bus adoption percentage rather than manually computing it for specific intervals.

The total CO_2 emissions from the provincial bus fleet can be estimated using a linear relationship between the number of electric buses (e-buses) introduced and the corresponding emission reductions. Given that a fully diesel-powered fleet (10,862 internal combustion engine buses) results in 2,784,355 t.CO_2e per year. In contrast, a fully electrified fleet reduces emissions to 205,334 t.CO_2e per year, a direct relationship can be established. The reduction in emissions per additional e-bus is calculated as 237.6 t.CO_2e per bus per year, leading to the formulation of a linear equation:

$$E = -237.6 \times N + 2,784,355$$

where **E** represents total CO_2 emissions (t.CO_2e/year) and **N** denotes the number of E-buses in operation. This equation enables the estimation of total CO_2 emissions for any given number of E-buses within the range of 0 to 10,862, facilitating a comprehensive assessment of emission reductions under varying levels of fleet electrification.

6 Route-Based Analysis

A Route-based emission analysis is conducted to assess emissions from internal combustion engine (ICE) buses and electric buses (E-buses), providing a practical understanding of their environmental impact. Among the 455 bus routes operated under the Road Passenger Authority of the Western Province, Sri Lanka, a specific route is selected for this analysis based on factors such as route characteristics, practicality of electric bus adoption, and data availability. Selected Route details are given in Table 4.

Table 4. The selected bus routes for the analysis

Route	341 (Maharagama–Piliyandala)
Route Grade	A
Daily operation	27
Number of Turns	111
Km	7.8
Daily Running Km	3153.6

Total Daily Running Distance = 111 turns × 7.8 km = 865.8 km/day.

This makes Route 341 a high-frequency urban route, where buses frequently stop for passenger boarding, leading to increased fuel consumption and emissions. Given these conditions, assessing the impact of transitioning from diesel-powered buses to electric buses is essential.

Using a fuel consumption rate of 3.25 L/km, the total daily fuel consumption for all buses operating on Route 341 is Daily Fuel Consumption = 865.8 km × 3.25 L/km = 2,813.85 L/day.

The annual fuel consumption is:
Annual Fuel Consumption = 2,813.85 L/day × 365 days = 1,026,056 L/year.

Based on standard emission factors, this fuel consumption results in significant CO_2 emissions, contributing to air pollution and climate change.

Replacing diesel buses with electric buses (E-buses) significantly reduces emissions. Assuming an energy consumption rate of 1.2 kWh/km, the total daily energy consumption for the fleet is:

Daily Energy Consumption = 865.8 km × 1.2 kWh/km = 1,038.96 kWh/day.

The annual energy consumption is:
Annual Energy Consumption = 1,038.96 kWh/day × 365 days = 379,214 kWh/year.

Even though Sri Lanka's electricity generation includes fossil fuels, E-buses still produce substantially lower emissions compared to diesel buses while improving energy efficiency and reducing operating costs.

6.1 Emission Analysis: ICE Buses vs. E-Buses for Route 341

1. Annual CO_2 Emissions from ICE Buses

Using the annual fuel consumption of 1,026,056 L per year and the emission factor for diesel:

Total Annual CO_2 Emissions from ICE Buses = 1,026,056 × 2.73 kg CO_2/L
= 2,801,945 kg CO_2/year = 2,802 t.CO_2/year

2. Annual CO_2 Emissions from E-Buses

Using the annual energy consumption of 379,214 kWh/year and Sri Lanka's grid emission factor (0.54 kg CO_2/kWh):

Total Annual CO_2 Emissions from E-Buses = 379,214 × 0.54 kg CO_2/kWh
= 204,775 kg CO_2/year = 205 t.CO_2/year

3. Emission Reduction & Percentage Comparison

Emission Reduction
= 2,802-205 = 2,597 t.CO_2/year
Percentage Reduction
= (2,597/2,802) × 100 = 92.7%

Thus, transitioning from ICE buses to E-buses on Route 341 results in a 92.7% reduction in CO_2 emissions, significantly improving environmental sustainability.

7 Discussion

The transition from internal combustion engine (ICE) buses to electric buses (E-buses) represents a pivotal strategy in reducing greenhouse gas emissions from the transport sector. This study quantitatively demonstrates the substantial environmental benefits of E-bus adoption, with findings indicating a 92.6% reduction in CO_2 emissions per bus when shifting from diesel to electric propulsion. Given the increasing urgency to mitigate climate change impacts, such reductions highlight the transformative potential of fleet electrification in achieving national and global sustainability targets.

A scenario-based assessment was employed to evaluate the effects of varying levels of E-bus integration within the provincial bus fleet. The results illustrate a clear linear relationship between the proportion of E-buses in operation and the corresponding decline in overall emissions. With full reliance on diesel-powered buses, the sector contributes significantly to national CO_2 emissions. However, as E-bus adoption increases, emissions decline proportionally, demonstrating that even partial electrification can yield substantial environmental benefits. A 50% E-bus transition, for instance, achieves a notable reduction in annual emissions, reinforcing the case for progressive electrification rather than an abrupt overhaul of the fleet. The findings further validate that a complete transition to E-buses would result in a drastic reduction in transport-related emissions, underscoring the feasibility of this approach for long-term sustainability.

A linear regression model was developed to quantify the emission reductions across different adoption levels systematically. This model enables policymakers, transport planners, and stakeholders to predict the environmental impact of varying degrees of fleet electrification, facilitating data-driven decision-making. The ability to estimate emissions based on fleet composition allows for strategic planning in alignment with national sustainability goals, ensuring a phased and economically viable transition.

It is also important to recognize that the carbon intensity of electricity varies throughout the day, depending on the mix of generation sources. Using a constant grid emission factor does not capture this variation and may overestimate benefits if charging occurs when the grid is more carbon-intensive. While this limitation does not undermine the current outcomes, future studies would benefit from scenario-based assessments that consider different charging times and renewable energy availability to provide a more accurate picture.

Beyond the macro-level assessment, a route-based analysis was conducted to understand the real-world implications of electrification. Given that route characteristics, operational efficiency, and local energy availability influence the feasibility of E-bus deployment, this approach provides practical insights into the transition process. The selection of a representative provincial route highlights the potential for emission reductions at the micro level, reinforcing the broader benefits observed in the fleet-wide analysis.

While the environmental advantages of E-buses are evident, challenges remain in implementation. Infrastructure readiness, particularly in terms of charging networks and energy grid capacity, poses logistical constraints that must be addressed to support widespread adoption. Additionally, the initial capital investment for electric bus procurement is higher than for diesel buses, necessitating financial incentives, public-private partnerships, and policy-driven support mechanisms. Despite these barriers, the long-term cost savings associated with reduced fuel dependency, lower maintenance expenses, and environmental benefits outweigh the short-term financial constraints.

8 Conclusions

This study provides clear evidence that transitioning from diesel-powered buses to electric buses offers a transformative solution for reducing greenhouse gas emissions in Sri Lanka's transport sector. The analysis shows that replacing internal combustion engine (ICE) buses with electric alternatives can lead to a significant reduction in carbon emissions, with each E-bus emitting only a fraction of what a diesel bus produces. As more electric buses are integrated into the fleet, emissions decrease predictably, reinforcing the environmental benefits of electrification.

The scenario-based analysis highlights that even a partial shift to electric buses can result in substantial emission reductions. A structured approach to fleet electrification allows policymakers and transit authorities to set realistic targets based on data-driven projections. Additionally, the route-based analysis provides practical insights into how E-buses perform under real-world conditions, further emphasizing the importance of region-specific planning and infrastructure development.

This study set out to address the pressing problem of high greenhouse gas emissions from Sri Lanka's diesel bus fleet, and the findings clearly demonstrate that transitioning to electric buses can reduce emissions by more than 90%, offering a practical and effective pathway toward cleaner public transport. By quantifying both fleet-wide and route-specific impacts, the analysis not only confirms the environmental benefits of electrification but also provides policymakers with data-driven insights for phased implementation. While the research focused primarily on operational emissions, future work should extend to lifecycle assessments encompassing vehicle manufacturing, battery management, and renewable energy integration to provide a more comprehensive understanding of sustainability outcomes.

Additionally, expanding the scope to nationwide fleet modeling, economic feasibility studies, and policy-driven frameworks such as incentives, subsidies, and public–private partnerships will be vital to accelerate adoption. Addressing these opportunities for future work will help ensure that electrification becomes a cornerstone of Sri Lanka's long-term sustainable mobility strategy.

While the environmental and economic benefits of E-buses are undeniable, the transition comes with challenges. Infrastructure readiness, investment in charging stations, and initial capital costs remain key barriers. Addressing these issues requires strong policy support, financial incentives, and collaboration between the government, private sector, and stakeholders in the transport industry.

Despite these challenges, the long-term advantages of E-bus adoption far outweigh the obstacles. Reduced fuel dependency, lower operational costs, and improved air quality are just a few of the lasting benefits. Moving forward, a well-structured, phased transition to electric public transport will be crucial in meeting Sri Lanka's sustainability goals. With the right policies, investments, and commitments, electrification of the bus fleet can become a cornerstone of a cleaner, more efficient, and sustainable transport system.

Acknowledgment. The authors would like to thank Mr. D. Varatharatnam for his input on greenhouse gas emission calculations.

References

1. United Nations: Kyoto Protocol to the United Nations Framework Convention on Climate Change (1997)
2. United Nations Framework Convention on Climate Change (UNFCCC). Paris Agreement (2015)
3. Centre for Intelligent Transport Systems: The final report: feasibility study and market research on light-duty electric vehicles. University of Moratuwa (2018)
4. Eckstein, D., Hutfilas, M.L., Winges, M.: Global Climate Risk Index 2019. Germanwatch E.V. (2019)
5. Thilakshan, T., Sugathapala, A.G.T., Bandara, J.M.S.J.: Implementation of electric buses into an existing public transit network: A Sri Lankan perspective. Moratuwa Engineering Research Conference (MERCon), Moratuwa, Sri Lanka, pp. 794–799 (2023). https://doi.org/10.1109/MERCon60487.2023.10355524
6. Ministry of Environment Sri Lanka: Nationally Determined Contributions Implementation Plan (2021–2030). Climate Change Secretariat, Government of Sri Lanka (2023)
7. US EPA. Overview of greenhouse gases. US Environmental Protection Agency (2025). Retrieved from https://www.epa.gov/ghgemissions/=overview-greenhouse-gases
8. Premasiri, H.D.S., Pemasiri, K.A.D.U., Athapattu, B.C.L., Navaratne, A.N.: Air pollutant exposure levels of passengers using public transport to reach Colombo from surrounding cities. NBRO Symposium (2015)

9. Sustainable Waterloo: Calculating GHG emissions from personal vehicle travel (2010)
10. KPMG: GHG Emissions Reporting Handbook. Greenhouse Gas Protocol (2024)
11. Intergovernmental Panel on Climate Change: IPCC guidelines for national greenhouse gas inventories. IPCC (2006). Retrieved from https://www.ipcc-nggip.iges.or.jp/public/2006gl/
12. National Transport Commission of Sri Lanka: National Transport Statistics (2024)

Concepts of Circular Economy to the Sustainable Development of the Sri Lankan Construction Industry

U. M. Samararathne[1(✉)], I. U. Atthanayake[2], and W. N. Kawmudi[1]

[1] General Sir John Kotelawala Defence University, Ratmalana, Sri Lanka
umeshasamararathne@gmail.com
[2] Open University of Sri Lanka, Nawala, Nugegoda, Sri Lanka

Abstract. Sri Lankan construction industry significantly contributes to environmental degradation through excessive use of resources and the generation of waste, necessitating a shift towards sustainability. Drawing on the use of Circular Economy (CE) principles to enhance sustainable development in the industry, this research examines the CE application. Through an examination of barriers to the adoption of CE, such as regulatory gaps, prohibitive initial investment costs, technology deficiency, attitude towards change, and knowledge gaps, the study identifies essential barriers to progress. Drawing on 20 semi-structured interviews with industry representatives like architects, engineers, and sustainability consultants, the study applies content analysis to nominate 28 barriers and recommend 15 solutions to address them. The overarching solutions include policy alterations, economic incentives like grants and low-interest loans, enhanced re-cycling facilities, education, and public awareness campaigns. These measures are intended to bridge the gap between CE theoretical principles and physical implementation, ensuring resource efficiency, sustainability, and economic resilience. The findings offer actionable insights to stakeholders, positioning the Sri Lankan construction sector in line with global sustainability targets while taking into account local resource limitations and environmental challenges, to-wards a regenerative, circular construction practice.

Keywords: Circular Economy · Construction Industry · Sri Lanka · Sustainability

1 Introduction

The construction industry not only accounts for a significant share of global resource consumption, waste generation, and greenhouse gas emissions but is also one of the most damaging sectors for the environment [1]. The industry's environmental footprint has grown as the industry has grown in response to urbanization and increased population [2]. Conventional construction practice often works on a linear model of harvesting resources and discarding waste with limits on available resources and landfill capacity [3]. The current unsustainable approach demands a transition to less sustainable, resource-intensive models; the circular economy is perhaps the latest star in this field [4].

M. Siriwardhana et al. (Eds.): IERS 2025, PTT, pp. 250–263, 2026.
https://doi.org/10.1007/978-981-95-7984-6_22

The construction industry has a considerable impact on the environment and operates based on limited resources, and thus it is not only critical but also necessary to embrace greener approaches in line with the global sustainability aims [5]. Examining the potential of circular economy concepts to revolutionize the Sri Lankan building sector, assisting in achieving this purpose [6]. This study goals to link the gap between theoretical potential and actual application by identifying implementation challenges and suggesting workable solutions, thereby promoting a more resource-efficient, environmentally accountable, and economically healthy building sector. The results will suggest insightful analysis for practitioners, legislators, and stakeholders that will help the sector toward a sustainable future, supporting Sri Lanka's development goals and supporting worldwide efforts in resource depletion and climate change prevention.

Although the use of Circular Economy principles, in theory, brings assumed bene-fits and now increasing acknowledgment that Circular Economy principles can and should be used in the construction industry, widespread usage and application remain rare and disconnected [7]. Furthermore, the absence of an articulated and implementable way to overcome the sector's specific dynamics further complicates the problem, leaving industry stakeholders uncertain about how to shift from linear to circular practices [8].

This is a gap of great challenge that restricts the industry's capability to greatly add to the sustainable development strategies globally [9]. Until new measures are put in place and a strategic framework is established on how change can be implemented despite the barriers, unsustainable habits can be repeated in the construction sector, further causing an increase in the number of environmental, economic, and social burdens [10].

Objectives

1. To identify barriers to Circular Economy adoption in the construction sec-tor.
2. To introduce strategies for overcoming barriers to Circular Economy adoption in promoting sustainable construction practices.

While the Circular Economy is studied a lot around the world, using it in the Sri Lankan construction industry is not very common and is often disorganized [11]. The study examines the Circular Economy from both a theoretical and an operational perspective. Making evidence-backed suggestions tied to the Sri Lankan situation helps to cover a major gap in what both experts and industries offer.

This research fills a gap in the construction industry by emerging with the need to transition toward a more sustainable, circular model. Apart from adding value to the academy of discourse on Circular Economy, it also presents pragmatic solutions to environmental conservation, economic resilience, and social well-being that can be implemented in industry surgeries.

2 Literature Review

2.1 Circular Economy (CE)

The construction industry consumes many resources while producing substantial waste, and it must adopt sustainable practices under growing pressure to change [12]. The Circular Economy provides a complete system design for sustainability, which extends resource usability while reducing waste production to maximize product value [1]. Circular Economy principles demand that builders relocate their thinking toward beneficial implementation methods that apply to real structures, and as a necessary step, there is a need to rethink all parts of building and infrastructure development from design through construction, usage, and termination [13]. Under the core principles of Circular Economy, reduce, reuse, and recycle, it can be understood how these aspects apply to construction while directly linking them to essential sustainability practices [1]. In construction, "reduce" operates as a principle that goes further than achieving bare material reduction [14]. The reuse practice within construction means utilizing existing materials and components in projects that have different purposes, and this practice enables materials to last longer, which results in a reduced need for virgin resources [15]. The recycling process in construction takes waste from building and demolition activities to make new substances that reintroduce waste back into supply chain systems, and the recycling process helps reduce waste and preserve resources, even though it does not match the environmental advantages of reuse or reduction [16]. The "reduce, reuse, recycle" sequence enables the construction industry to implement Circular Economy principles through a practical system of implementation [14].

2.2 Sustainable Development (SD)

Sustainable Development (SD) represents the essential principle through which people address multiple difficulties happening on Earth as our planet faces complex obstacles [17]. From the 1987 Brundtland Report, "Our Common Future," "sustainable development is development that meets the needs of the present without compromising the ability of future generations to meet their own needs." Sustainable development promotes a unified framework for balancing social equity with economic growth and environmental protection [17]. Sustainable development shows clear importance within the construction field because this sector significantly consumes resources, produces waste, and creates environmental impacts [18]. Construction activities directly impact community environments, economies, and social welfare structures, so sustainable development is a vital mechanism for industry future planning [18]. Incorporating environmental, social, and economic concerns into construction practices has been known as "sustainable construction" [19].

In the long run, sustainable construction is a required principal step in developing a physical environment that meets current requirements without jeopardizing those of the next generations [20]. Making use of the three pillars of sustainability and the ideas behind the circular economy can support the building industry in becoming friendlier to the environment, stronger financially, and more beneficial to society [3]. To reach these goals, the Circular Economy uses strategies that remain useful through the lifetime of the building [21].

2.3 Contribution of Circular Economy (CE) to Sustainable Construction

The circular economy provides a comprehensive system for sustainable construction because it strengthens all three pillars of sustainability [22]. The environmental sustainability goals of Construction are directly achieved through the application of Circular

Economy principles centered on reduce-reuse-recycle [23]. Material reduction leads to lower customary resource demands, which subsequently leads to decreased environmental side effects from resource extraction and manufacturing processes [24]. Both the reduction of waste material flow to landfills and the decrease of pollution, together with resource preservation, can be achieved through material reuse and recycling [24]. Energy efficiency improves through the Circular Economy because it minimizes the energy embedded in materials and supports builders using renewable energy during construction [21].

In the long run, sustainable construction is a required principal step in developing a physical environment that gratifies current requirements without jeopardizing those of the next generations [20]. Making use of the three pillars of sustainability and the ideas behind the circular economy can support the building industry in becoming friendlier to the environment, stronger financially, and more beneficial to society [3]. To reach these goals, the Circular Economy uses strategies that remain useful through the lifetime of the building [21].

2.4 The Sri Lankan Context: Demand for Building Circular Economies

An important part of Sri Lanka's economic growth, the building sector helps to create infrastructure, jobs, and urbanization by means of considerable contribution, but the sector is also one of the biggest waste creators and consumers of natural resources, which causes great social, financial, and environmental problems [25]. Though the world is moving toward sustainable practices, Sri Lanka still has a limited adoption of Circular Economy ideas, including energy efficiency, waste minimization, and material reuse [26].

Adopting circular economy practices would have helped to elude economic inadequacies, including increased raw material prices, waste discarding, and energy usage. Furthermore, limiting prospects for cost savings and income generation through reutilizing and material reuse is the reliance of the construction sector on virgin resources and poor waste management methods [25]. Employment of unsustainable building practices results in environmental degradation, compromises the economy, endangers the health of the community, and renders people more unable to access decent and affordable homes [27]. In addition, the development of a skilled labour force in sustainable construction is triggered by disregard and dismissal of circular economy practices.

3 Research Methodology

A research philosophy is a viewpoint on the best ways to collect, analyze, and use data. While this research used qualitative data, the philosophy of this research is interpretivism. The study population should be flexible in qualitative research to ensure the ability to gather diverse and comprehensive data in any given situation [28]. The professionals who are involved in construction projects and practice sustainable construction methods and circular economy principles in consulting, main contracting, and sub-contracting firms, and related professionals involved in the building projects in the Sri Lankan construction industry are the target population for this study. This study collected data through semi-structured interviews with the construction stakeholders as the target population.

The construction specialists were interviewed to obtain further insights into the Circular Economy concept. Every interviewee has prior Circular Economy experience in their respective positions. The main goal of the interviews was to learn more about the challenges to the construction industry's adoption of the Circular Economy.

Sampling involves selecting a few from a larger group, with quantitative research aiming for unbiased, representative samples, and qualitative research considering various factors affecting selection [29]. For the semi-structured interviews purposive sampling method is used [29]. This technique is a nonrandom method commonly employed in qualitative research to gather information from participants with knowledge and experience [30]. Table 1 provides Profiles of the Interviewees.

Table 1. Profiles of the Interviewees

Reference No	Profession	Experience in the construction industry
R01	Senior Architect	15 Years
R02	Senior Construction Project Manager	12 Years
R03	Sustainability Consultant	10 Years
R04	Senior Civil Engineer	15 Years
R05	Sustainability Consultant	12 Years
R06	Construction Materials Supplier	18 Years
R07	Construction Project Manager	10 Years
R08	Architect	12 Years
R09	Senior Mechanical Engineer	13 Years
R10	Civil Engineer	08 Years
R11	Construction Project Manager	15 Years
R12	Construction Sustainability Consultant	12 Years
R13	Sustainability Manager	08 Years
R14	Construction Waste Management Consultant	10 Years
R15	Senior Civil Engineer And Sustainability Auditor	15 Years
R16	Sustainability Coordinator	08 Years
R17	Construction Project Manager	07 Years
R18	Senior Quantity Surveyor	14 years
R19	Mechanical Engineer	09 Years
R20	Architect	08 Years

Content analysis was used to examine the interview data. This technique allowed for the identification of factors reflecting the proposed objective and the weight of each stated factor, making it a valuable tool for comprehensive data analysis. Content analysis is a widely used technique in qualitative research. Key topics, trends, and insights about the obstacles and tactics for implementing CE in the construction industry may be found using this method. The following Table 2 provides a summary of the Methodology.

Table 2. Summary of the Methodology

Objective	Sample	Sample technique	Data Collection Method	Data Analysis Method
To identify barriers to CE adoption in the construction sector	20	Purposive Sampling	Semi-Structured Interviews	Content analysis
To introduce strategies for overcoming barriers to CE adoption on promoting sustainable construction practices	20	Purposive Sampling	Semi-Structured Interviews	Content analysis

4 Results and Discussion

4.1 Challenges to Circular Economy Adoptions in the Construction Sector

The interview asked about the barriers to Circular Economy adoption, mainly to achieve the fourth objective of the research. According to the respondents, interviewees stated that there is a broad range of perceived barriers to adopting Circular Economy principles in the construction sector, particularly within the Sri Lankan context.

Several challenges affect the implementation of the Circular Economy (CE) in the construction sector of Sri Lanka, as highlighted by various professionals interviewed, such as architects, engineers, project managers, and sustainability consultants. There are also serious regulatory barriers; despite building codes focusing on safety and cost rather than sustainability, they do not have specific requirements regarding CE targeting waste management or recycling, and certification standards on recycled materials do not exist, restraining their use in official projects. The issue of financial difficulty lies in the fact that CE technologies can be very expensive in the short term, i.e., modular building and waste sorting systems, and all this lacks incentives such as tax breaks or grants, which would make CE adoption especially unavailable for small and medium-sized enterprises (SMEs). The industry technologically faces a lack of infrastructure to recycle complex materials such as insulation or composite, insufficient use of digital technologies, including Building Information Modeling (BIM), and a lack of skills in modularization or reverse logistics to rescue material. Cultural challenges are high, as there is a popular tendency towards using new materials because it is believed to be the best in terms of quality and prestige, an inability to accept the change due to the accumulation of experience with the traditional and simple patterns, and high levels of the stakeholders' awareness of the benefits of CE. There are also systemic challenges holding it back, such as a lack of knowledge about CE because of the lack of CE training among professionals and laborers, the lack of continuity in the supply chains to ensure the delivery of recycled material, and the demands of projects, which are frequently time-constrained, discouraging time-culture-inherent CE practices. Other challenges include under established markets of circular products, the lacking workforce experienced in circular production techniques such as deconstruction, a lack of data regarding the life cycle of material, mismatched priorities among the stakeholder groups, issues that arise regarding logistics such as the lack of land area where waste can be variously sorted through, doubts about the structural integrity of recycled materials, and the digital divide that inhibits the adoption of CE technologies.

Table 3. Summary of the barriers to implementing the CE concept

Barrier	No. of Responses
Regulatory Gaps	18
High initial cost	17
Technological limitations	15
Cultural resistance	14
Knowledge deficit	13
Supply chain fragmentation	12
Time pressures	08
Market readiness	08
Skilled labour shortages	07
Data gaps	07
Infrastructure defects	06
Lack of CE certifications	05
Client misconceptions	05
Stakeholders' misalignments	04
Public perceptions	04
Design limitations	04
Lack of local standards	04
Insurance resistance	04
Land availability	04
Short-term mindset	04
Energy grid dependency	03

(*continued*)

Table 3. (*continued*)

Barrier	No. of Responses
Land use conflicts	03
Informal waste sector	03
Financial risks	03
Waste sorting inefficiency	03
Climate adaptation conflicts	03
Contractual limitations	02
Reverse logistics complexity	02

Regulatory Gaps (18 responses), High Initial Cost (17 responses), Technological Limitations (15 responses), Cultural Resistance (14 responses), and Knowledge Deficit (13 responses) are the top five barriers to the implementation of Circular Economy (CE) concepts in the Sri Lanka construction industry based on the response by professionals in the industry such as the architects and engineers. The gaps in regulation in the form of missing CE-tailored policies, waste decree, or use of recycled materials hamper adoption because the regulatory focus is on safety and affordability rather than on sustainability. Technologies such as the ones that involve constructing in modules are very expensive to set up, which makes them unattractive to firms, and in particular SMEs, since there are no incentives to take them up, such as subsidies. The reuse of materials is inhibited by technology (lack of recycling facilities and low utilization of tools such as BIM) and the culture (preference for new products over the recycled ones), motivated by such misperceptions as inferiority and lack of CE awareness. The lack of expertise of professionals and the clients has been perpetuated since most of them are not adequately trained to get out of the linear practice justified by a lack of knowledge. Addressing these barriers in a comprehensive way, Sri Lanka will be able not only to make the leap to a sustainable, circular construction sector that will change the face of global sustainability but also to overcome local resource issues. According to the interview results, Table 3 shows the identified barriers and the most effective barrier to Circular Economy adoption in the construction industry.

4.2 Strategies for Overcoming Barriers to Circular Economy Adoption in Promoting Sustainable Construction Practices.

After identifying the barriers, then identified the strategies to overcome the identified barriers. The interviewees stated several strategies to overcome the identified barriers. According to that, policy reforms and financial incentives were the most frequently mentioned. To list the promising approaches to removing major barriers to implementing the principles of Circular Economy in the construction sector of Sri Lanka, the respondents put forward the following strategies: Financial Incentives (16 responses), Policy Reforms/CE Building Codes (13 responses), Recycling Infrastructure and Tech Tools (11 responses), Training and Education Programs (10 responses), and Public Awareness Campaigns (8 responses). The cost of obtaining CE technologies as part of modular construction or waste recycling systems is very high initially, and therefore, financial incentives, in the form of grants and low-interest loans, are direct attempts to address this financial barrier, which can further be made accessible, especially to small and medium-sized enterprises (SMEs) that are predominant in the sector. By offering an articulated policy, such as waste audit requirements and recycling estimates, as well as recycled material certification systems, regulatory gaps are closed because of the motivating environment of sustainable practices and control of safety. The reforms lead to an enabling environment through which the CE practices can prosper, thus causing the reduction of uncertainty and fostering adoption throughout the industry. These strategies also combine top financial assistance with the strategic regulatory requirements and establish a strong basis to overcome any economic and structural impediments so that firms can invest in long-term sustainable solutions.

To supplement these initiatives, investments in the recycling infrastructure and digitization, such as with BIM, focus on addressing technological constraints, thus allowing the effective processing of complex construction waste and improving the optimization of resource tracking with the aim of reusing them. The academic training and education programs to develop capacity on high-tech CE processes, which include deconstruction and modular design among others, are designed and incorporated into undergraduate and postgraduate academic curricula and in the industry-based workshops, which address the knowledge gap among professionals and laborers. The need to publicize awareness of cultural resistance to ensure that stakeholders, clients and contractors, and the general populace are educated about the environmental and economic value of CE through the utilization of pilot projects to testify on its quality and viability of recycled materials is an important aspect in overcoming such resistance. Other initiatives, including online markets of salvaged materials, community material banks, and integration of supply chains, increase the availability of resources and the collaboration among stakeholders. According to the interview results, the following Table 4 shows the identified strategies to overcome barriers to Circular Economy adoption in the construction industry.

Table 4. Strategies to overcome identified barriers

Suggested Strategies	No. of Responses
Policy reforms/CE building codes	13
Financial incentives (grants, tax breaks, loans)	16
Recycling infrastructure and tech tools (BIM)	11
Training and education programs	10
Public awareness campaigns	08
CE material certification systems	04
Digital marketplaces for reused materials	07
Pilot/demonstration projects	03
Material lifecycle data platforms	03
Supply chain/industry integration	02

(*continued*)

Table 4. (*continued*)

Suggested Strategies	No. of Responses
CE award and recognition programs	04
Energy efficiency bonds or funding	03
Waste zoning or sorting	04
Regional CE labs for testing designs	02
Community materials banks	03

4.3 Theoretical/Practical Implications

Also used semi-structured interviews to get some practical implementation and outlook. According to the interviewees stated that, CE implementation in Sri Lana's construction industry hinges on an integrated strategy encompassing training, funding, technology, policy, and collaborative networks. The necessity of skill improvement is a recurring issue in the suggestions for online courses, CE workshops, and certification programs for labourers, engineers, and architects. With goals like certifying 1,000 CE practitioners in five years, these programs seek to increase the talent pool and integrate CE ideas into professional practice. The sector's demand for both technical ability and creativity is reflected in the subsidized training provided by trade groups and collaborations with universities for CE research and technology development.

R07, R11, and R19 stated that "another recurrent aspect is financial accessibility. To assist CE experimental projects and businesses, interviewees recommend low-interest loans, seed money, and microgrants. For small and medium-sized businesses (SMEs), which frequently lack the funds to test out CE strategies, these financing tools are especially essential. The industry-wide capacity can be increased by matching SMEs with bigger, CE-experienced companies for mentorship, while proposals for government-backed microfinance programs and shared recycling facilities with pooled equipment can lower entrance barriers for SMEs".

With demands for easily available 3D printing equipment for recycled materials, drones for mapping site trash, and digital monitoring systems like material passports, the significance of technology deployment is further emphasized. To democratize access to these technologies, especially for smaller businesses, shared technology centers that rent out equipment are being advocated. As buildings start to be planned as "material banks," saving components for later reuse, a notion that is gaining favor worldwide, digitally enabled technologies are anticipated to optimize material flows.

The necessity of a national CE roadmap with quantifiable goals, standards, and procedures for incorporating CE into building regulations and public tenders is emphasized in policy proposals. Mandating CE in government projects, rewarding innovative work with national CE awards, and making CE certification easier for SMEs are some of the suggestions. Scaling CE beyond pilot programs and into general practice is said to need strong government commitment. Additionally, it is believed that the establishment of regional recycling centers by 2030 and the scaling of pilot projects are essential to constructing the physical foundation needed for widespread CE adoption.

Collaboration and knowledge sharing are also seen as crucial. The interviewee envisions forums, CE networks, and awareness campaigns to disseminate best practices and drive behavioral change among contractors and clients. These would be instrumental in shifting industry norms, especially if paired with public campaigns that highlight the environmental and economic benefits of circular construction.

The construction industry, as one of the world's largest consumers of natural resources and generators of waste, holds a pivotal role in advancing sustainability through the adoption of CE principles. By reducing operational and embodied carbon emissions, the industry may make a significant contribution to mitigating climate change. This is directly in line with Sustainable Development Goal (SDG) 13 on climate action and may be accomplished using ultra-low carbon materials, energy-efficient building designs, and the integration of renewable energy sources. Furthermore, by lowering reliance on limited virgin resources and lowering the geopolitical risks associated with raw material extraction, CE techniques like life-cycle costing (LCC) over traditional first-cost budgeting and urban mining from decommissioned buildings can greatly improve resource security.

By forecasting closed-loop systems and reducing reliance on polluting industries, CE also supports pollution reduction across air, water, and soil, while biodiversity protection is promoted by limiting land degradation and ecosystem disruption from raw material extraction. With the adoption of innovation, the establishment of green jobs, and the establishment of novel business models, such as the use of performance-based contracts, where sustainability measures, such as energy and material efficiency, are core key performance indicators (KPIs), and the transformation of the construction industry to circularity, which helps in facilitating an economic transition. These modifications display how CE might be operationalized at scale and suggest a model for further industries to adopt.

The social dimensions of CE are also evident. By encouraging waste-free buildings and low-waste infrastructure, circular architecture improves urban resilience (SDG 11), making cities more habitable while also advancing social fairness. Urban circular hubs and decentralized material reuse centers may boost local economies, particularly when combined with cross-sector partnerships. One example would be repurposing waste streams from industrial or agricultural sources in construction. In addition to decreasing material waste, these facilities provide access to reasonably priced building materials and create jobs at the local level.

R15 stated that "Use energy-efficient buildings to cut emissions" (SDG 13). Encourage material reuse to encourage responsible consumption (SDG 12). Low-waste infrastructure can increase urban resilience (SDG 11). Moreover, CE in construction catalyzes innovation, pushing advancements in design, material science, digital technologies, and logistics. Adoption of methods like life-cycle assessment, material passports, and digital twins, which maximize resource utilization across a structure's life span, is encouraged by the need to rethink conventional construction systems. All things considered, the construction sector is in a unique position to spearhead the shift to a circular, inclusive, and sustainable economy while also directly advancing important global goals like SDGs 11, 12, and 13.

5 Conclusions and Recommendations

5.1 Conclusions

Achievement of objective One-To identify barriers to Circular Economy adoption in the construction sector.

Determining the obstacles to implementing the Circular Economy in the Sri Lankan construction industry is the fourth objective in this study. Semi-structured interviews were used to collect data, and content analysis was used to analyze the data. According to the collected data, 28 barriers were identified, and among them, most stated that regulatory gaps are the most affecting barrier to implementing the Circular Economy in the Sri Lankan construction industry.

Achievement of objective Two-To introduce strategies for overcoming barriers to CE adoption in promoting sustainable construction practices.

After identifying the barriers, then identified the strategies that can be used to overcome them identified barriers were identified. To achieve the final objective also semi-structured interviews were also used, and content analysis was used to analyze data. According to the interviewees' responses, 15 strategies were identified, and among the most interviewees stated that giving financial incentives (grants, tax breaks, and loans) is the most suitable way to overcome the identified barriers.

5.2 Recommendations and Future Research

The responses identified some recommendations to implement in the Sri Lankan construction industry. Responses point to a more systems-oriented and strategic approach to promoting the Circular Economy in Sri Lanka's construction industry, with a focus on financial innovation, infrastructure, technology, education, cooperation, and regulatory change. The development of regional hubs and mobile units for material recovery, the introduction of various financing instruments like green bonds and micro grants, and the integration of the Circular Economy into national codes and procurement are some of the main recommendations. Stakeholders also emphasize the value of cross-sector cooperation, extensive education initiatives, and digital technologies like BIM and material passports. Notably, forward-thinking concepts like digital twins, product-as-a-service models, and climate-resilient Circular Economy solutions indicate a growing comprehension of Circular Economy as a comprehensive, long-term change rather than a collection of discrete actions.

Circular Economy provides a revolutionary way of solving issues related to the construction industry in Sri Lanka, as well as moving in accordance with the global sustainability objectives, along with solving the lack of resources and environmental destruction locally. CE can foster economic flexibility, environmental stewardship, and social fairness by highlighting energy efficiency, material reuse, and waste minimization. The suggested future research opportunities, the creation of CE indicators, the analysis of material flows, online infrastructure, policy experiments, peer-to-peer model at the community level, behavior change, and the inclusion of circular design in education, give guidance to the next steps towards the prevalence of CE. These spheres will contribute to the delivery of the knowledge gap between theory and practice, as well as develop a sustainable infrastructural environment that is applicable to the Sri Lankan context. A regenerative mode of operation in the construction sector is achievable through concerted efforts on the part of policymakers, industry players, and academic institutions that can ensure a sustainable built environment for future generations, besides promoting sustainability goals at the national and global levels.

References

1. Ahmed, W., Siva, V., Bäckstrand, J., Sarius. and Sundberg, H.Å.: Circular economy: extending end-of-life strategies. Sustain. Prod. Consumption **51**, 67–78 (2024). https://doi.org/10.1016/j.spc.2024.09.003
2. Alcalde-Calonge, A., Sáez-Martínez, F.J., Ruiz-Palomino, P.: Evolution of research on circular economy and related trends and topics. A thirteen-year review. Ecol. Inform. **70**, 101716 (2022). https://doi.org/10.1016/j.ecoinf.2022.101716
3. IIllankoon, M.C.S., Vithanage, S.C.: Closing the loop in the construction industry: a systematic literature review on the development of circular economy. J. Build. Eng. **76**, 107362–107362 (2023). https://doi.org/10.1016/j.jobe.2023.107362
4. Benachio, F., Freitas, M.C.D., Tavares, S.F.: Circular economy in the construction industry: a systematic literature review. J. Clean. Prod. **260**(1), 121046 (2020)
5. Mukalula, P., Chisumbe, S., Kabaso, K., Aigbavboa, C., Akinradewo, O.: Benefits of adopting sustainability practices by construction companies: developing country's perspective. Baltic J. Real Estate Econ. Constr. Manage. **12**(1), 55–73 (2024). https://doi.org/10.2478/bjreecm-2024-0004
6. Jayamanna, J.M.D.S.D., Denagama, T.D., Dias, S.N.C.M.: A study on adaptations of circular economic concepts in construction industry of Sri Lanka. The 14th International Conference on Sustainable Built Environment (ICSBE) (2023)
7. Yu, Y., Yazan, D.M., Junjan, V., Iacob, M.E.: Circular economy in the construction industry: a review of decision support tools based on information & communication technologies. J. Clean. Prod. **349**, 131335 (2022). https://doi.org/10.1016/j.jclepro.2022.131335
8. Yang, Y., Guan, J., Nwaogu, J.M., Chan, A.P.C., Chi, H., Luk, C.W.H.: Attaining higher levels of circularity in construction: scientometric review and cross-industry exploration. J. Clean. Prod. **375**, 133934 (2023). https://doi.org/10.1016/j.jclepro.2022.133934
9. Ghufran, M., et al.: Circular economy in the construction industry: a step towards sustainable development. Buildings **12**(7), 1004 (2022). https://doi.org/10.3390/buildings12071004
10. Khan, S.A.R., Shah, A.S.A., Yu, Z., Tanveer, M.: A systematic literature review on circular economy practices: challenges, opportunities and future trends. J. Entrepreneurship Emerg.Econ. **14**(5) (2022). https://doi.org/10.1108/jeee-09-2021-0349
11. Samaraweera, T., Gunawardhana, T.: The adaptability of sustainable construction practices: an analysis of issues faced by Sri Lankan Construction Industry. 3rd International Conference on Real Estate Management and Valuation (ICREMV 2019) (2019)
12. Liu, Z.J., Pypłacz, P., Ermakova, M., Konev, P.: Sustainable construction as a competitive advantage. Sustainability **12**(15), 5946 (2020). MDPI. https://doi.org/10.3390/su12155946
13. Kochanski, M., Lotz, M.T., Korczak, K.: Benchmarking circular economy measures in buildings along the 11R framework: a systematic review of quantified impacts on material use, energy consumption, GHG emissions, and costs. J. Clean. Prod. **485**, 144337 (2024). https://doi.org/10.1016/j.jclepro.2024.144337
14. Norouzi, M., Chàfer, M., Cabeza, L.F., Jiménez, L., Boer, D.: Circular economy in the building and construction sector: a scientific evolution analysis. J. Build. Eng. **44**(1), 102704 (2021). https://doi.org/10.1016/j.jobe.2021.102704

15. Citraningrum, A., Iyati, W., Wardoyo, J.: Sustainability through reuse of materials and components in buildings: an Indonesian. Perspective (2023). https://doi.org/10.2991/978-94-6463-140-1_24
16. Gherman, I.E., et al.: Circularity outlines in the construction and demolition waste management: a literature review. Recycling **8**(5), 69 (2023). https://doi.org/10.3390/recycling8050069
17. Mangukiya, R., Sklarew, D.: Analyzing three pillars of sustainable development goals at sub-national scales within the USA. World Dev. Sustain. **2**, 100058 (2023). https://doi.org/10.1016/j.wds.2023.100058
18. Lima, L., Trindade, E., Alencar, L., Alencar, M., Silva, L.: Sustainability in the construction industry: a systematic review of the literature. J. Cleaner Prod. **289**, 125730 (2021)
19. Moghadam, M.A., Bagherpour, M., Ghannadpour, S.F.: Sustainability assessment in construction projects: a sustainable earned value management model under uncertain and unreliable conditions. Environ. Syst. Decisions (2023). https://doi.org/10.1007/s10669-023-09913-2
20. Walker, A.M., Opferkuch, K., Lindgreen, E.R., Simboli, A., Vermeulen, W.J.V., Raggi, A.: Assessing the social sustainability of circular economy practices: industry perspectives from Italy and the Netherlands. Sustain. Prod. Consumption (2021). https://doi.org/10.1016/j.spc.2021.01.030
21. Hafez, F.S., et al.: Energy efficiency in sustainable buildings: a systematic review with taxonomy, challenges, motivations, methodological aspects, recommendations, and pathways for future research. Energy Strategy Rev. **45**(101013), 101013 (2023)
22. Rodríguez-Espíndola, O., et al.: The role of circular economy principles and sustainable-oriented innovation to enhance social, economic and environmental performance: evidence from Mexican SMEs. Int. J. Prod. Econ. **248**(248), 108495 (2022). https://doi.org/10.1016/j.ijpe.2022.108495
23. Çimen, Ö.: Construction and built environment in circular economy: a comprehensive literature review. J. Clean. Prod. **305**, 127180 (2021). https://doi.org/10.1016/j.jclepro.2021.127180
24. Alsheyab, M.A.T.: Recycling of construction and demolition waste and its impact on climate change and sustainable development. Int. J. Environ. Sci. Technol. **19** (2022). https://doi.org/10.1007/s13762-021-03217-1
25. Jeewananda, B.J., Bandara, S.: A study on the impact of the Construction industry on Sri Lanka's National Economic Growth. Symp. Int. Build. Econ. Res. (2023)
26. Mudalige, D.: Knowledge on resilience and brand image of Sri Lankan business organizations, Rajarata University of Sri Lanka (2022). https://www.rjt.ac.lk/wp-content/uploads/2024/08/RUJ-2022_07_02_005.pdf
27. Dissanayake, N., Withanawasam, A., Sarjoon, A.: Achieving sustainable development goals by adopting the blue-green economic provisions: an analysis of opportunities and challenges of Sri Lanka. J. Hum. Soc. Sci. Stud. **3**(6), 18–29 (2021). https://doi.org/10.32996/jhsss.2021.3.6.3
28. Willie, M.M.: Distinguishing between population and target population: a mini review. Surg. Res. J. **3**(2) 2021
29. Kumar, B.: A practical guide to adopting BIM in construction projects. Whittles Publishing (2015). ISBN: 978-1-84995-146-3
30. Saunders, M., Lewis, P., Thornhill, A.: Understanding research philosophies and approaches. 4, Chapter 4: pp. 106–135 (2009). https://www.researchgate.net/publication/309102603

Design Methodology for Shoreline Protection in Sarawak Shoreline Using Rubble-Mound Breakwater

P. A. Ratnasinghe[1(✉)] and K. K. Kuok[2]

[1] Department of Civil Engineering, Swinburne University of Technology Sarawak Campus, 93350 Kuching, Sarawak, Malaysia
paratnasinghe321@outlook.com

[2] Faculty of Engineering, Computing and Science, Swinburne University of Technology Sarawak Campus, 93350 Kuching, Sarawak, Malaysia

Abstract. The Sarawak shoreline in East Malaysia has been exposed to erosion in the long term and has experienced severe damage in recent years. This specific event occurred mainly due to the influence of global warming, wave energy, and marine currents that transfer sediments distantly. Strategic implementation of geotextile tubes, rock revetment, and dune stabilization were visible across the Sarawak shoreline. However, breakwaters are considered a profoundly feasible method compared to the current applications for the preservation of the shoreline. The main objective of this research is to achieve an appropriate design template for the preservation of seven different regions of Sarawak's shoreline. After adequate research, it was revealed that a rubble mound breakwater is suitable as the structural development for the coastline. The research is focused on calculating the different components of a breakwater through Microsoft Excel. Hudson's formula and the Shoreline Protection Manual were reviewed when performing the calculation for the design of armor units. A Cubipod calculator was used to calculate the size of the armor units. This paper presents a series of calculations done through obtaining relevant data and inputs from reliable sources regarding Sarawak shoreline. The results disclosed that the average weight of an armor unit is about 3.7 t, which is a Cubipod. Region 2 and 3, which include Sarikei, and Sibu, require the largest armor unit due to their higher water depth. The estimated overtopping of the breakwater in regions 1 and 2 is slightly lower than the rest of the regions. The south side of the Sarawak shoreline necessitates a large total structure height compared to the north side. The final results obtained through the Cubipod calculator and spreadsheet calculations provide the parameters to create the design template for the breakwater. Furthermore, the utilization of Cubipod as the armor unit increases the overall hydraulic stability of the structure and its structural strength, lowers run-up and overtopping rates, and provides higher friction with the filter layer.

Keywords: Rubble mound breakwater · Cubipod · Cubipod Calculator · Shoreline preservation · Coastal Protection Manual

M. Siriwardhana et al. (Eds.): IERS 2025, PTT, pp. 264–279, 2026.
https://doi.org/10.1007/978-981-95-7984-6_23

1 Introduction

1.1 Background

Coastal erosion is one of the significant problems experienced in Malaysia, causing damage to infrastructure and the environment. It was regarded as a national issue because many socioeconomic activities were threatened and the situation was deteriorating over time potentially causing damages to the local coastal landscape [1]. This phenomena gradually occurs in Sarawak and is one of the prime issues that has been getting more severe over the last few years as a result of natural forces, primarily global warming, wave energy, and ocean currents [2]. According to the National Coastal Erosion Study (2015), 17% of the Sarawak coastline is currently eroding, and as a result of this, an ocean-level escalation of 0.67–0.74 mm on average annually occurs, as well as an increase of wave energy. Waves are created by wind and the wave height is decided on three major factors of wind speed, duration of the wind, and the fetch that the size of the storm. Energy of the wave is the major factor that creates coastal erosion [3]. Waves crash in opposition to the shores and flush out the sediments along the coastline, causing erosion. The impact of the bigger waves is that they easily obliterate the sediments, causing more coastal erosion. The understanding of the wave conditions like wave height, fetch length, wind speed, and duration is essential in achieving a satisfactory design for the breakwater. Understanding these dynamics is crucial for effective coastal management and conservation efforts [4].

1.2 Shoreline Protection in Sarawak

Shoreline preservation is paramount, as it directly affects the marine environment and the livelihoods of the residents along the coastline. The restoration of geological stability, which has been deteriorated by coastal erosion, is regarded as a primary goal of shoreline protection. Many methods have been used by the authorities in order to prevent shoreline erosion in Sarawak including armor rocks, geotextile tubes, and rock revetment [1]. Revetments function as shoreline protection structures with core flexibility. Revetments depend on the surface for support of their protection while they are assembled out of armor rock. These structures, being comparatively insubstantial, are appropriate to be located in a lightwave-conditioned atmosphere. Furthermore, geotextile tubes are usually considered to be massive bags made out of permeable woven geotextiles that are heavier than average sandbags. These have been utilized to preserve the coastline, recover the land area, and reconstruct the coasts in the long run [5].

However, there is an absence of a convenient design template for a breakwater along the Sarawak shoreline. Breakwaters are identified to decrease the outcomes of storm outpouring and shoreline erosion with their productive methodology that enthralls high tidal energy during storm events. There are various types of breakwaters that can be utilized to preserve the shoreline in Sarawak. The experts have not yet identified the appropriate breakwater as each type shows differences in the ability to resist wave energy due to the alterations in the local coastal environment and geotechnical conditions [6].

The research will be coordinated to find the most appropriate breakwater type and design for the preservation of the Sarawak shoreline and to propose a design template that can be used in all seven different regions in Sarawak. The study is accompanied through a series of calculations using an Excel spreadsheet and using Cubipod calculator to calculate different design parameters of the breakwater. The Hudson's formula, data from the Department of Drainage and Irrigation and Shoreline Protection Manual are used for the design process of the Rubble Mound breakwater.

The importance of this research is that coastal erosion around Sarawak shoreline is experiencing severe damage and this situation is out of the professionals' consideration. The design template that is gained as result of this research can be used for the future reference of coastal engineers as it delivers all the necessary information like type of breakwater, type of rock material, and calculation of wave conditions including run-up. This study is useful for the environment as it preserves the coast-line from erosion. The government and society will benefit from this research because it will provide information about coastal erosion and how breakwater can be used to protect the shoreline in Sarawak.

1.3 Existing Shoreline Protection Using Breakwater in Malaysia

According to the Department of Irrigation and Drainage Management [5], the adaptations of breakwater for shoreline preservation can be seen in different regions of Malaysia. These hard structures are constructed in Haji Dorani, Kuantan Port, and Sungai Marang river mouth in Terengganu. The breakwater found near Sungai river mouth is constructed with combining a perpendicular groyne to enhance the stability of the structure. The emerged breakwater restoration in Haji Dorani represents an incorporated approach to shoreline restoration by reducing water flow and severe ecological outcomes. An attached breakwater was constructed in Kuatan port and reef ball structures which are considered as porous submerged breakwaters can be seen in Sabah park at Selingan Island in Sabah.

2 Methodology

The area of the study is focused on the Sarawak shoreline in east part of Malaysia which borders with the South China sea. The Coastal Engineering Manual, which is published by the United States of Army Corps is used to obtain guidelines for the design [7]. An excel spreadsheet was developed in order to get the results for the design template and the below steps illustrate the procedure of obtaining the design template accordingly, that can be used for every region of the Sarawak Shoreline.

2.1 Identifying and Studying the Area

The study will focus on the shoreline of Sarawak, which spans approximately 1235 km, starting from Telok Melano and extending to Lawas and Tawau, encompassing seven distinct regions (Fig. 1) which are Kuching, Sarikei, Sibu, Mukah, Bintulu, Miri, and Limbang [2]. The Sarawak Sea, situated on the continental shelf of the South China Sea, exhibits significant bathymetric variation with predominantly shallow waters, where depths gradually increase towards the outer shelf, beginning at approximately seven meters. The northern side of the Sarawak Sea, bordering Sabah, primarily consists of minor marine unadorned sequences. In contrast, the southern side features mudflats and potentially submerged gorges [8]. The seven regions include ports with a considerable amount of vessel motion. The data compilation regarding the shoreline highlighted the importance of offshore structures being resilient to environmental factors and their installation in relation to coastal waters. The standard ports mostly attracted cargo ships that exhibited translational movement, defined as the ship moving at a consistent average speed at all points while experiencing variations in buoyancy and weight [9].

Fig. 1. Seven Regions of Sarawak Shoreline [10]

2.2 Selection of Breakwater and Collection of Data

Table 1. Collection of Data for run-up calculation [10]

Region	Standard Port	Water Depth (m)	Design High Water (m)	Hs (m)	Tm (sec)	Still water Level (m)
1	Kuching	9.7	2.22	4.4	6	2.37
2	Sarikei	7	2.86	4.5	6	2.54
3	Sibu	7	1.69	4.5	6	1.48
4	Mukah	9.7	1.36	3	6	0.96
5	Bintulu	9.7	1.42	3	6	3.63
6	Miri	8.1	1.47	3.1	6	2.23
7	Limbang	8.1	1.49	3.1	6	2.31

The primary function of the breakwater is to provide erosion protection and control wave action. Consequently, an ideal solution would be a sea-backed structure that requires minimal human intervention. A detached rubble mound breakwater was selected, given that the Sarawak shoreline features sandy beaches alongside a few rocky sea beds. This option aligns well with the metocean conditions, which include variations in waves and currents that change daily, alongside seldom extreme events. Additionally, the nearshore hydrodynamics and the site's constructability were considered [9]. Nearshore water depth, design wave height (H_s), still water level, wave period (T_m) and design high water are among the data that were obtained for the calculations which are presented in Table 1 [10].

2.3 Design Conditions and Parameters

The design life of a rubble mound breakwater can extend up to 40 years or more, depending upon the stability and its resilience to wave forces. Several assumptions factored as the design conditions during the calculations. The structure is considered to be symmetrical, therefore front and back slopes are equal. Allowable overtopping (q) for port protection is 0.03 L/s/m [11]. The structure slope is 1:1.5 [12]. Breakwater crest height relative to design high water and freeboard are equal as a crown wall is not considered during the design [13]. The breakwater length approximately runs up to 1km, whereas variations can occur due to the location and the breakwater is built detached to the shoreline [14]. The governing design parameters calculated are as follows.

Environmental Conditions

Estimated overtopping and run-up (Fig. 2) were calculated using Eqs. 1 and 2 [14].

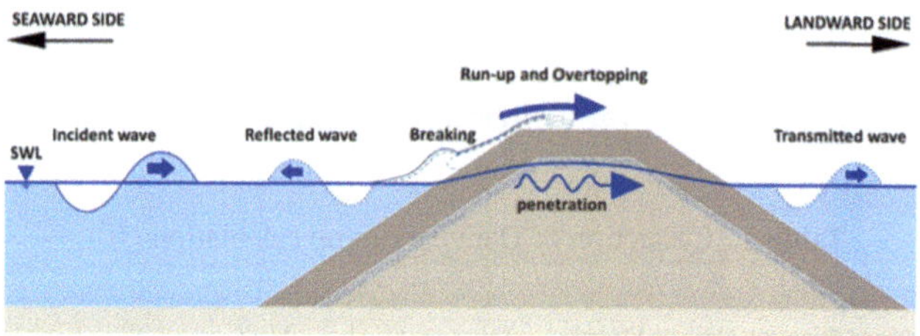

Fig. 2. Illustration of Run-up and Overtopping [14]

$$\frac{q}{gH_sT_M} = aexp\left(-b\frac{R_m^*}{\gamma_r}\right); \quad \text{Owen Model Equation for overtopping} \tag{1}$$

$$\frac{R_{ui\%}}{H_s} = B(\xi_{om})^C; \quad \text{Wave Run}-\text{up} \tag{2}$$

Dimensions of the Structural Components of the Breakwater-Design of Layers

Figure 3 illustrates an adequate example of the components of a rubble mound breakwater. The computation of crest height, toe-berm width, maximum water depth, and the width and length of both the structure and trunk effectively defines the viable structural geometry of the breakwater [e]. Both the breakwater and its components are designed as a trapezoid, as the particular shape enhances stability and effectively dissipates wave energy.

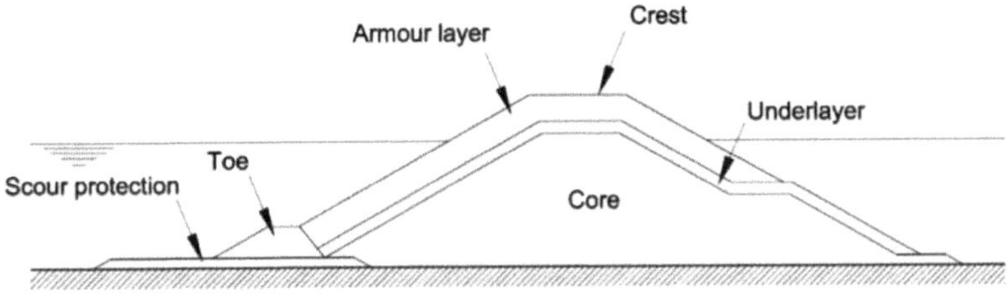

Fig. 3. Cross-section of a Rubble mound breakwater [e] [14]

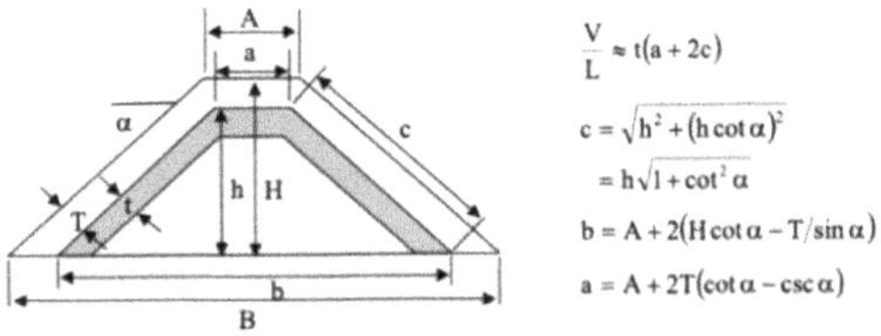

Fig. 4. Equations for the dimension calculations [13]

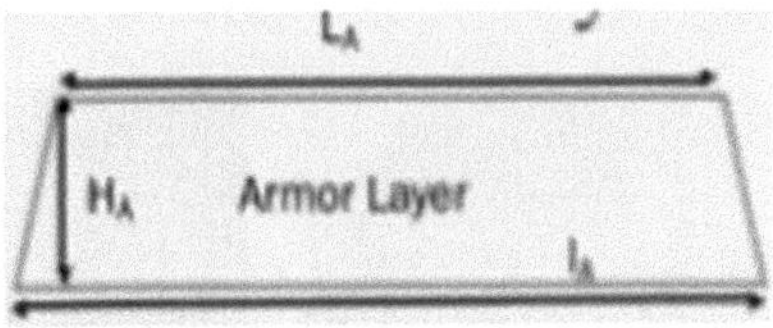

Fig. 5. Width and height of armor layer [13]

The cross-sectional dimensions, given in Figs. 4 and 5, thickness, height, and width of each layer will be calculated in this particular stage of the design. The cross-sectional dimensions are computed for the three main components of Armor layer, filter layer, and core. The dimensions can be computed using the below trigonometric equations and the layer thickness obtained from the cubipod calculations [13].

Toe Berm, Bed, Crest Width, and Trunk Design

The manufacture of the toe berm considerably influences the proper positioning of the first row of concrete armor units, which interrupts the upper rows of units, because armor units in a row are positioned on units of the row underneath. Figure 3 indicates the cross section of a mound breakwater with an armor sustained by a toe berm located on scour protection apron. The objective of having a bed is to protect the foundation material from discharging, avoid pore compression build-up in the base material, and keep the structure away from excessive settlement. The bedding does not require a filter layer if the base is made out of consistent material and a coating of quarry stone will be positioned as the bedding coat to minimize settlement. The crest width refers to the distance from the ocean side to the shore side of the breakwater at its peak point, which corresponds to the width of the armor layer [13]. All the above described components belong to the trunk design of the breakwater. The design of this segment is complex and it requires the ideal placement of Cubipods under the conditions of crawler cranes, heavy winds, waves, and underwater blind placement.

$$t = nk\Delta\left(\frac{W}{\gamma_a}\right)^{\frac{1}{3}} \quad \text{Crest width} \tag{3}$$

The Material Properties of the Breakwater

A rubble mound breakwater consists of bedding, a core, a filter layer and an armor layer. The armor unit used in this design will be a Cubipod (shown in Fig. 6) and the rest of the layers will be occupied with different kinds of quarry stone. The material properties of the breakwater including unit weight of armor, core, armor, and filter layer thicknesses were obtained using the Cubipod calculator [15]. Thickness, porosity, and unit geometry are some of the characteristics that should be attenuated when selecting an armor unit. The availability of the casting forms and the quality of concrete were considered when selecting a proper armor unit. The cubipods will be placed in a single-layer armor with homogeneous armor porosity. These units have the ability to arrange themselves in random orientations (Fig. 7).

Fig. 6. Cubipod [13]

Fig. 7. Orientation of the Cubipods in the Breakwater [13]

In this specific design process, the Cubipod calculator, created by the OHLA Group in Spain, is used to determine the standard dimensions and weight of the armor units, as well as the layer thicknesses, thereby ensuring enhanced accuracy. After calculating the armor unit's weight, the filter and core material sizes can be optimized. The armor unit weight equals to 20 times of filter quarry and 200 times core quarry material weight. After the initial breakwater design had been completed, the capacity of the different material sizes should be compared to the allocation of the rock size at the quarry. Stone sizes or layer thickness may require modification to competently manipulate the quarry [13].

3 Results and Discussion

All the design parameters were obtained through spreadsheet calculations and using Cubipod calculator. These results include all the seven different regions of the Sarawak shoreline. Further analysis and discussion related to the obtained results were conducted to acquire the design template of the breakwater.

Rubble mound breakwaters protect the shoreline by breaking the waves on the slope of the rigid structures. These massive configurations require huge volumes of quarry material. The armor units should have a specific weight and excellent resistance to abrasion and weathering. The primary armor layer of the breakwater, constructed with precast concrete units, should be able to resist the forces originated by the waves breaking on the slope during storm events. This particular design will be focused on mound breakwaters that are constructed using the primary concrete armor unit of Cubipods. Rubble mound breakwaters are considered the most durable and best suited for extreme wave conditions, and reinforcement is not being utilized in mound breakwaters.

3.1 Overtopping and Run-Up

The estimation of overtopping and run-up determines the crest elevation. Run-up and overtopping in a rubble mound breakwater rely on several factors, such as armor characteristics, the permeability of the core, filter layers, and crest geometry. An armor layer with a higher roughness reduces overtopping and accepts the design for a lower crest elevation.

Table 2. Overtopping and run-up

Region	Standard Ports	$R_{u10\%}$ (Run-up) (m)	q (Estimated Overtopping) (L/s/m)
1	Kuching	4.41	0.02
2	Sarikei	4.48	0.02
3	Sibu	4.48	0.02
4	Mukah	3.70	0.03
5	Bintulu	3.05	0.03
6	Miri	3.13	0.03
7	Limbang	3.13	0.03

According to Table 2, it is evident that both overtopping and run-up are acceptable for the design, as the values obtained fall within the range of port protection. The south side of the Sarawak shoreline exhibited moderately high run up values and low overtopping. In contrast, the north side showed low run-up and high overtopping, satisfying the design conditions. Wave run-up, caused by waves breaking on the structure slope, mainly decreases with escalating water depth at the toe of the structure. Conversely, overtopping tends to increase with greater water depth. This is evident when comparing Sarikei and Bintulu. As the water depth increases, Bintulu, situated on the northern side of the Sarawak shoreline, displays reduced wave run-up and heightened overtopping in comparison to Sarikei. The latter has a slightly lower water depth and overtopping, yet experiences a significantly higher wave run-up. An increase in wave run-up can lead to elevated crest levels (Table 3).

Table 3. Comparison between Sarikei and Bintulu

Region	Water Depth h_t (m)	Overtopping (L/s/m)	Wave Run-up(m)
Sarikei	7	0.02	4.48
Bintulu	9.7	0.03	3.05

The high homogeneity of the Cubipods used in the breakwater allows for a reduction in run-up and a decrease in pressure compared to other armor units [13]. The estimated overtopping is particularly significant for a rubble mound breakwater, given that these structures are low-crested. However, estimating wave run-up is also considerable, as it provides a prediction of the number of waves that will reach the crest of the breakwater [11]. Access to the breakwater crest must be limited during extreme annual storm events in order to avoid injuries due to breaking off the allowable overtopping. This can occur during the service time and will sometimes result in severe injuries to vessels, infrastructure, and people if the rear slope is not secured enough [13].

3.2 Analysis of Material Properties, Design Layers and Cross-Sectional Dimensions of the Breakwater

Material Properties

A rubble mound breakwater consists of three main layers. The armor layer, filter, and core. For the armour layer, Cubipod armor units are used, while quarry stones are used for the filter layer and core. The accuracy is relatively high compared to using the traditional Hudson's equation given in Eq. 4 [12]. The Cubipod is made out of concrete. Therefore, density of concrete is 2.35 tm^3. The structure has a slope of 1:1.5.The filter layer quarry stone size and core quarry stone size are computed with respect to the armor unit weight. Table 4 provides the sizes of the Cubipod armor units with respect to the parameters and the sizes of the filter layer and core quarry stones with respect to the armor weight.

Table 4. Material Sizes of the Breakwater calculated according to the Design Conditions

Standard Port	Weight of the armor unit (t) **W**	Weight of the filter layer quarry stone (t) **Wf**	Weight of the core layer quarry stone (t) **Wc**
Kuching	3.56	0.18	0.02
Sarikei	5.62	0.28	0.03
Sibu	5.62	0.28	0.03
Mukah	3.66	0.18	0.02
Bintulu	1.67	0.08	0.01
Miri	1.84	0.09	0.01
Limbang	1.84	0.09	0.01

$$W(t) > W_f \frac{W}{20}(t) > W_c \frac{W}{200}(t) \quad \text{relation between material sizes [12]} \tag{4}$$

The results show that Sarikei and Sibu, located on the southern side of Sarawak's coastline require heavy armor, filter and core material. The minimum required armor unit weight, recorded at 1.67 t, ensuring the structure's stability against extreme wave action, despite being moderately small compared to the standard size of 6 t [13].

Design Layers

The breakwater consists of three main layers. The armor layer is the most important part of the design, as it is the first to encounter the ocean's high wave energy. Figure 8 shows a comparison of the wave height and armour layer thickness according to the seven different regions of the Sarawak shoreline. It shows that with a greater wave height, the thickness of the armour layer also increases. Port Sarikei and Sibu, located in the southern region of the Sarawak shoreline, require the largest armour unit, specifically a Cubipod weighing 5.62 t. This requirement leads to an increased armour layer thickness of 1.78 m, which corresponds to the areas that experience the highest wave height of 4.5 m. The filter layer thickness and core were evidently the highest, corresponding to greater wave heights, with the respective lowest values recorded for Bintulu being 0.77 and 0.63.

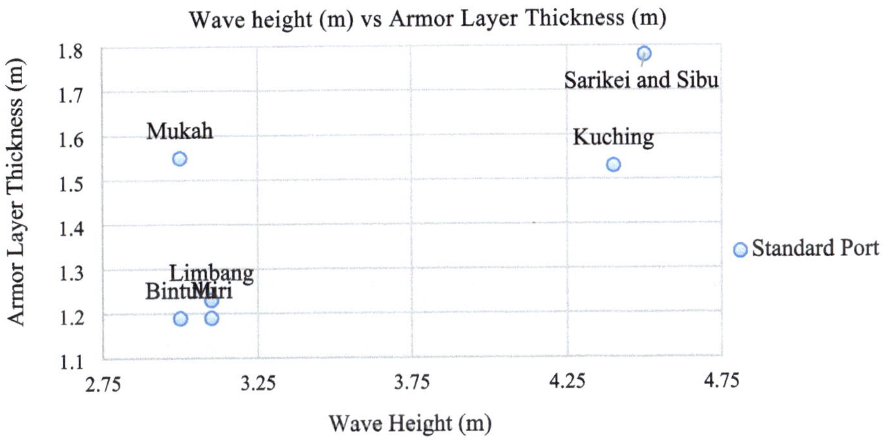

Fig. 8. Wave Height vs Armour Layer Thickness

Analysis of the Round Head Design and Trunk Cross-Section

The breakwater structure is designed as a trapezoidal prism, incorporating armour, a core, a filter layer, and a trunk (shown in Figs. 9 and 10). This structure functions as a sea-backed detached breakwater; consequently, both the round head and trunk are constructed from the same material, allowing the trunk to be viewed as an extension of the round head, although with increased length as same cross-section. As per the design conditions specified in Sect. 2.3, the breakwater's length may extend up to 1–1.5 km, contingent upon the length of the region's shoreline and the positioning of the breakwater, which is intentionally constructed to be detached from the shoreline to enhance its overall effectiveness [14].

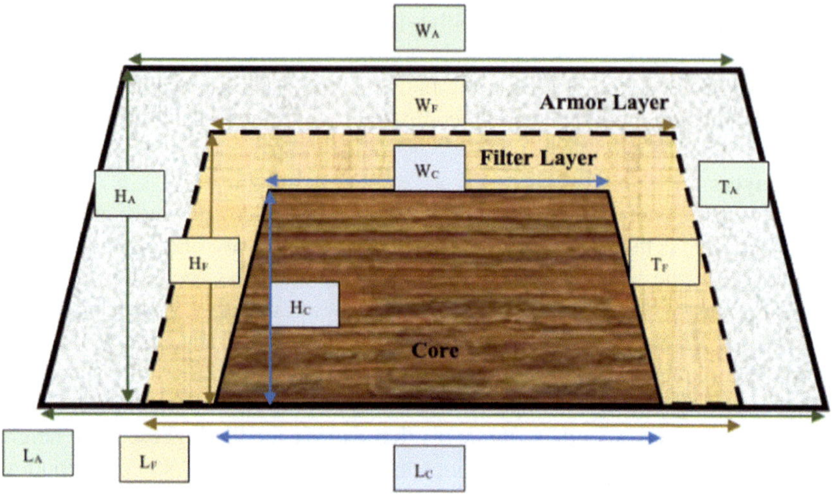

Fig. 9. Cross-Section Dimensions of the Round Head and Trunk Design

Fig. 10. Visual Layout of the Round Head and Trunk [5]

Cubipod has a higher hydraulic stability coefficient value of 5 (K_D) compared to other armor units, including Tetrapods and Dolos [15]. Therefore, the obtained crest width resulted in a higher value and is acceptable for the large structure of a rubble mound breakwater. Figure 9, Fig. 10 and Table 5 displays a typical cross section of the design and its dimensions. The values of different parameters differ according to the varying water depth, design high water, still water level, wave period and design wave height of the seven regions of the Sarawak shoreline. The structure slope is around 33.69° for both the front and back slopes of the structure due to its symmetrical shape. The crest height and toe width were also calculated and resulted in moderately significant values according to the seven different regions (Table 6). The toe berm should be positioned on the sea bottom even though the Cubipod units are larger than expected even at the deepest points. The sea bottom experiences a dynamic equilibrium with waves and sea currents when the breakwater is constructed on a sandy ocean floor with necessary scour protection. The positioning of the massive structure makes notable changes to the current pattern around it and modifies the motion of sand while producing scour around the toe of the structure [13]. The horizontal width of the bedding layer extends further than the width of the cross-section of the breakwater.

Table 5. Design Calculation for cross-section dimensions for all the regions

Region 1 – Kuching										
Armor Layer (m)				Filter Layer (m)				Core (m)		
TA	**W_A**	**L_A**	**H_A**	**TF**	**W_F**	**L_F**	**H_F**	**WC**	**LC**	**HC**
1.53	8.60	17.20	9	1.42	5.67	8.50	4.59	1.83	5.48	2.37
Region 2–Sarikei										
1.78	10	20	9.88	1.27	6.56	9.83	5.4	2.09	6.27	2.54
Region 3–Sibu										
1.78	10	20	7.65	1.17	6.56	9.83	3.17	2.09	6.27	1.48
Region 4–Mukah										
1.55	8.68	17.36	6.02	1.02	5.67	8.50	2.32	1.83	5.49	1.36
Region 5–Bintulu										
1.19	6.70	13.40	8.1	1.06	4.34	6.50	5.05	1.46	4.37	2.23
Region 6–Miri										
1.23	6.92	13.84	6.83	1.22	4.51	6.76	3.7	1.46	4.37	1.47
Region 7–Limbang										
1.23	6.92	13.84	6.93	1.17	4.51	6.76	3.8	2.18	6.55	1.48

Table 6. Details of Crest Height and Toe Design

Standard Port	Crest Width-B (m)	Toe			
		Width from bottom of the structure (m)	Material Weight (t)	Thickness (m)	Height (m)
Kuching	8.60	7.23	0.18	0.99	1.25
Sarikei	10	10.78	0.28	1.08	1.11
Sibu	10	9.75	0.28	0.75	1.03
Mukah	8.68	8.21	0.18	0.86	0.95
Bintulu	6.70	7.34	0.08	1.23	0.98
Miri	6.92	8.42	0.09	0.88	1.08
Limbang	6.92	8.11	0.09	1.11	1.08

A Cubipod, made out of pre-cast concrete and belongs to the cube family and the design is robust with much higher structural integrity. Wave energy and the influence of global warming can impact the structure long-term. Therefore, the great hydraulic stability combined with self-positioning and self-rehabilitating implementation places the Cubipod in a higher position compared to the other armor units. It allows the breakwater towards lower overtopping and run-up rates. These massive units have the ability to self-position when arranged randomly on the slope with homogeneous armor porosity. They tend to withstand rough sea conditions and design ambiguities like discrepancy settlements. The Cubipod armor unit is an appropriate design substitute for the adverse conditions and limitations in construction.

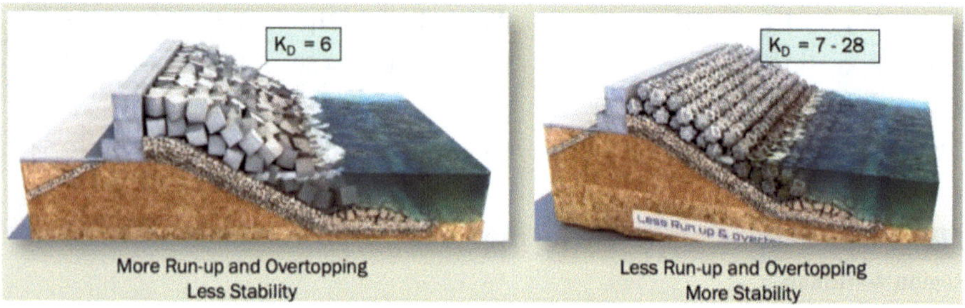

Fig. 11. Difference between using Cube and Cubipod in Breakwater [13]

A single Cubipod layer is used in the trunk of this design, which will reduce the amount of concrete used during the manufacturing process and further reduce the construction cost of the breakwater. The Cubipod layer carries higher armor randomness, has higher friction with the underlayer, and produces lower run-up and overtopping rates (Fig. 11). It has much higher structural strength compared to other units. The casting of Cubipod is easy using vertical molds. The production rate is higher, and the handling of this massive object is safe and efficient, which will improve the quality of the breakwater design.

3.3 Final Design Template for the Breakwater

Table 7 outlines the design parameters for the breakwater template for Bintulu in region 5 of Sarawak shoreline.

The economic cost of the mound breakwaters is determined by several environmental conditions and one major factor. That is the precast concrete unit used for the armor layer construction and the geometrical configuration of the unit. Construction of mound breakwater includes several complex logistical procedures, which can easily be solved with organized and managed equipment and personnel. Therefore, using a single-layer breakwater can be both cost-effective and impeccable for this environment. For this research, there are some suggestions to improve the calculations. It is recommended to use a collection of updated data to enhance the accuracy of the design. Engineers must also take into account environmental factors and the potential impacts of climate change when planning for future resilience. This will be useful for future use when designing coastal protection structures for the Sarawak shoreline.

Table 7. Design Template for Region 5 - Bintulu

Design Parameter	Symbol	Values
General Design Parameters		
1. Actual Overtopping	q	0.03 (m^3/s/m)
2. Run- up	$R_{u10\%}$	3.05m
3. Water Depth of Structure relative to design high water (DHW)	h	9.7 m
4. Depth of structure toe relative to still water level (SWL)	h_t	7.34 m
5. Crest Width	B	6.70 m
Toe design		
6. Toe height	$2r_f$	0.98 m
Layer Thickness		
7. Thickness of layers	T- armor layer	1.19 m
	t- filter layer	0.77 m
	Toe armor	1.23 m
Unit weight		
10. Armor unit weight	W_A	1.67 t
11. Filter material unit weight	w_f	0.08 t
12. Core material unit weight	w_c	0.01 t
13. Toe material unit weight	w_t	0.08 t
Cross-sectional Parameters		
Armor Layer		
14. Armor layer height	H_A	8.1 m
15. Armor layer upper width	W_A	6.70 m
16. Armor layer bottom width	L_A	13.40 m
Filter Layer		
17. Filter layer height	H_F	5.05 m
18. Filter layer upper width	W_F	4.34 m
19. Filter layer bottom width	L_F	6.70 m
Core Layer		
20. Core layer height	H_c	2.23 m
21. Core layer upper width	W_c	1.46 m
22. Core layer bottom width	L_c	6.55 m

4 Conclusions

Coastal erosion in Sarawak has been getting severe in the past few years and is commonly caused by global warming, wave energy and, marine currents on the coastline. The construction of a breakwater has been suggested, but there is currently an absence of a design template for a breakwater that is appropriate to all seven different regions in Sarawak. This research project was able to achieve its main aims and objectives by proposing an appropriate design template for the preservation of the Sarawak shoreline using breakwater. The results obtained from the calculations present the suitable sizes of the armor unit, filter, and core material for the breakwater. The structure dimensions, such as depth of the structure, toe width, crest width, and thickness of layers were obtained. From the numerical analysis, it is proven that with greater structure depth, the armor thickness and unit size increases. Design wave height mainly depends on the water depth. The northern region, which includes Sarikei, showed larger values for the dimensions and properties of the armor unit. Furthermore, it elaborates that using Cubipod as the armor unit increases the hydraulic stability of the structure and strengthens the structural dynamics of the breakwater. A rubble mound breakwater is the most commonly used breakwater for coastline protection. These breakwaters are stable structures made up of several layers of stone and gravel that are protected on the outside by a layer of randomly located or positioned artificial armour units. The most prominent advantage of this breakwater type is the simplicity of construction, function, and maintenance. This design serves to protect the coastline from erosion and ensures that the structure can endure the forces of nature over time. By optimizing the geometry, engineers can significantly enhance the durability and performance of coastal defenses.

5 Future Research Path

Hydraulic stability can be achieved by developing a prototype to investigate the performance of the breakwater. The settlement and the bearing capacity of the breakwater can also be calculated for further classifications. A comparison between the other methods of shoreline preservation and breakwater can be done to acquire the most effective and appropriate method for the Sarawak shoreline.

References

1. Ghazali, N.H.: Coastal erosion and reclamation in Malaysia. Aquat. Ecosyst. Health Manage. **9**(2), 237–247 (2006)
2. Muller, J.: Coastal Length of Malaysia 2019, Statista, 201, viewed 7 April 2022, <https://www.statista.com/statistics/947844/malaysia-coast-length-by-state/>
3. Luo, Y., Shi, H., Wang, W.: Response of extreme significant wave height to climate change in the South China sea and Northern Indian sea. Aquat. Ecosyst. Health Manage. **24**(3), 103–110 (2021)
4. Chang, G., Jones, C., Roberts, J., Neary, V.: A comprehensive evaluation of factors affecting the levelized cost of wave energy conversion projects. Renew. Energy **127**(4), 344–354 (2018)

5. Department of Irrigation and Drainage, Integrated coastal zone management. Sarawak Government (2022). viewed 17 Apr 2022. <https://did.sarawak.gov.my/page-0-123-476-INTEGRATED-COASTAL-ZONE-MANAGEMENT.html>
6. Department of Irrigation and Drainage, Coastal management. Ministry of Environment and Water (2022) viewed 17 Apr 2022. <https://www.water.gov.my/index.php/pages/view/1296?mid=305>
7. Mokhtar, A., Ghazali, N.M., Isnin, A., Sivioli, J., Lee, V.: Planning the Malaysian coastline Integrated shoreline management plan. 10th International conference on Asian and Pacific Coasts (APAC 2019), 25-28 Sep 2019, Hanoi, Vietnam, pp. 4-8. APAC 2019. viewed 31 Mar 2022. <https://docs.jpa.gov.my/sme/penulisan/kejuruteraan/1.pdf>
8. Vannak, D., Liew, M., Yew, G.: Time domain and frequency domain analyses of measured metocean data for Malaysian waters. Int. J. Geol. Environ. Eng. **7**(8), 549–554 (2013)
9. Manson, A., Saleh, E., Mujahid, A., Juneng, L.: Temperature and salinity analysis off Sarawak waters. Malays. Trans. Sci. Technol. **5**(3), 216–220 (2018)
10. Kuok, K.K., Ying, S.T., Chiu, P.C.: Integrated coastal zone management to protect the Sarawak shoreline. J. Coast. Conserv. **25**(48), 25–35 (2021)
11. Van der Meer, J.W., et al.: Manual on wave overtopping of sea defences and related structures. EurOtop (2018). www.overtopping-manual.com
12. U.S. Army Corps of Engineers: Coastal Engineering Manual. U.S Army Corps of Engineers, Washington D.C. (2006)
13. Mediana, J.R., Gomez-Martin, M.E.: Cubipod Manual. University of Valencia, Spain (2016)
14. Van de Meer, J., Sigurdarson, S.: Geometrical design of berm breakwaters. Coast. Eng. Proc. **1**(34), 1–15 (2014)
15. Sato C.: Cubipod Calculator, Cubipod (2022). viewed 18 Apr 2022. https://www.cubipod.com/en/design/cubipod-calculator/

Developing a National Safety Standard in Using Flammable/Toxic Refrigerants in HVAC/R Applications in Sri Lanka

C. H. Kannangara(✉), J. B. Senarath, and W. P. D. Fernando

Department of Mechanical Engineering, Faculty of Engineering, University of Peradeniya, Peradeniya, Sri Lanka

charith.kannangara@eng.pdn.ac.lk

Abstract. The necessity of a global shift toward low-GWP and no ODP refrigerants is forcing countries like Sri Lanka to revise and strengthen their refrigerant safety regulations since they have not yet fully implemented those regulations. The global trend in the refrigeration industry is to move towards natural refrigerants to replace the existing common refrigerants that are mostly man-made. However, most of the natural refrigerants used in the HVAC/R sector are flammable, toxic, or both, leading to safety concerns. Due to the lack of localized safety standards to support the safe adoption of these alternatives, they are mostly overlooked. The critical challenges spring from how to avoid, minimize the flammability issues, toxicity issues, and improper end-of-life handling of refrigerants currently used in the refrigeration industry. These concerns have become very significant since the usage of low-GWP refrigerants expands without a comprehensive regulatory framework. Going through a detailed analysis, it was identified that the existing Sri Lankan Standard, SLS 1262:2004, does not sufficiently address the safety and environmental risks and international regulations associated with new refrigeration technologies. This paper presents the development of SLS 1262:2025, a revised standard created by referring to internationally recognized guidelines such as ISO 817, ISO 5149, IEC 60335, EN 378, and ASHRAE 15/34. With a focus on adapting global best practices to local climatic, economic, and infrastructural realities, this paper outlines the most suitable regulatory provisions to ensure the safe use of flammable and toxic refrigerants in Sri Lanka.

Keywords: Refrigerants · Flammability · Toxicity · Safety Standards · Low-GWP

1 Introduction

The Heating, Ventilation, Air Conditioning, and Refrigeration (HVAC/R) sector is undergoing a fundamental transformation driven by the dual imperatives of environmental sustainability and operational safety globally. As nations accelerate their transition toward low global warming potential (GWP) and no ozone depletion potential (ODP) refrigerants, concerns have increased regarding the flammability and toxicity of potential

M. Siriwardhana et al. (Eds.): IERS 2025, PTT, pp. 280–289, 2026.
https://doi.org/10.1007/978-981-95-7984-6_24

replacements, and the need for robust safety standards has become increasingly urgent. Developed countries have responded with updated regulatory frameworks grounded in international benchmarks such as ISO [1], ASHRAE [2, 3], and BS EN [4–6]. However, direct adoption of these standards remains impractical for developing nations due to variations in climate conditions, economic capacity, infrastructure, and workforce training.

In Sri Lanka, the existing national standard for refrigerant safety is SLS 1262:2004 [7] and was developed at a time when the sector was less complex and dominated by high-GWP refrigerants like CFCs and HFCs. This standard laid the foundation for basic safety compliance. However, it lacks technical depth, classification rigor, and environmental provisions required to manage the emerging risks associated with low-GWP alternatives such as ammonia (R717) and propane (R290) as refrigerants. The urgency of modernizing this framework has been amplified by the country's international commitments under the Kigali Amendment and its increasing adoption of natural refrigerants in new commercial and industrial applications.

The study was undertaken with the objectives,

- to identify gaps between SLS1262:2004 and internationally recognized refrigerant safety standards.
- to adapt international best practices to Sri Lanka's climatic, infrastructural, and socio-economic context.
- to propose and validate a revised national standard (SLS 1262:2025) that ensures both safety and environmental compliance for the handling of flammable/toxic refrigerants in Sri Lanka's HVAC/R sector.

Referring to globally accepted good practices and deeply contextualized through extensive engagement with local stakeholders, technical feasibility studies, and local infrastructure evaluations, the revised standard offers a comprehensive, actionable, and resilient approach to refrigerant safety. This paper outlines a literature review, gap analysis, and contextual adaptation approach to show how Sri Lanka can advance in both environmental goals and occupational safety by establishing a modernized and context-specific regulatory framework for refrigerants use refrigeration industry.

2 Methodology

This study followed a document-based research methodology structured in three key phases as Literature Review, Gap Analysis, and Contextual Adaptation. Each phase helped to build a comprehensive and locally relevant standard for the safe use of flammable and toxic refrigerants in Sri Lanka's HVAC/R sector.

2.1 Literature Review

A detailed review of well-established international safety standards was conducted to understand the globally recognized approaches to handling flammable and toxic refrigerants in the refrigeration sector. These standards were selected based on the relevance to the flammable and toxic refrigerants, global impact and recognition and flexibility for adaptation. The adaptability of these standards to local climatic conditions and implementation requirements was systematically evaluated through a comparative analysis with the Indian National Standards [8], which are themselves based on the reviewed international standards. The following standards were identified as the foundation for the literature review, based on the evaluation criteria discussed above.

- ISO 817-Classifies refrigerants by safety grouping and toxicity levels.
- EN 378-Provides guidance on the design, construction, installation, and operation of refrigerating systems, focusing on safety and environmental protection.
- ISO 5149-Details of safety and environmental requirements for refrigerating systems and heat pumps.
- ASHRAE 15 and 34-Defines safety procedures for refrigerant-containing systems and classifications.

These standards were studied to recognize the critical safety provisions, the limitations of flammable/toxic refrigerant quantities used in given systems, the refrigeration system design requirements, the refrigerant leak detection requirements, and the emergency procedures that are relevant to refrigeration systems using A2L, A3, and B class refrigerants.

2.2 Gap Analysis

A structured gap analysis was conducted to evaluate the extent to which the existing national standard (SLS 1262:2004) aligns with the identified international benchmarks. This analysis was performed by comparing the international standards with the national standards to identify deficiencies, omissions, or outdated provisions. Special attention was given to risk mitigation strategies for flammable and toxic refrigerants, as well as evolving global practices related to low-GWP alternatives.

2.3 Contextual Adaptation

In relation to ensure the feasibility to local conditions, the research adapted the identified international best practices to the Sri Lankan context through the following steps.

- **Technical Adjustments-**Adaptations were made to inline typical refrigeration system sizes commonly used in Sri Lanka.
- **Economic Feasibility-**High-cost provisions were adjusted, allowing cost-effective alternatives where appropriate.
- **Climatic and Infrastructure Considerations-**This standard was adapted for tropical climates and naturally ventilated environments that are common in Sri Lankan architecture
- **Skill Level and Workforce Capacity-**A draft standard was prepared to formalize structured technician training, certification, and continuous safety education, reflecting the varying levels of industry expertise.

The contextualization process was validated through stakeholder consultations involving mechanical engineers, HVAC/R contractors, environmental regulators, and relevant industry bodies. A Technical Advisory Committee comprising representatives from relevant industry bodies was established, and a series of meetings were conducted to evaluate the content. The draft standard document was published on the SLS website to invite public comments before finalizing the content. A questionnaire was sent to several relevant organizations through the stakeholders. The feedback from the stakeholders and answers received through the questionnaire helped ensure that the revised standard is both technically sound and locally applicable.

3 Comparison of the Previous (SLS 1262:2004) and the Revised (SLS 1262:2025) Standards

A comparative analysis between the previous standard (SLS 1262:2004) and the revised standard (SLS 1262:2025) was conducted to better understand the advancements introduced in the revised national standard. The key differences between the main sections of the two standards are shown in Table 1. This comparison outlines the significant shifts from the generalized approach to a more rigorous, application-specific framework that aligns with the best international practices and addresses contemporary challenges related to the use of flammable and toxic refrigerants.

Table 1. Comparative review of the previous (SLS 1262:2004) and the revised standard (SLS 1262:2025).

SLS 1262:2004	SLS 1262:2025
Primary Focus and Scope	
General safety requirements for all mechanical refrigerating systems used for cooling and heating	Specific focus on the safety, handling, and environmental aspects of systems containing flammable and toxic refrigerants
Refrigerant Classification According to Safety	
Classified only into 3 groups based on the combined effect of flammability and toxicity. (e.g., Group 1 - Non-flammable refrigerants which are not substantially injurious to the health of people. [7])	Refrigerant matrix with every possible scenario, with LFL and toxicity thresholds. Providing more detail on toxicity (A/B) and flammability (1, 2L, 2, 3) (e.g., A2L, A3, B2L). Aligned with ISO 817, EN378–1:2026, and ASHRAE 34 [3]
System Classification According to Access Categories	
Classify occupancies into five detailed categories, A, B, C, D, and E	Simplifies classification into three Access Categories as General access, Supervised access, and Authorized access
Topology (arrangement) Classification	
Basic classification without a comprehensive comparison (Direct, Indirect)	Highly detailed classification based on **Topology** (e.g., Direct Ducted, Indirect Vented) with clearer diagrams [4]
Refrigerant Charge Limit Calculation	

(*continued*)

Table 1. *(continued)*

SLS 1262:2004	SLS 1262:2025
Relied on "practical limits" derived from an appendix, based on the volume of the smallest occupied space	Provides detailed formulae and modern methodologies for calculating allowable charge limits, especially for flammable refrigerants. Introduces concepts like minimum room area and maximum charge [9]
Recovery, Reuse, Recycle, and Disposal of Refrigerants	
Included as an appendix (Appendix D) with general requirements	A comprehensive and dedicated Section. Explicitly forbids the release of CFCs, HCFCs, HFCs, PFCs, HFOs, and HCs to the atmosphere
Control of Ignition Sources and Fire Danger	
Contained general requirements and warnings regarding flammable refrigerants and ignition sources	Detailed sections and appendices have been given (e.g., Ignition source control, hazardous zone classification, specific requirements for detectors, alarms, and ventilation systems)
Structure of the Standards	
Organized by general topics such as Design, Construction, Installation, and Operation	More comprehensive distribution of content, as shown in Sect. 4

4 Structure of the Revised Standard (SLS 1262:2025)

The revised Sri Lankan national standard for the safe use of flammable and toxic refrigerants in HVAC/R systems is structured into five comprehensive sections. This format progresses from refrigeration system classifications and refrigeration system design to operational safety and life cycle management. In comparison to the previous version (SLS 1262:2004), the updated standard integrates international best practices while responding to Sri Lanka's specific environmental and technical conditions.

4.1 Refrigerant Classifications and Systems Containing Refrigerants

This section focuses on refrigerant classification and system types, laying the foundation for the safety requirements.

Classification of Refrigerants and Systems. Refrigerants are categorized by toxicity and flammability, enabling clear identification of hazard levels [1, 3]. This classification is given in Table 2.

Table 2. Classification of Refrigerants according to Their Toxicity and Flammability

According to Toxicity	According to Flammability	Code	Description
Class A Toxicity has not been identified when its concentration is less than or equal to 400 ppm by volume	Group 1	A1	Lower toxicity and no flame propagation (least hazardous)
	Group 2	A2L	Lower toxicity and lower flammability
		A2	Lower toxicity and medium flammability
	Group 3	A3	Lower toxicity and higher flammability
Class B Produces evidence of toxicity when workers are exposed to a concentration below 400 ppm by volume	Group 1	B1	Higher toxicity and no flame propagation
	Group 2	B2L	Higher toxicity and lower flammability
		B2	Higher toxicity and medium flammability
	Group 3	B3	Higher toxicity and higher flammability (most hazardous)

The classification of refrigeration systems according to the access category has been updated to the current standard, as shown in Table 3 [4].

Table 3. Classification according to the access categories.

Categories	General characteristics
General access **a**	Rooms, parts of buildings, and buildings where sleeping facilities are provided. People's movements are limited. An uncontrolled number of people are present. Any person has access without being personally acquainted with the necessary safety precautions
Supervised access **b**	Rooms, parts of buildings, buildings where only a limited number of people may assemble, some being necessarily acquainted with the general safety precautions of the establishment
Authorized access **c**	Rooms, parts of buildings, buildings where only authorized people have access, who are acquainted with general and special safety precautions of the establishment, and where manufacturing, processing, or storage of material or products takes place

Refrigerant Charge Limits and Space Volume. This presents the maximum allowable refrigerant charge limits in an occupied space to mitigate exposure risks. These limits are determined based on access category and space volume, with detailed methods provided for volume calculations, including considerations for interconnected spaces, false ceilings, and ventilation systems [9].

4.2 Safety Requirements During Manufacturing of System Components

This subsection explains the critical safety requirements for governing the manufacturing of individual components and piping used specifically in refrigeration systems that use flammable and toxic refrigerants. These requirements ensure the integrity, compatibility, and safe operation of components before system assembly. It specifies the relevant national and international standards that components must comply with, covering items such as heat exchangers, compressors, pumps, valves, and pipe joints [5]. The standard also explains additional safety criteria, including tightness and strength pressure testing to ensure leak-free and structurally sound components.

Material compatibility is emphasized, requirements that ferrous, non-ferrous, and non-metallic materials used in components are suitable for the intended refrigerants, lubricants, pressures, and temperatures, while minimizing the risks related to corrosion, chemical reactions, and mechanical failure. Furthermore, the standard mandates thorough testing procedures and detailed documentation for traceability and compliance verification. For electrical and refrigerating appliances, it references relevant International Electrotechnical Commission (IEC) safety standards, ensuring components meet necessary safety levels according to their risk category.

4.3 Safety Requirements for Design, Assembly, and Installation

The revised standard, SLS 1262:2025, introduces rigorous safety requirements for the designing, assembling, and installing of refrigerating systems to ensure structural integrity, safe operation, and compliance with environmental and operational regulations. Key provisions emphasize robust mechanical support, bases, and piping to endure anticipated loads, vibrations, and environmental factors such as wind, seismic activity, and thermal expansion [10].

Detailed specifications are given for piping materials, joints, and fittings, requiring adherence to type-approved standards and proper installation to prevent damage due to freezing, vibration, or mechanical stress [5]. Safety devices such as pressure relief valves and safety switching devices are mandated to prevent overpressure hazards. These must be appropriately set, tested, and maintained to protect all system sections. Isolation and overflow protections are outlined, along with requirements for devices indicating valve discharge and for safe discharge piping arrangements.

Additional provisions address safety during transport, prevention of fire and explosion hazards, especially with flammable refrigerants. The standard SLS 1262:2025 also includes specifications for noise, electrical design, protection against hot surfaces and moving parts, and the use of ventilated enclosures. Comprehensive system testing protocols, including strength, tightness, and functional conformity tests, must be performed before commissioning, ensuring operational readiness and regulatory compliance.

4.4 Safety Requirements for Operation and Maintenance

The updated standard SLS 1262:2025 introduces strengthened provisions to ensure the safety of personnel, systems, and the environment during the operation, maintenance, and repair of refrigerating systems that include toxic and flammable refrigerants. Installation sites must include visible warning notices, emergency procedures, and accessible protective equipment [6].

Regular inspections, leak tests, and preventive maintenance are required, with detailed procedures for oil draining, vacuum evacuation, and refrigerant charging. Only competent personnel should perform repairs, welding, or system modifications, and all activities must be properly documented. Refrigerant replacements, refilling, adding, removing, and disposals must follow strict planning and safety protocols.

4.5 Recovery, Reuse, and Disposal of Refrigerants

The revised standard SLS 1262:2025 introduces strict protocols for recovery, reuse, recycling, and disposal of refrigerants at the end of their service life. These measures are taken to minimize atmospheric emissions, ensure operational safety, and align with both national and international environmental regulations.

Key additions include clearly defined responsibilities, requiring that all recovery and disposal operations to be carried out by certified personnel using approved equipment. The standard SLS 1262:2025 also specifies technical specifications for recovery equipment, including minimum pressure thresholds and performance validation protocols [11]. Specific attention has been given to the separation of refrigerant mixtures, proper labeling and storage of refrigerant containers, and prohibition of single-use containers to prevent unintentional emissions. Furthermore, strict requirements have been provided for reclaiming and reusing refrigerants, including quality assurance benchmarks to match new refrigerant specifications. The standard SLS 1262:2025 outlines environmentally friendly disposal methods for refrigerants that are not suitable for reuse, including procedures for dealing with ammonia-water mixtures, used refrigeration oils, and associated system components. It also discourages unsafe retrofitting of existing systems to certain low-GWP refrigerants without proper compatibility and safety assessments.

5 Adaptation Challenges and Justifications

The adoption of the revised standard SLS 1262:2025 provides a range of adaptation challenges, particularly in the Sri Lankan context. Technical difficulties include the unavailability of compliant components, such as leak detection systems and ventilated enclosures, which are essential for systems using flammable refrigerants. Furthermore, a lack of training among technicians for handling Class A2L or B2 refrigerants poses operational safety concerns. Economic barriers also exist, as small-scale operators may struggle with the increased costs of equipment upgrades and system redesign. Institutional challenges are compounded by the absence of robust regulatory frameworks and enforcement mechanisms aligned with the revised standard SLS 1262:2025.

However, the justification for adapting to the standard SLS 1262:2025 remains strong. It significantly enhances safety by addressing the fire and toxicity risks associated with commonly used refrigerants like ammonia. Moreover, it supports Sri Lanka's environmental goals by promoting the use of low-GWP alternatives, contributing to global climate mitigation efforts. Although the transition may incur initial costs, the long-term benefits, such as improved energy efficiency, reduced accident risks, and compliance with global best practices, make the adaptation both necessary and justifiable. Therefore, a phased implementation approach with capacity building and policy support is recommended to overcome these barriers effectively.

6 Industry Implications

The revised standard SLS 1262:2025 introduces significant changes that will reshape the way refrigerants that will be selected, refrigeration systems designed, and safety managed. The industrial sectors that have recently been focusing on sustainable and energy-efficient refrigeration systems are the most affected by this. The range of these changes is not limited to technical operations alone but also includes regulatory compliance and workforce development. The industries are now obliged to implement stricter classification and location-based safety protocols due to the use of flammable or toxic refrigerants.

Moreover, the standard drives the workforce to gain new technical skills in handling flammable refrigerants like R290 (propane) and R717 (ammonia), which correspond to A3 and B2 classes, respectively. This gives both a challenge and an opportunity for industry-wide capacity building and the modernization of vocational training curricula. From a compliance standpoint, manufacturers and service providers will be required to maintain traceability records and safety certifications, aligning with global best practices.

The adaptation may incur higher initial investment however, the industry stands to benefit in the long term from improved energy efficiency, reduced operational risks, and enhanced environmental sustainability. The sectors that comply with the revised standard SLS 1262:2025 and move early are those that could position themselves for gaining a decisive advantage in international markets and being ready for the next regulatory frameworks, for example, the Kigali Amendment and future carbon-price schemes [12].

7 Conclusion and Future Directions

This comprehensive study has provided a complete revision of an existing national standard of Sri Lanka for the safe use of flammable and toxic refrigerants in the HVAC/R sector. The enhanced standard bridges the gap between the best global practices and local contextual realities, addressing critical safety, environmental, and technical challenges associated with the use of toxic and flammable refrigerants. Since aligning with internationally recognized frameworks and adapting to Sri Lanka's infrastructure, economic capacity, and workforce skill level, the revised standard sets a robust foundation for safer and more sustainable refrigeration practices. Key changes include modernized refrigerant and system classifications, detailed charge limit methodologies, mandatory documentation, and clear requirements for operation, maintenance, and end-of-life refrigerant management. The three objectives mentioned above have been achieved by,

- identifying critical deficiencies in SLS 1262:2004, including outdated classifications and weak disposal protocols,
- adapting global best practices to Sri Lanka's climatic, economic, and workforce realities, and
- validating the proposed revisions through structured stakeholder consultations and questionnaire results received from the industry, ensuring technical feasibility and local acceptance respectively.

Continued collaboration among regulators, industry professionals, and training institutions will be critical for the effective implementation of the standard. The publishing of the document would not be sufficient, as there must be awareness of the newly adopted regulations. Consequently, conducting targeted workshops is essential for updating engineers, designers, and operators. Simplified guidelines and fact sheets can be provided instead of the whole standard, as it is not a convenient document to be handled in the industrial environment. The finalization of this standard SLS 1262:2025 should be regarded not as a terminal event, but as the initiation of an iterative process of review and refinement. The standard's sustained relevance and effectiveness are contingent upon periodic revisions, which must be informed by empirical feedback gathered from its extensive implementation across the HVAC/R sector. The standard is only productive if the workforce is competent to apply it. The technical curriculum has to be revised according to the new requirements, and vocational training centers can be arranged to enhance understanding.

References

1. ISO 817: Refrigerants-Designation and Safety Classification (2014)
2. ASHRAE 15: Safety Standard for Refrigeration Systems (2016)
3. ASHRAE 34: Designation and Safety Classification of Refrigerants (2019)
4. EN 378-1: Refrigerating systems and heat Pumps-Safety and environmental requirements-Part 1: basic requirements, definitions, Classification and Selection Criteria (2016)
5. EN 378-2: Refrigerating systems and heat Pumps-Safety and environmental requirements-Part 2: design, construction, testing, Marking and documentation (2016)
6. EN 378–3: Refrigerating systems and heat pumps-Safety and environmental requirements-Part 3: installation site and personal protection (2016)
7. SLS 1262: Mechanical refrigerating systems used for cooling and heating-Safety requirements (2004)
8. IS 16678: Refrigerating Systems and Heat Pumps (2022)
9. ISO 5149-1: Refrigerating systems and heat pumps-Safety and environmental requirements Part 1: definitions, classification and selection criteria (2014)
10. ISO 12100: Safety of machinery-General principles for design-Risk assessment and risk reduction (2010)
11. ISO 5149-4: Refrigerating systems and heat pumps-Safety and environmental requirements Part 4: operation, maintenance, repair and recovery (2014)
12. Kigali Amendment to the Montreal Protocol on Substances of the Deplete Ozone Layer (2016)

Development of an Oil-in-Water Emulsion Model for Ocean Oil Spills in Tropical Conditions

Lalani Kaluarachchi[1](✉), Ajith de Alwis[2], and Indrajith D. Nissanka[1]

[1] Department of Mechanical Engineering, University of Moratuwa, Katubedda, Sri Lanka
{kaluarachchil.24,nissankai}@uom.lk

[2] Department of Chemical and Process Engineering, University of Moratuwa, Katubedda, Sri Lanka
ajith@uom.lk

Abstract. This study develops a numerical model to examine the weathering of oil, with a focus on emulsification in the context of marine oil spills. In marine spills, oil weathering plays a significant role as it affects the density and viscosity of the oil in water. The emulsification process affects the density and viscosity of oil, forming a water-in-oil emulsion that has a negative impact on the environment and cleanup operations. There are only limited studies that have been conducted in the literature related to oil spills in tropical oceanic conditions. Thus, current research focused on studying the oil spill behavior in tropical oceanic conditions. During this study, existing models in the literature were extensively studied to establish the different empirical and semi-empirical formulas to model the emulsification process of spilled oil. The simulation results were validated using the experimental data available in the literature. The sensitivity of the emulsification process to environmental factors was investigated using the developed model. The findings demonstrated that the ambient wind speed significantly affects the process of oil emulsion. The analysis showed that viscosity reduces when sea surface temperature increases; however, viscosity significantly increases over time as a result of evaporation and emulsification processes. Having proper numerical models to assess the oil viscosity variations and the factors influencing those changes will be key in accurately predicting the oil fate in a spill, which is the future direction of this research.

Keywords: Oil Weathering · Emulsification · Viscosity Modeling · Ocean Oil Spills

1 Introduction

Due to increases in oil and gas demand around the world, new oil types, which vary from heavy crude oil to light crude oil and condensate, continue to be produced in bulk. The behavior and fate of various crude oils could alter significantly if they are spilled at sea due to numerous variances in both their chemical and physical properties [1]. Oil weathering processes, which include spreading, evaporation, dispersion,

M. Siriwardhana et al. (Eds.): IERS 2025, PTT, pp. 290–299, 2026.
https://doi.org/10.1007/978-981-95-7984-6_25

emulsification, and dissolution, which act in the early stages and sedimentation, photo-oxidation, and biodegradation in the later stages, cause crude oil to gradually deteriorate after an accidental spill. Oil weathering mechanisms lead to changes in the physical and chemical properties of oil over time [2]. One of the natural mechanisms that assists removing spilled oil from the sea surface is evaporation. As soon as the oil spill occurs, the evaporation process begins, and over the course of the spill, the rate of evaporation drops exponentially. The chemical nature of the oil, along with the weather, sea surface temperature, and oil film thickness, determines how much of oil evaporates [1]. When sea surface temperature is high, the evaporation process becomes accelerated [2]. During the evaporation phase, the volatile components of oil separate from the slick and enter the atmosphere as a gaseous phase, leaving the high molecular weight components at sea. As a result, density and viscosity of the slick increase, leading to increased emulsification of oils [3].

When oil spills at sea, it spreads on the ocean surface. In the early phases of an oil spill, spreading is often the most prominent process; however, as the viscosity and density of the residual oil increase, spreading becomes less significant. Water-in-oil (W/O) emulsion formation has a major impact on fate of oil spills at sea and their cleanup strategies [4]. Depending on the oil characteristics, emulsification can cause the entire emulsion volume to increase by up to six times the initial spilled amount, which is another factor that keeps oil on the sea surface when w/o emulsion occurs [5].

1.1 Significance of the Study

Sri Lanka is crucial to international maritime lines that connect the East and West. Sri Lanka's strategic location allows for swift and effective global and regional connections, facilitating freight movement between Asia and the world [2]. Therefore, there is a potential for an oil spill. Hence, the study of behavior of oil spilled in tropical conditions is vital in understanding the oil behavior during accidental spills and subsequent cleanup processes. However, most of the studies for oil spills reported in the literature were done for the temperature of approximately 10 °C, 15 °C, and 20 °C and wind conditions also at lower values such as 2.5 m/s [6, 7].

Over tropical waters, deep convection is typically seen above a sea surface temperature threshold, which is around 26 °C to 28 °C [8]. Wind speed over Indian ocean around Sri Lanka varies from 10 m/s to 17.5 m/s [8]. Thus, there is a clear research gap in the literature to develop a comprehensive oil spill model considering the oil weathering process and other physico-chemical processes oil undergoes in the tropical oceanic conditions. This study aims to contribute toward addressing this research gap by examining the various oil weathering models and their sensitivity to ambient conditions in tropical climate conditions.

1.2 Emulsification Process

Water in oil emulsion is the result of emulsion formation in oil spills. The physical characteristics of oil and oil cleanup procedures can be negatively impacted by the formation of water-in-oil emulsion. This brown emulsion resembles chocolate mousse in appearance [9]. The volume of the contaminated material is significantly increased by the addition of water. Because mousse is viscous, it produces a cohesive slick that is resistant to both natural and artificial dispersion that makes it challenging for the majority of mechanical recovery equipment to work effectively. The effective window of opportunity for particular countermeasures is determined by the rate at which spilled oil emulsifies [4]. According to Fingas and Fieldhouse [5], crude oil and water are able to create four different kinds of emulsions: entrained, unstable, meso-stable, and stable water-in-oil emulsions.

Stable water-in-emulsion appears as reddish-brown colored semisolid substances, and it remains for at least 4 weeks. When forming the emulsion viscosity is increased. In stable water in emulsion, the viscosity increases by an average of 400 times the original viscosity [5]. The average water content of the stable emulsion is about 70 to 80%. In meso-stable water-in-oil emulsion is a viscous liquid and its color is reddish-brown, which is a stable emulsion. It breaks down into water about 20% within one week. This type of emulsion increases the viscosity, and it is about 7 times of initial viscosity on the day of formation [5]. Entrained water-in-oil emulsion is a black color viscous liquid. This emulsion type could be identified in viscous oils and petroleum products however, not in extremely viscous products. Here, the factor of viscosity increase of the emulsion is less than that of the stable and meso-stable types, and it is about two times of the viscosity of day of formation of the oil slick. The average water content of the entrained emulsion is about 40 to 50% [5]. Unstable emulsion decomposes within few hours and it does not hold a significant amount of water [5].

The fundamental mechanism of emulsification is the formation of a high-strength viscoelastic asphaltene layer covering water droplets in oil, which stabilizes the water in oil emulsion. The role of the resin content of the oil in the formation of emulsion is important as it increases the solvation of asphaltenes. According to the Fingas laboratory experiment [5], the ratio between resin and asphaltenes 2:1 could increase the stability of the emulsion.

There are numerous models for identifying water in oil emulsion. The majority of the parameters related to the emulsification process were reported in the works of Fingas and Mackay [10]. In 1980, Mackay developed the relationship between water uptake due to the emulsification process [10], which is given in (1).

$$\ln\left[(1 - K_2 W)\exp(-2.5W/(1 - K_1 W))\right] = -K_1 St/D\mu_0 = -K_3 t \tag{1}$$

where, t is time, K_1 is a rate constant for water incorporation. K_2 is the rate constant regarding coalescing tendency and K_3 is given by Eq. (2), here. W_s is wind speed at 10 m height.

$$K_3 = 3.43 \times 10^{-5} \times W_s^2 \tag{2}$$

1.3 Change of Oil Viscosity

The SIMAP (Spill Impact Model Application Package) oil spill model has used a mix of components, such as temperature changes, evaporation, and the emulsification process, to calculate the net viscosity changes (μ_{net}) of spilled oil [4].

Viscosity changes due to variation of the temperature can be calculated by (3).

$$\mu_{temp} = \mu_o \exp\left(-63.16\left(\frac{1}{T_{oil}} - \frac{1}{T_{ref}}\right)\right) \tag{3}$$

here, μ_o is the viscosity of the fresh oil at the reference temperature (T_{ref}) and T_{oil} is the temperature of the surrounding water in Kelvin. Therefore, net viscosity could be found by using eq. (4) as,

$$\mu_{net} = \mu_{temp} \times \mu_{evap} \times \mu_{wateruptake} \tag{4}$$

Mooney's equation is commonly used in oil weathering models to describe the viscosity modification caused by emulsion formation [9]. Viscosity changes due to emulsification process and evaporation process can be calculated by Eqs. (5) and (6) respectively.

$$\mu_{wateruptake} = \mu_o \exp\left(\frac{2.5F_w}{1 - 0.654F_w}\right) \tag{5}$$

$$\mu_{evap} = \mu_o \exp\left(\frac{C_{evap} \times F_v}{F_{bp380}}\right) \tag{6}$$

where, C_{evap} is the empirical constant and it is defined as 4.6077 by Mackay [4]. F_{bp380} is the fraction of oil that distills below the temperature 380 °C [4]. F_w is defined as the fraction of water in oil [10]. The fraction of water in oil emulsification can be calculated by Eq. (7),

$$\frac{dF_w}{dt} = C_1(W + 1)^2 \times \left(1 - \frac{F_w}{F_w \max}\right) \tag{7}$$

where, C_1 is an empirical constant and it is 2×10^{-6} for stable, meso-stable, and entrained water-in-oil emulsion and zero for non-emulsifying oils, $F_w max$ is the maximum water fraction of the fully emulsified oil and it is 0.7 for typical crude oil and 0.25 for heavy fuel oils [9].

1.4 Stability of Emulsion

In order to determine the emulsion stability (phase equilibrium), Mackay and Zagorski (1982). proposed a stability index S_A that is computed from the asphaltene, wax content, and temperature [10].

$$S_A = X_a \Lambda_a \exp\left[K_{ao}(1 - X_a - X_w)^2 + K_{aw}{X_w}^2\right]\exp[-0.04(T - 293)] \tag{8}$$

here, Λ_a activity of asphaltene, K_{ao} =3.3 at 293 K, K_{aw} = 200 at 293 K, X_a is the fraction of asphaltene, and X_w is the fraction of wax [10].

Stable emulsion has asphaltene content over 7% in weight and both the amount of asphaltene and resin is over 3% in meso-stable emulsion. If these conditions are not fulfilled, then the emulsion is unstable [10]. However, when the asphaltene content typically over 10%, the stable emulsion will not form [5]. A new parameter S is introduced [10] and represented by Eq. (9).

$$S = S_A / \Lambda_a \tag{9}$$

According to the [10],
when $S \geq 1.22$, emulsion is stable
when $0.67 \leq S \leq 1.22$, formed emulsion is meso-stable
when $S < 0.67$, emulsion is unstable.

In the current study, the previously mentioned equations were employed to develop an oil weathering/fate and transport model with the most appropriate formulas for effectively modeling the oil weathering process in tropical ocean settings for use in an oil spill fate and transport model. Furthermore, the research study has been extended to evaluate how ambient conditions such as sea surface temperature, wind velocity, and ambient temperature affect the weathering process, thereby determining the most important parameters influencing oil weathering and the spread of oil slicks in the ocean.

2 Model Formulation

In the current study, an oil weathering model was developed by combining the existing oil weathering formulation in the literature. All variations and oil weathering processes were analyzed by using relevant formulas, which were already discussed above, and the most up-to-date models were used. The model was run for the Arabian Medium crude oil with API (American Petroleum Institute) gravity 29.5. The evaporation has been simulated by using ADIOS2 oil spill model. The emulsification process was modelled, and the water-in-oil percentage was calculated using Eq. (7). The water-in-oil percentage was studied at different wind velocities. The variation of viscosity of the spilled oil scenario was studied by varying water in oil fraction according to (5) and varying the sea surface temperature. Furthermore, viscosity changes were studied by varying evaporation fraction of spilled oil by using (6). To determine the type of emulsion, the stability index was calculated using Eqs. (8) and (9). The viscosity changes due to variation of temperature, evaporation and emulsification processes were calculated and analyzed with time considering a reference temperature of 15 °C. Combining all the above equations resulted in a set of differential equations that were solved by using the Runge-Kutta 4th order method to model the emulsification process because it produces higher level accuracy and by achieving convergence with a larger time step. Since oil spills should be simulated for a duration of few days the existing oil spill models also use this numerical method to solve the ordinary differential equations [11].

The results obtained from the model were plotted using the MATLAB and the results are presented in Fig. 1,2, 3,Fig. 4. All these models have been previously validated with experimental data in the literature; however, an additional validation step was conducted by comparing the calculated results with the laboratory experimental results [6] for the emulsification. The water percentage of oil in the water emulsion was calculated at a wind speed 0.25 m/s. The viscosity of the spilled oil was calculated by using similar values which were used in the experiment for calculating evaporation fraction and initial viscosity. Required chemical and physical properties of the oil were obtained from the ADIOS Oil Database [12].

3 Model Simulation

The evaporation fraction, viscosity change, and water in oil emulsion were calculated in scenario case simulations with selected types of oil. Due to its widespread use and the availability of the necessary oil properties, Arabian Medium crude oil was chosen for the simulations. For weathering models, the model computations employed the input parameters listed below.

The simulation was done for Arabian Medium crude oil and instantaneous release was selected with the amount of spilled was considered as $2000 m^3$. Sea surface temperature between 15 °C and 25 °C were considered separately for each simulation while other conditions were kept as constant. To analyze the water in oil percentage in emulsion, Eq. (7) was used. Here, input constant values for C_1 and $F_w\ max$ were selected as 2×10^{-6} and 0.7 following existing studies in the literature. Further, analysis was carried out by varying the wind speed for 1 m/s, 2.5 m/s, 5 m/s, 7.5 m/s and 10 m/s, simulating the tropical wind conditions.

Changes in viscosity due to emulsification and evaporation were studied. Simulated output from ADIOS2 was used as the required evaporation fractions. For the emulsification simulation, the reference oil temperature was selected as 15 °C while varying the sea surface temperature as 20 °C, 25 °C and 10 °C for the sensitivity analysis. Calculated water in oil fractions by using Eq. (7) were used as the input data in this analysis.

To determine the type of the emulsion formed, both Eqs. (8) and (9) were used. Here fraction of asphaltene of Arabian Medium crude oil and the fraction of wax content of Arabian Medium crude oil were equal to 0.06 and 0.06, respectively. $K_{ao} = 3.3$ at 293 K and $K_{aw} = 200$ at 293 K were selected from literature.

4 Results and Discussion

Using the developed model, the water-in-oil percentage was calculated for the wind speed at 0.25 m/s. The comparison of the values obtained for the water percentage in oil and viscosity is provided in Table 1. The viscosity changes due to the emulsification process were calculated under the same experimental conditions the initial viscosity, and evaporation fraction. Viscosity was calculated for 7 days after the oil spilled. It is evident that the model's computations closely resemble the findings of the experiment. The sea surface temperature was selected as 20 °C for this simulation, resembling the experimental conditions.

Table 1. Comparison of model results and experimental results of emulsion formed by Arabian Medium crude oil.

Parameter	Results obtained from the model	Experimental results [6]
Water in oil percentage - F_W	65.32	65
Viscosity (mPa.s)	45367.0526	46000

The water percentage in oil was determined using the developed model, and the distribution of the water percentages in the oil over time is shown in Fig. 1. Five distinct wind speeds 1 m/s, 2.5 m/s, 5 m/s,7.5 m/s, 10 m/s were used in the computations. As the emulsification process begins, the water content in oil gradually changes. However, the time for the oil to achieve a constant water in oil percentage drastically varied with wind speed.

From Fig. 1, it can be observed that with higher wind speed, such as 5 m/s and greater, water in oil percentage achieves a constant value (maximum water- in-oil fraction) within 12 hrs of formation of emulsion. However, at lower wind speeds, it is possible to achieve a stable water content in oil over several days. These results show that the emulsification process is strongly dependent on the ambient conditions, especially the wind speed.

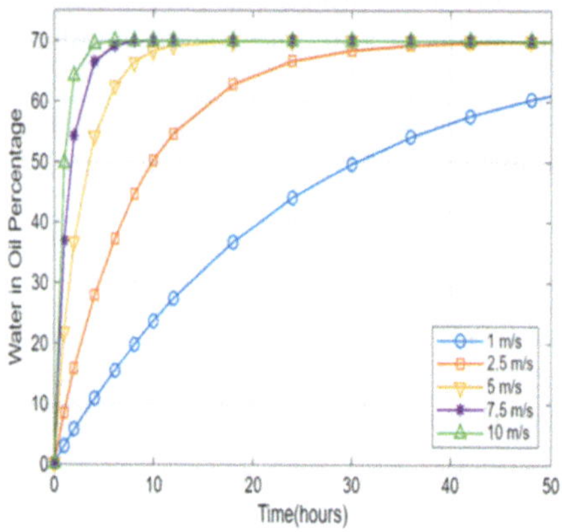

Fig. 1. Variation of water percentage in oil with time for different wind conditions.

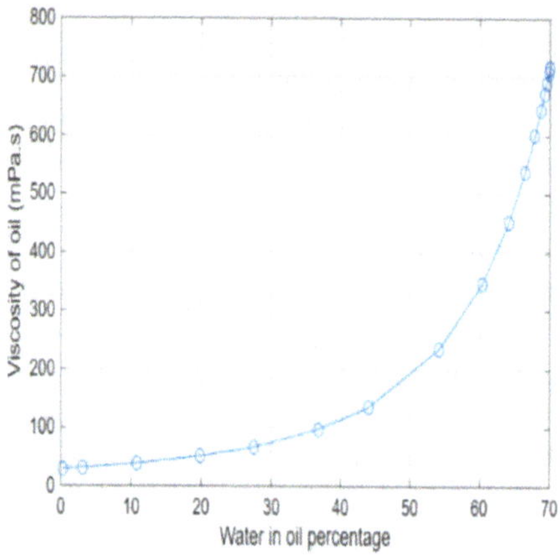

Fig. 2. Viscosity variation with water-in-oil percentage

According to the literature, the viscosity of the oil due to the emulsification process increases with time. Therefore, this study was extended to analyze how the viscosity varies with the water percentage in oil. Fig. 2 shows the variation of viscosity with water percentage in oil. This illustrates the impact of the emulsification process on viscosity. It is gradually increased but initial rate of change of viscosity is low. However, after some time it achieves an approximately constant value. Thus, rate of change of water-in-oil fraction becomes zero. Fig. 3 shows the variation of evaporation percentage with time. Initially rate of evaporation is high. However, after 1 day, rate of evaporation gradually decreases and achieves constant value after 3 days.

Fig. 4 illustrates the variation of viscosity due to both evaporation and emulsification processes with time. It can be observed that within 2 days after the oil spill, changing viscosity due to evaporation is higher than the viscosity due to emulsification. The evaporation completely removes the high volatile components while the heavier components remain on the sea surface, resulting in an increase in the oil slick viscosity. After 2 days of spillage, contribution of emulsification process for viscosity increment is much greater

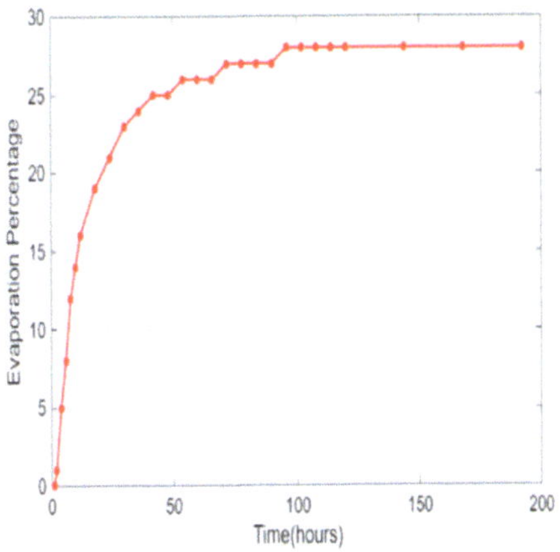

Fig. 3. Evaporation variation with time

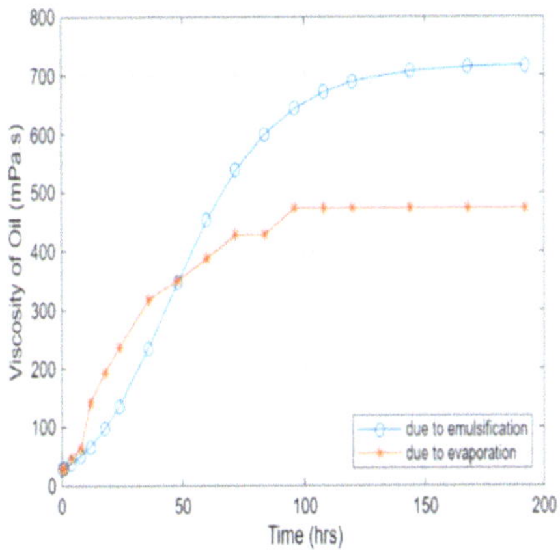

Fig. 4. Variation of viscosity due to emulsification and evaporation at 20 °C constant temperature

than evaporation. It can be clearly identified that viscosity variation due to emulsification is higher than that of evaporation process after the initial phase. It can be seen from Fig. 4. that the slope of the viscosity variation is greater between 2 days to 4 days after spills this is where the major change in oil happens due to the emulsification process.

Table 2 demonstrates the results obtained by calculating the viscosity due to changes of temperature. Here reference temperature was set as 15 °C. By comparing the viscosity of the fresh oil at 15 °C, it can be clearly identified that the viscosity has changed with temperature. However, the percentage increase due to temperature changes, is not significant compared to the initial value. Also, it was observed that viscosity also reduces with increases in sea surface temperature. According to Table 3. The model result for stability index is greater than 1.22. Therefore, the emulsion is stable. This was also verified by the experimental results in literature.

Table 2. Comparison of model results and experimental data of the viscosity of the emulsion formed by Arabian Medium crude oil at different temperatures

Sea Surface Temperature	Results obtained from the model	Initial viscosity at reference temperature [12]
10 °C	30.39 mPa.s	29 mPa.s
20 °C	28.13 mPa.s	29 mPa.s
25 °C	27.63 mPa.s	29 mPa.s

Table 3. Comparison of model result for stability index with experimental results

Stability Index from model	Emulsion Stability [6]
1.5874 (S > 1.22)	Stable

5 Summary and Conclusion

This paper presents a numerical analysis to study emulsification oil weathering process in ocean oil spills under tropical climate condition. The study identified many oil weathering processes that undergoes in the water. The short-term oil weathering process consists mostly of evaporation, spreading, dissolution, and emulsification. The limitations and applicability of existing models in tropical ocean conditions were investigated. By merging existing models, an oil weathering model for the emulsification process was built and verified using experimental data from the literature.

The model was then utilized to investigate the sensitivity of the emulsification process to environmental variables. The analysis of emulsification of spilled oil by varying the temperature and wind speed was investigated for four distinct types of water-in-oil emulsification processes reported in the literature. Arabian Medium crude oil was utilized for the analysis. It can be concluded that after one day of oil spilling at a greater wind speed, the water content in the emulsion reaches its maximum. However, at lower wind speeds emulsification process takes more time where even after one week, the water content in oil increases. As a result, wind speed determines the ability to achieve a stable emulsion in a short amount of time. Also, the evaporation proportion of spilled oil increases with sea surface temperature. The viscosity of spilled oil varies due to oil evaporation, temperature, and emulsification.

This is continuing work, and research is expected to extend to further improve the formulas to calculate the oil weathering process through experimental and numerical analysis and integrate with a hydrodynamic model to calculate the fate and transport of ocean oil spills.

6 Acknowledgement

This research was funded by NRC Investigator Driven Grant (Grant number: NRC-22-070)

References

1. Sørheim, K.R., Hellstrøm, K.C.: Aasta Hansteen Blend– Weathering Properties and Behaviour at Sea. SINTEF, Norway (2020)
2. Kaluarachchi, L., De. Alwis, A., Nissanka, I.D.: Analysis of factors effecting oil weathering of Accidental Ocean spills. In: ERU Symposium 2024, Colombo (2024)

3. Barker, C.H., et al.: Progress in operational Modeling in support of oil spill response. Mar. Sci. Eng. **8**(31), 668 (2020)
4. French-McCay, D.P., Frediani, M., Gloekler, M.D.: Modeling emulsification influence on oil properties and fate to inform effective spill response, journal of Frotiers in environmental. Science. **10**(26), 19 (2022)
5. Fingas, M.: Water-in-Oil Emulsions: Formation and Prediction, in Handbook of Oil Spill Science and Technology., Alberta, Canada, p. 692. John Wiley & Sons, Inc., Hoboken, New Jersey (2015)
6. Fingas, F., Fieldhouse, B.: Formulation of water-in-oil emulsions and application to oil spill modeling. J. hazardous Mater. **107**, 37–50 (2004)
7. Wang, L., Chu, S., Zhu, D., Liu, C., Luan, G., Gao, S.: Experimental study on the synergistic effect between evaporation weathering and emulsification of oil spills. ACS Omega. **6**, 20865–20876 (2021)
8. NARA, Sri Lanka Ocean Weather Forecast, Coastal Oceanography, (2023). Available: https://boreas.mywire.org/juraj/uwa/index.html?lang=english. [Accessed 8 Apl. 2025].
9. Sergey, K.A., Arlamov, M.V.: Oil Spill Initial Stage Spreading and Physical Properties Model. Kyushu University, Research Institute for Applied Mechanics (2000)
10. Xie, H., Yapa, P.D., Nakata, K.: Modeling emulsification after an oil spill in the sea. J. Mar. Syst. **68**, 489–506 (2007)
11. Zhen-Gang Ji, Johnson. W. R., Price, J M., Marshall, C. F.: Oil Spill Risk Analysis For Assessing Environmental Impacts, in International Oil Spill Conference, (2003).
12. N. O. A. A. Administration, ADIOS Oil Database, March 2021. [Online]. Available: https://adios.orr.noaa.gov/oils. [Accessed 26 Mar. 2025].

Investigate the Volumetric Heat Transfer Coefficient for the Rotary Drying of Coir Pith

S. A. P. S. Silva[1](✉), H. K. A. Mahela[2], S. Sarmilan[2], A. D. U. S. Amarasinghe[2], and M. Narayana[2]

[1] National Engineering Research and Development Center, Ja-Ela, Sri Lanka
shalinda@nerdc.lk

[2] Department of Chemical and Process Engineering, University of Moratuwa, Moratuwa, Sri Lanka

Abstract. The effects of drying parameters on the volumetric heat transfer coefficient during rotary drying of coir pith were experimentally investigated using a pilot-scale rotary dryer. Experiments were conducted at a constant inlet air temperature of 55 °C, varying the inlet air velocity (0.2 m/s, 0.3 m/s, and 0.4 m/s) and feed rate (0.67 L/min, 1 L/min, and 2 L/min), while maintaining constant rotational speed and relative humidity. To assess the effect of temperature, addi-tional tests were performed by varying inlet air temperature (55 °C, 70 °C, and 80 °C) at a constant air velocity (0.3 m/s) and feed rate (1 L/min). The results confirm that the inlet air temperature is the dominant factor influencing the vol-umetric heat transfer coefficient, with its impact becoming more significant at elevated temperatures. An empirical model was developed to predict the volu-metric heat transfer coefficient as a function of inlet air temperature, air veloci-ty, and feed rate. Statistical analysis showed that inlet air temperature had a sig-nificant effect ($p = 0.029$), whereas air velocity and feed rate were less influen-tial. The model demonstrated a strong predictive capability, with good agree-ment between experimental and predicted values ($R2 = 0.894$), further support-ed by residual analysis. These results highlight the potential applicability of the developed model for the design and optimization of rotary drying processes in industrial applications.

Keywords: Volumetric Heat Transfer Coefficient · Mass-dependent Heat Transfer Coefficient · Rotary Drying

1 Introduction

Coir pith is a light brown, spongy lignocellulosic material obtained from the non-fibrous tissues of coconut husks during fiber extraction [1, 2]. It is widely used in compressed products such as briquettes, discs, grow bags, and bales [1], and serves as an effective soil amendment and substrate, enhancing crop productivity and soil structure [3]. The global demand for coir pith has grown significantly due to rising environmental awareness and the move toward organic and sustainable agriculture [4]. Its high porosity, excellent water retention, and rich nutrient content make it a preferred growth medium in horticulture and agriculture [1].

M. Siriwardhana et al. (Eds.): IERS 2025, PTT, pp. 300–310, 2026.
https://doi.org/10.1007/978-981-95-7984-6_26

To use coir pith effectively as a substrate, it must be converted into value-added products through proper drying and treatment to improve quality and consistency. Drying is a critical step, as the high initial moisture content poses a major challenge for processing [4]. Common drying methods include sun drying, hot air drying, vacuum drying, and rotary drum drying [5, 6]. Although sun drying is widely practiced in tropical regions like Sri Lanka due to its low cost, it is unsuitable for large-scale production because of slow drying rates, weather dependence, space requirements, labor costs, and contamination risks [7]. Thus, more advanced and controlled drying methods are essential to meet market demands.

Rotary drying is a promising technique for large-scale coir pith processing, providing uniform drying and effective mixing [8–12]. In rotary dryers, solids tumble inside a rotating drum while heated by direct or indirect hot gases [13]. These dryers handle various material types, including sticky and heterogeneous feed-stocks, making them suitable for coir pith [14–16]. A key design parameter is the volumetric heat transfer coefficient, which governs drying rates and energy efficiency [17]. While correlations have been developed for materials like soymilk powder [18], cassava [8], sawdust [9], and fertilizers [19, 20], coir pith's unique properties, such as higher moisture content and distinct physical characteristics limit the direct application of existing models [4]. Therefore, developing correlations for the volumetric heat transfer coefficient based on key operating parameters of rotary dryers is essential for accurate modeling and simulation. This study experimentally investigates the effects of inlet air temperature, air velocity, and feed rate on the volumetric heat transfer coefficient during the rotary drying of coir pith and establishes empirical correlations among these parameters.

2 Theoretical Models for Volumetric Heat Transfer Coefficient for Rotary Drying of Coir Pith

The volumetric heat transfer coefficient can be calculated as Eq. (1),

$$q_T = U_a V \Delta T_{LM} \tag{1}$$

where q_T = rate of heat transfer, U_a = volumetric heat transfer coefficient, V = volume of the cylindrical drum and ΔT_{LM} = logarithmic mean temperature difference between inlet and outlet flow streams in cocurrent configuration. The concept of the volumetric heat transfer coefficient has been introduced due to the uncertainty of the actual area available for heat transfer and the difficulty in measuring the exposed surface area of particles in rotary dryers [21].

Alternatively, the rate of heat transfer can be expressed using mass-dependent heat transfer coefficient, as given by Eq. (2) [21].

$$q_T = m_f h \Delta T_{LM} \tag{2}$$

where h = mass-dependent heat transfer coefficient or specific heat transfer coefficient and m_f = mass of the dry feed.

By combining (1) and (2), volumetric heat transfer coefficient can be obtained as (3),

$$U_a = \frac{h m_f}{\mathrm{V}} \tag{3}$$

The amount of heat transferred per unit mass of material during drying in a rotary dryer is known as the heat load, Q, and can be calculated using energy balance equation considering the thermal properties of coir pith and moisture as in Eq. (4) [22].

$$Q = C_{ps}\left(T_P - T_f\right) + X_f C_{PL}\left(T_v - T_f\right) + \lambda\left(X_f - X_P\right) + X_P C_{PL}(T_P - T_v) + C_{PV}\left(X_f - X_P\right)(T_{Vb} - T_v)) \quad (4)$$

where, C_{PS} denotes the specific heat capacity of the solid material, C_{PL} and C_{PV} stand for the specific heat capacities of liquid water and water vapor, respectively. T_v is the water's vaporization temperature and T_{Vb} is the maximum vapor temperature. T_f and T_P are the initial and final temperatures of the material, respectively. X_f and X_P are the initial and final moisture contents on dry basis, respectively. λ is the latent heat of water vaporization. Thermal energy needed for effective drying of coir pith in a rotary dryer is determined by combining these factors. The energy accumulation was considered to be negligible since the rotary dryer was operated at steady state during experiments.

By using heat load, rate of heat transfer can be given by Eq. (5),

$$Q = \frac{q_T}{\dot{m}_s} \quad (5)$$

where $\dot{m}_s$ = mass flow rate of dry solids.

For cocurrent material drying in a rotary dryer, the logarithmic mean temperature difference can be calculated as Eq. (6) [22],

$$\Delta T_{LM} = \frac{(T_{ai} - T_v) - (T_{ao} - T_v)}{\frac{(T_{ai} - T_v)}{(T_{ao} - T_v)}} \quad (6)$$

T_{ai} = inlet air temperature and T_{a0} = outlet air temperature.

T_v is derived from the turning point of the curves where the outlet temperature varies over time for a constant inlet temperature as described by Ademiluyi et al. [8]. The coir pith specific heat capacity was taken as 4.02 kJ/kg.K. The latent heat of vaporization (λ) at the vaporization temperature is taken from engineering standard steam tables [23]. The specific heat capacity of pure water (C_{PL}) is assumed to be 4.2 kJ/kg.K.

3 Material and Methods

3.1 Sample Preparation

Fresh coir pith samples were collected from a coconut fiber manufacturing facility located in Wennappuwa, Sri Lanka. The collected samples were sieved to remove coir fibers and other impurities to ensure uniformity of the mixture. The initial moisture content was measured using an A&D MX-50 Moisture Analyzer (accuracy: ± 0.1), and the average moisture content of the samples was 140% (w/w dry basis).. Initial feed temperature was considered to be same as the atmospheric temperature at the time of experiment.

3.2 Details of Pilot-Scale Rotary Dryer Used in the Study

A pilot-scale rotary dryer was designed and fabricated at the National Engineering Research & Development Center (NERDC), Sri Lanka. The dryer consists of a cylindrical drum with an effective length of 1.20 m and an internal diameter of 0.250 m. It is horizontally inclined at 5° downward to facilitate continuous gravitational flow and uniform axial movement of the material.

The drying process utilizes direct hot air as the heat transfer medium. Inside the drum, three longitudinal lifters are incorporated to facilitate effective lifting, cascading, and thorough mixing of coir pith, thereby improving heat and mass transfer efficiency. Hot air is supplied by a centrifugal blower fitted with an orifice plate, allowing precise adjustment of inlet air velocity to suit experimental requirements.

An electric heating system preheats the inlet air to a maximum temperature of 90 °C. Inlet air temperature and velocity were monitored using a thermocouple and a thermo-anemometer, respectively. The drum is driven by an electric motor with a variable-speed controller, enabling adjustments in rotational speed to optimize residence time and material flow behavior (Fig. 1).

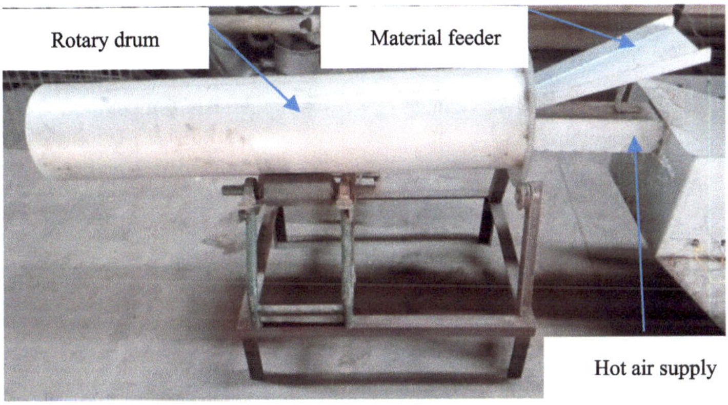

Fig. 1. Pilot scale rotary dryer fabricated for experiment

Continuous feeding of coir pith is achieved through a controlled mechanical feeder, ensuring a consistent feed rate. Dried product is discharged at the lower end of the drum and collected for further analysis. This pilot-scale configuration enables systematic investigation of drying performance, and heat transfer characteristics.

3.3 Experimental Procedure

Three experimental sets were systematically designed to investigate the effects of inlet air temperature, inlet air velocity, and feed rate on the volumetric heat transfer coefficient in a pilot-scale rotary dryer.

In the first and second experiment sets, the inlet air temperature was maintained at a constant 55 °C. This temperature was selected based on coir pith drying data from the existing rotary dryer developed by NERDC, which operates an exhaust air temperature of approximately 55 °C under steady-state conditions. The choice of 55 °C as the inlet air temperature ensured alignment with the thermal profile of industrial-scale systems while allowing controlled experimental validation [24].

With the air temperature fixed at 55 °C, the effects of inlet air velocity and feed rate were investigated independently. In rotary drying systems, an inlet air velocity of around 0.2 m/s is generally considered optimal for maximizing convective heat transfer while avoiding excessive particle entrainment [25]. Due to system constraints, velocities below 0.1 m/s were not achievable; therefore, experiments were conducted at 0.2 m/s, 0.3 m/s, and 0.4 m/s. The feed rate was controlled by varying the time required to introduce a fixed volume of pre-processed coir pith through the feeder, enabling precise adjustment of the material loading. The maximum feed rate of 2 L/min was chosen to achieve a material hold-up of approximately 15% of the drum's internal volume, which is critical for maintaining sufficient residence time and ensuring uniform thermal exposure and drying efficiency [23, 24].

Table 1. Experiment Data Set used to evaluate the Heat Transfer Coefficient at different drying conditions

Experiment	Variable	Constant Parameters
Experiment Set 1	**Inlet Air Velocity** 0.2 m/s 0.3 m/s 0.4 m/s	Inlet Air Temperature = 55 °C Feed Rate = 1 L/min Drum Drive Speed = 35 rpm
Experiment Set 2	**Feed Rate** 0.67 L/min 1 L/min 2 L/min	Inlet Air Temperature = 55 °C Inlet Air Velocity = 0.3 m/s Drum Drive Speed = 35 rpm
Experiment Set 3	**Inlet Air Temperature** 55 °C 70 °C 80 °C	Inlet Air Velocity = 0.3 m/s Feed Rate = 1 L/min Drum Drive Speed = 35 rpm

In the third experiment set, the effect of inlet air temperature on the volumetric heat transfer coefficient was examined. Experimental trials were conducted at inlet air temperatures of 55 °C, 70 °C, and 80 °C, while keeping inlet air velocity 0.3 m/s, and feed rate 1 L/min. For all experiments, drum rotational speed was kept constant at 35 rpm to isolate the temperature effect on drying performance and heat transfer behavior. Ambient relative humidity, ranging from 75% to 85% in Ja-Ela, Sri Lanka [26], was assumed to remain constant throughout the experiments.

The third experiment set focused on combined effects and cross-interactions among parameters to validate the empirical model under various operational scenarios.

Table 1 summarizes the experimental conditions used for determining the volumetric heat transfer coefficient across different parameter sets.

Before each experimental run, the rotary dryer was brought to steady-state operating conditions by stabilizing the inlet air temperature, drum rotational speed, and inlet air velocity at their predefined constant values. To confirm the attainment of steady-state, the outlet air temperature was continuously monitored using an Extech CFM thermo-anemometer (accuracy: ± 0.1 °C for temperature and ± 0.01 m/s for velocity). Measurements were recorded until the outlet temperature reached a constant value, indicating thermal equilibrium within the system.

Once steady-state conditions were established, coir pith samples were manually and continuously fed into the drum via the feeder. During each experimental pass, key process parameters including outlet air temperature, bulk product temperature, moisture content, and residence time for each pass were measured to characterize the drying kinetics.

After completion of each pass, the partially dried coir pith was reintroduced into the system through the feed hopper, and the cycle was repeated until the target final moisture content of approximately 40% (w/w dry basis) was achieved. This typically required six to seven successive passes. Between each pass, steady-state conditions were re-established to minimize transient effects and ensure consistent drying performance across cycles.

4 Results and Discussion

The Fig. 2 illustrates the variation in outlet air temperature across multiple drying passes for different inlet air temperatures (55 °C, 70 °C, and 80 °C). A sharp drop in outlet temperature is observed during the first pass following the feeding of coir pith, attributed to its high initial moisture content and the corresponding latent heat required for evaporation. In the subsequent passes, the outlet temperature gradually declines, reflecting a reduction in evaporative demand. Beyond the fourth pass, the outlet temperature begins to rise, indicating a transition into the falling-rate drying period and the approach toward quasi-equilibrium conditions.

The minimum point on each curve was assumed as the T_v value according to the method described by Ademiluyi et al. [8]. The T_v values are required to calculate the volumetric heat transfer coefficient using Eqs. (1) to (6) given in Sect. 2. Table 2 represents the experimental mass-dependent heat transfer coefficients for different drying conditions at constant temperature.

Table 2 depicts that the heat transfer coefficient decreases with increasing inlet air velocity and feed rate. Higher air velocities reduce residence time and remove heat from the system, thereby reducing effective heat transfer to the material. Similarly, enhanced feed rates increase the thermal load and reduce air–solid contact efficiency, resulting in lower overall heat transfer performance.

Multivariate linear regression analysis conducted for experiment set 1 and 2, where inlet air temperature was kept constant at 55 °C, yielded a coefficient of determination (R2) of 0.844. When the temperature effect was also included with the experiment set 3 data, the coefficient of determination (R2) improved to 0.894. Therefore, all three parameters were included in developing the linear empirical model, which was derived in Eq. (7):

$$h = 451.714 - 3.049T_{ai} - 405.776V_{ai} - 27.053F_i \quad (7)$$

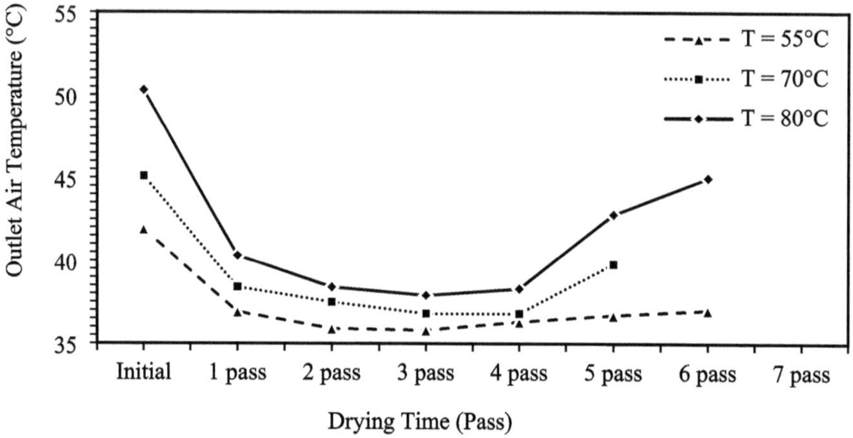

Fig. 2. Outlet air temperature variation over drying time

Table 2. Calculated mass-dependent heat transfer coefficient at different drying conditions

Experiment	Variable	Mass-dependent Heat Transfer Coefficient (W/kg°C)
Experiment Set 1	**Inlet Air Velocity** 0.2 m/s 0.3 m/s 0.4 m/s	160.0128 146.3832 78.8575
Experiment Set 2	**Feed Rate** 0.67 L/min 1 L/min 2 L/min	154.7776 146.3832 111.6828
Experiment Set 3	**Inlet Air Temperature** 55 °C 70 °C 80 °C	146.3832 105.3050 49.5233

Table 3 summarizes the regression analysis of drying parameters affecting the specific heat transfer coefficient. Inlet air temperature had a significant negative effect (B = -3.05, p = 0.029), while inlet air velocity showed a pronounced but marginally significant negative influence (B = -405.78, p = 0.057). Feed rate did not significantly affect the coefficient (p = 0.247).

Table 4 is the ANOVA analysis, which summarizes the overall significance of the regression model. The model demonstrates marginal significance (F = 8.457, p = 0.056), which may be due to the limited temperature range used in these experiments. Expanding the temperature range in future analyses could improve the model's statistical significance.

Table 3. Statistical Analysis of the heat transfer coefficient

Parameter	Unstandardized Coefficients		t value	significance	95.0% Confidence Interval for B	
	B	Std. Error			Upper	Lower
Constant	451.714	67.849	6.658	0.007	235.788	667.639
Inlet Air Temperature (°C)	-3.049	0.773	-3.945	0.029	-5.508	-0.590
Inlet Air Velocity (m/s)	-405.776	134.678	-3.013	0.057	-834.381	22.828
Feed Rate (L/min)	-27.053	18.848	-1.435	0.247	-87.035	32.928

Table 4. ANOVA analysis of the model

Model	Sum of Squares	Degree of freedom	Mean Square	F	Significance
Regression	9203.166	3	3067.722	8.457	0.056
Residual	1088.285	3	362.762		
Total	10291.450	6			

Figure 3 presents a normal probability scatter plot of residuals comparing the experimental and model-predicted data. The data points closely follow the reference line, indicating that the residuals are approximately normally distributed and that the model provides a good fit to the observed data. This alignment suggests that the regression model reliably predicts the experimental outcomes within the studied range, supporting the adequacy and validity of the model assumptions.

Using Eq. (3) and the proposed model given in Eq. (7), volumetric heat transfer coefficient can be derived as in Eq. (8).

$$U_a = \frac{(451.714 - 3.049T_{ai} - 405.776V_{ai} - 27.053F_i) * m_f}{V} \tag{8}$$

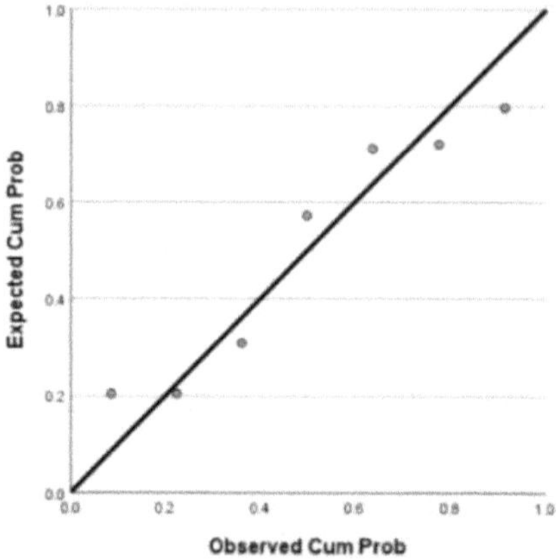

Fig. 3. The comparison between experimental and model-predicted data

5 Conclusions

This study systematically investigated the effects of inlet air temperature (T_{ai}), inlet air velocity (V_{ai}), and feed rate (F_i) on the volumetric heat transfer coefficient during rotary drying of coir pith. Experimental results demonstrated that increases in all T_{ai}, V_{ai}, and F_i resulted in a decrease in the volumetric heat transfer coefficient. Statistical analysis confirmed that inlet air temperature had a significant negative influence, inlet air velocity exhibited a marginal effect, and feed rate had no significant impact. A linear regression model was developed to predict the volumetric heat transfer coefficient as a function of T_{ai}, V_{ai}, and F_i. . The model exhibited good agreement with experimental data (R2 = 0.894), further supported by the normal probability plot of residuals, indicating the model's adequacy within the tested range. Future work should consider a broader temperature range to further enhance model robustness. The proposed model provides a practical tool for the design, analysis, and optimization of rotary drying processes for coir pith in industrial applications.

References

1. Fernando, J.A.K.M.: Flash Drying of Coir Pith as an Alternative to Sun Drying. Moratuwa Engineering Research Conference (MERCon), pp. 199–204 (2019)
2. Neethi Manickam, I., Subramanian, P.: Study of physical properties of coir pith. Int. J. Green Energy **3**(4), 397–406 (2006). https://doi.org/10.1080/01971520600873384
3. Sahoo, C.R., Kar, B.B.: Use of Coir Pith as a Soil Amendment Material. Asian J. Water Environ. Pollut. **15**(1), 97–100 (2018). https://doi.org/10.3233/AJW-180011
4. Fernando, J.A.K.M., Amarasinghe, A.D.U.S.: Drying kinetics and mathematical modeling of hot air drying of coconut coir pith. Springerplus **5**(1) (2016). https://doi.org/10.1186/s40064-016-2387-y
5. Doymaz, I., Smail, O.: Drying characteristics of sweet cherry. Food Bioprod. Process. **89**(1), 31–38 (2011). https://doi.org/10.1016/j.fbp.2010.03.006

6. Lee, S.H., et al.: Drying Characteristics of Agricultural Products under Different Drying Methods: A Review. J. Biosys. Eng. **41**(4), 389–395 (2016). https://doi.org/10.5307/jbe.2016.41.4.389
7. Fernando, J.A.K.M., Amarasinghe, A.D.U.S.: Effects of retting and drying on quality of coir pith and coco discs. J. Natl. Sci. Found. **45**(1), 3–12 (2017). https://doi.org/10.4038/jnsfsr.v45i1.8032
8. Ademiluyi, F.T., Abowei, M.F.N., Puyate, Y.T., Achinewhu, S.C.: Effects of drying parameters on heat transfer during drying of fermented ground cassava in a rotary dryer. Drying Technol. **28**(4), 550–561 (2010). https://doi.org/10.1080/07373931003622669
9. Alvarez, P.I., Shene, C.: Experimental Determination Of Volumetric Heat Transfer Coefficient In A Rotary Dryer. Drying Technol. **12**(7), 1605–1627 (2007). https://doi.org/10.1080/07373939408962189
10. Balakrishnan, M., et al.: Design, development, and evaluation of rotary drum dryer for turmeric rhizomes (Curcuma longa L.). J. Food Process Eng. **45**(6), 1–16 (2022). https://doi.org/10.1111/jfpe.14052
11. A'yuni, D.Q., Subagio, A., Prasetyaningrum, A., Sasongko, S.B., Djaeni, M.: The optimization of paddy drying in the rotary dryer: energy efficiency and product quality aspects analysis. Food Res **8**, 125–135 (2024). https://doi.org/10.26656/fr.2017.8(S1).17
12. Pelegrina, A.H., Elustondo, M.P., Urbicain, M.J.: Design of a semi-continuous rotary drier for vegetables. J. Food Eng. **37**(3), 293–304 (1998). https://doi.org/10.1016/S0260-8774(98)00082-X
13. Marinos-Kouris, D., Krokida, M., Mujumdar, A.: Rotary Drying, Handbook of Industrial Drying, Third Edition. CRC Press (2006). https://doi.org/10.1201/9781420017618.ch7
14. Iguaz, A., Esnoz, A., Martínez, G., López, A., Vírseda, P.: Mathematical modelling and simulation for the drying process of vegetable wholesale by-products in a rotary dryer. J. Food Eng. **59**(2–3), 151–160 (2003). https://doi.org/10.1016/S0260-8774(02)00451-X
15. Adapa, P.K., Schoenau, G.J., Arinze, E.A.: Fractionation of alfalfa into leaves and stems using a three pass rotary drum dryer. Biosyst. Eng. **91**(4), 455–463 (2005). https://doi.org/10.1016/j.biosystemseng.2004.12.003
16. Mohseni, M., Peters, B.: Surface Area Effects On Particulate Drying in a Rotary Dryer. 12th International Conference on Heat Transfer, Fluid Mechanics and Thermodynamics, p. 1439–1446 (2016). https://repository.up.ac.za/server/api/core/bitstreams/40aff788-663c-47d3-857a-ba882139919a/content
17. Poós, T., Szabó, V.: Application of volumetric heat transfer coefficient on fluidized bed dryers **2**, 1–4 (2016)
18. Nguyen, D.Q., Nguyen, T.H., Allaf, K.: Volumetric heat transfer coefficient in spray drying of soymilk powder. Drying Technol. **40**(6), 1146–1152 (2022). https://doi.org/10.1080/07373937.2020.1857768
19. Lisboa, M., Hills, D.: Study of the performance of the rotary dryer with fluidization. School of Chemic (2004)
20. Arruda, E.B., et al.: Modeling of Fertilizer Drying in Roto-Aerated and Conventional Rotary Dryers Modeling of Fertilizer Drying. Drying Technol. **27**(11), 1192–1198 (2009). https://doi.org/10.1080/07373930903263129
21. Audu, T.O.K., Ikhu-Omoregbe, D.I.O.: Drying characteristic of fermented ground cassava. Nigerian J. Eng. Technol. Develop. **7**, 2–20 (1982)
22. McCabe, W.L., Smith, J.C., Harriott, P.: Unit Operations of Chemical Engineering, 5th-edition. McGraw Hill International Editions (1993)
23. The Engineering Toolbox, Water Properties: Vaporization Heat vs. Temperature - Charts and Calculator. Accessed: 02 Jun 2025. https://www.engineeringtoolbox.com/water-properties-d_1573.html

24. Silva, S.A.P.S., Amarasinghe, A.D.U.S., Narayana, M.: Drying of coir pith under changing humidity conditions. Int. Eng. Res. Symposium (2023)
25. Zhang, H., Liu, C., Zhang, X., Li, B., Peng, G.: Experimental Study on Variation Characteristics of Pressure Drop and Process Optimization for Corn Kernels During the Hot Air Convective Drying Process. Processes **13**(4) (2025). https://doi.org/10.3390/pr13041180
26. "Ja-ela, Western, LK Climate Zone, Monthly Averages, Historical Weather Data." Accessed: 02 Jun. 2025. https://weatherandclimate.com/sri-lanka/western-sri-lanka/ja-ela

Modeling of Thermal Zones and Heat Transfer in BFB/CFB Incineration of Sludge

K. A. C. G. Siriwardhana[1(✉)], K. R. R. Mahanama[2], I. U. Atthanayake[3], and D. H. G. S. R. Somasundara[3]

[1] Faculty of Graduate Studies, University of Colombo, 35/30, Colombo 07, Sri Lanka
gaya@maxtherm.lk

[2] Department of Chemistry, University of Colombo, PO 1490, Colombo 07, Sri Lanka

[3] Department of Mechanical Engineering, The Open University of Sri Lanka, Nugegoda, Sri Lanka

Abstract. Escalating sludge generation and enforced environmental regulations dictates advanced thermal treatment methods for sustainable energy recovery and zero carbon. This study develops a comprehensive Euler-Euler multiphase model for sludge incineration in a bubbling fluidized bed (BFB) integrated into a circulating fluidized bed (CFB). The modular reactor (0.6 m x 14 m) comprises five sequential thermal zones: drying, devolatilization, volatile combustion, char oxidation, and burnout. Governing equations are derived for gas and solid phases using the interpenetrating continuum approach in the CFB riser, incorporating conservation of mass, momentum, species, and energy for each phase. Euler-Euler formulation with Kinetic Theory of Granular Flow (KTGF), volumetric reactions, species transport, and interphase heat exchange is applied. Gas phase is modeled as Newtonian, while sludge follows non-Newtonian granular behavior via kinetic theory, improving particle drying and fast fluidization. Convective, conductive, and radiative heat transfer are incorporated. Moisture evaporation and multi-step pyrolysis kinetics for volatile release are included using semi-empirical and literature models (EDM-FRC). For preliminary assessment, simplified zone-wise finite difference simulations are implemented in Python to evaluate axial temperature, pressure drop, and species concentration. Final transient 3D CFD simulation is conducted in Ansys Fluent using PISO and finite volume method, with detailed species transport and heat transfer with UDF based boundary conditions. Results capture low-temperature drying, sharp combustion rise, and nearly linear pressure drop along the freeboard. Model predictions of temperature distribution, heat-flux, conversion rates, and flow dynamics are validated against pilot-scale experimental data. The model pro-vides realistic insights for sludge to energy incineration.

Keywords: Circulated Fluidized Bed (CFB) · Bubbling Fluidized Bed (BFB) · Computational Fluid Dynamics (CFD) · Sludge Incineration · Euler-Euler Multiphase Model

M. Siriwardhana et al. (Eds.): IERS 2025, PTT, pp. 311–321, 2026.
https://doi.org/10.1007/978-981-95-7984-6_27

1 Introduction

We can pave the way for sustainable future by embracing innovative technologies for sustainable waste management. Sludge incineration is used as a waste management technology. The cost implications, emissions and public perception are concerns meeting sustainability. It requires innovative approach like advanced incineration focusing on energy recovery and adopting best practices. Continuous improvement and adaption will be the key achieving future waste management challenges.

Fluidized bed combustion offers excellent mixing and heat transfer for waste incineration [1]. Integrating BFB with CFB enhances combustion efficiency, energy re-covery, and sludge volume reduction. Sludge, rich in moisture and organics like cellulose, lignin, proteins, and lipids, can be thermochemically processed for volume reduction, pathogen destruction, and energy conversion.

Sludge is typically fed as moist solids. Therefore, its thermal treatments generally follow the five sequential stages drying, devolatilization, volatile combustion, char oxidation, and burnout [2]. It is also important to note that these stages may overlap, particularly when the processes are examined at the particle scale. As far as the emission control is concerned, the CFB reactor is virtually divided into primary and secondary chambers, the latter is turbulent dominant zone to ensure complete combustion of unburnt hydrocarbons and thereby achieving low emissions.

In the drying zone, water evaporates from dewatered sludge with typical moisture content of 73–84% [3]. The sludge used in this study contains 67% moisture. The free water, surface water and certain interstitial water has been removed at early stages by mechanical de-watering. The remaining interstitial water and vicinal water, which is made up of multiple layers of water molecules held tightly to particle surface by hydrogen bonding, can be removed through the application of heat [3]. The removal of bound water requires changes in the structure or composition of the material.

Sludge drying differentiates from other materials, showing a constant rate followed by two falling-rate periods for interstitial and vicinal water removal. In the granular zone, sludge becomes crumbly, mixes well then release gases like H_2, CO, CO_2, CH_4, and light hydrocarbons while devolatilization. Even though the thermal zones be able to overlapped within a sludge particle, spatial separation is possible across the reactor due to temperature gradients. Accurate modeling of these complex processes needs a multiphase CFD approach that captures solid-gas interactions, mixing, thermal decomposition, homogeneous and heterogeneous reactions, and heat and mass transfer under quasi-steady state conditions.

In light of these complexities and gaps, the present study focusses on thermal performance and energy recovery potential of a modular CFB sludge incinerator, investigated using a validated CFD approach. The specific objectives are to develop a CFD model integrating sludge drying, devolatilization, and combustion in a BFB-CFB system, characterize thermal zones, evaluate energy recovery potential, and validate the model predictions against pilot-scale data for temperature profile and pressure drop.

The Euler-Euler model treats gas and solids as interpenetrating continua, applying mass, momentum, species and energy conservation to each stage. In this study, the Euler-Euler multiphase framework coupled with the KTGF is used to capture gas-solid interactions, particle-particle collisions, and hydrodynamic behavior. This approach predicts key parameters such as particle mass loss, granular temperature, pressure drop, velocity profiles, and species transport in sludge incineration. Table 1 summarizes the thermal zones based on temperature and key reactions.

Table 1. Typical thermal zones of sludge incineration

Drying Zone	Temp. Range (°C)	Primary Process	Key Reactions	Species Behavior
Drying zone	80 - 150	Moisture evaporation	Physical phase change	Moisture rapidly decreases
Devolatilization Zone	150 - 350	Release of light volatiles	Pre-pyrolysis gas evolution	H_2, CO, CH_4, light HC increase, O_2 consumption begins slowly
Pyrolysis Zone	350 -500	Breakdown of heavy hydrocarbons	Complex thermal decomposition	Volatile (H_2, CO, CH_4, C_xH_y) peak, tar formation starts
Volatile Combustion Zone	500- 900	Combustion of volatiles	Exothermic oxidation	Sharp O_2 consumption, rapid temperature rise, CO_2 and H_2O formation
Char Combustion and Cooling Zone	900 - 1100	Oxidation of fixed carbon	Exothermic char oxidation	Gradual O_2 consumption, temperature stabilizes or slowly decreases, CO_2 increases

The model incorporates sludge rheology, considering both Newtonian and non-Newtonian behavior with shear-dependent viscosity across the bed and riser. Using a Euler-Euler approach with KTGF, it captures particle stress through granular temperature and gas-solid interactions via drag forces based on slip velocity and Reynolds number. This detailed method, offers a more accurate particle and source term representation than macroscopic models, providing a robust framework for sludge incineration.

Single particle pyrolysis and combustion [4] or used empirical heat and mass balances been modeled on sludge incineration, but a full 2D/3D multiphase development model is limited. Detailed sub-models like moisture evaporation kinetics, devolatilization reactions, and heterogeneous combustion into a Eulerian CFD frame-work are integrated in this paper and the results are validated against experiment data [1].

2 Model Development and Analysis

This study models two primary phases: gas (oxidizer and combustion products) and solids, consisting of sand (200–250 μm), sludge particles (67% moisture), and char. The liquid phase is negligible at typical operating temperatures (>850 °C) except during diesel-assisted ignition at startup. Therefore, it is not explicitly modeled. The continuity equations for each phase are as follows Eq. (1) – Eq. (3) [5].

$$\frac{\partial}{\partial t}(\varepsilon_g \rho_g) + \nabla(\varepsilon_g \rho_g U_g) = S_{H2O} + S_{vol} + S_r + S_{sec} \tag{1}$$

$$\frac{\partial}{\partial t}(\varepsilon_p \rho_p) + \nabla(\varepsilon_p \rho_p U_p) = -(S_p + S_{vol} + S_r) - (1 - \eta_{ash})m_{char} \tag{2}$$

$$\frac{\partial}{\partial t}(\varepsilon_s \rho_s Y_{ash}) + \nabla(varepsilon_s \rho_s U_s Y_{ash}) = dotm_{ash} \tag{3}$$

The conservation of momentum for the solid and gas phases is expressed as in Eq. (4) [5].

$$\frac{\partial}{\partial t}(\varepsilon_q \rho_q U_q) + \nabla(\varepsilon_q \rho_q U_q U_q) = -\varepsilon_q \nabla P + \nabla \tau_g + \rho_q \varepsilon_q g + M_q \tag{4}$$

where q = g, s denotes gas or solid, ε_q is the volume fraction, ρ_q is the density, U_q is the velocity vector, and τ_g is the stress tensor. The interphase momentum exchange term M_q is derived from drag correlation [5]. The Syamlal-O'Brien drag model is applied to evaluate gas-solid interaction. To capture the particle-particle interaction, the KTGF is incorporated. The key variable is the granular temperature (θ), defined as the kinetic energy of random particle motion per unit mass.

2.1 Species and Relation Modeling

Key species in both phases are tracked. in the gas phase, species such as CO, H_2, CH_4, CO_2, H_2O, N_2 evolve by convection, diffusion, and reaction where in the solid phase, the composition of the sludge such as moisture, volatile precursor, fixed carbon are tracked. The species conservation formula is given in Eq. (5).

$$\frac{\partial}{\partial t}(\varepsilon_g \rho_g Y_i) + \nabla(\varepsilon_g \rho_g Y_i U_g) = \nabla(D_i \nabla Y_i) + R_{vol} + R_r \tag{5}$$

where R_{vol} is rate of reaction for release of volatiles from devolatilization and R_r is rate of reaction for release of volatiles from pyrolysis.

Gas species such as CO, CO_2, CH_4, H_2, water vapor, and N_2 are transported with convection, diffusion, and reaction. Solid species such as moisture, volatile precursors, char, and ash are tracked via source/sink terms.

$$\frac{\partial}{\partial t}(\varepsilon_p \rho_p Y_{char}) + \nabla(\varepsilon_p \rho_p U_p Y_{char}) = -R_{char} \tag{6}$$

where R_{char} is reaction rate in kg/m^2s of char conversion, can be expressed as in Eq. (6).

These solids undergo drying and devolatilization. Moisture evaporates to water vapor. Volatiles are released according to the pyrolysis kinetics. A multi-step kinetic scheme: first order arrhenius rates or four step model as shown in Table 2, is used for sludge pyrolysis and char oxidation [4]. This four-step model helps differentiate complex processes into manageable sub-models.

Table 2. Four step sub-model.

Step	Process	Control mechanism	Kinetic model
1	Drying	Heat and mass transfer	Evaporation
2	Pyrolysis	Chemical kinetics	First order Arrhenius
3	Volatile combustion	Chemical/turbulent Interaction	FRC/EDM hybrid
4	Char combustion	Diffusion/surface reaction	Shrinking core or surface model

Moisture evaporation from sludge particles has been modeled using a lumped heat balance, presumed uniform particle temperature. The Stefan flow term accounts for vapor convection from the particle surface, driven by phase change, influencing mass transfer and gaseous species transport. Volatile release, thermally activated at higher temperatures, follows arrhenius-based kinetics with temperature-dependent reaction rates. Char combustion dominates, proceeding more slowly than volatile combustion at lower temperatures, but providing sustained heat. Lower flame temperature is affected by moisture evaporation and dilution effects, decelerating pyrolysis and causing some pyrolysis products to condense on the particle surface. This multi-stage thermal process couples drying, devolatilization, pyrolysis, volatile combustion, and char combustion, occurring sequentially across time and space within the CFB. The reaction kinetics used in the model as shown in Table 3 [6, 7].

Table 3. Reaction Kinetics

Process	Reaction	Model/Expression	Kinetic parameter
Drying	$H_2O_{(l)} \rightarrow H_2O_{(g)}$	Evaporation rate (T = 100°C)	Empirical correlation
Pyrolysis	$C_{10}H_{16}.NO_3S_{0.2} \rightarrow$ Volatile + Char	Arrhenius $K = Ae^{-E/RT}$	A = 1 x 107 S−1 E = 85 kJ/mol
Volatile combustion	$CH_4 + 2O_2 \rightarrow CO_2 + 2H_2O$	EDM/FRC	Global, Magnussen constants, A = 4, B = 0.5
Char combustion	$C + O_2 \rightarrow CO_2$	Shrinking core	E = 35 kJ/mol $A = 1 \times 10^5\ S^{-1}$

The char combustion is predominantly a heterogeneous reaction occurs at the surface of the particles. The primary surface reaction is represented by Eq. 7. This reaction is surface limited and typically dominates in the dense phase region in the CFB. Oxygen is consumed at the char surface to produce Carbon Monoxide (CO) as the primary product. CO produced is further oxidized in the gas phase through homogeneous.

$$C_{(s)} + \frac{1}{2}O_2 \rightarrow CO \tag{7}$$

Surface limited char oxidation is a heterogeneous reaction at the particle surface, producing CO as the primary intermediate. The char oxidation can be modeled either as heterogeneous reaction itself or converted surface reaction into volumetric reaction by UDFs in Ansys Fluent. The CO formed is subsequently oxidized in the gas phase via homogeneous reaction with O_2. Source terms for O_2, CO, and CO_2 are coupled homogeneous-homogeneous reaction pathways.

2.2 Energy and Heat Transfer

Separate energy equations are solved for gas and solids in differential form. The overall energy transport equation after volatile combustion and char oxidation in gas phase is as shown in Eq. 8.

$$\frac{\partial}{\partial t}\left(\varepsilon_g \rho_g h_{tot}\right) + \nabla\left(\varepsilon_g \rho_g U_g h_{tot}\right) = \Delta\left(K_{eff} \Delta T\right) + S_h + S_{rec} + S_{rad} + S_\theta \tag{8}$$

S_h is heat release source terms by volatile and char.

Convection, conduction, and radiation heat transfer modes are incorporated in the model. Convective and conductive terms are solved directly by the energy balance, while radiation heat transfer is modeled using P1 radiation Model. Both the solid and gas phases are considered as participating media, which is critical in high temperature CFB environments to accurately capture radiation heat exchange. The typical operating temperature is 850–950 ºC but it may reach 1100–1200 ºC in the burn out region to ensure compliance with environmental regulations.

The interphase convection, Q_{gs} is introduced in the energy transport Eq. (9).

$$Q_{gs} = h_{gs}A_{gs}\left(T_g - T_s\right) \tag{9}$$

It is computed using Nusselt number correlation for particle bed. The resulting energy equations capture the temperature profile across drying, devolatilization and combustion zones. In the drying zone, most of heat goes as latent heat of vaporization, capping particle temperature (T_s) near 100 °C, till moisture is depleted [4]. In the volatile combustion zone, intense exothermic gas phase reaction increases (T_g) rapidly, as seen in Fig. 1 by the results of this study. Beyond 1 m bed height, the drying, volatile release, and part of combustion are completed, and since the burnout stage was not considered, no further heat release and oxygen consumption is observed. It is consistent with the scope, which is to isolate the early thermal conversion zones with burnout.

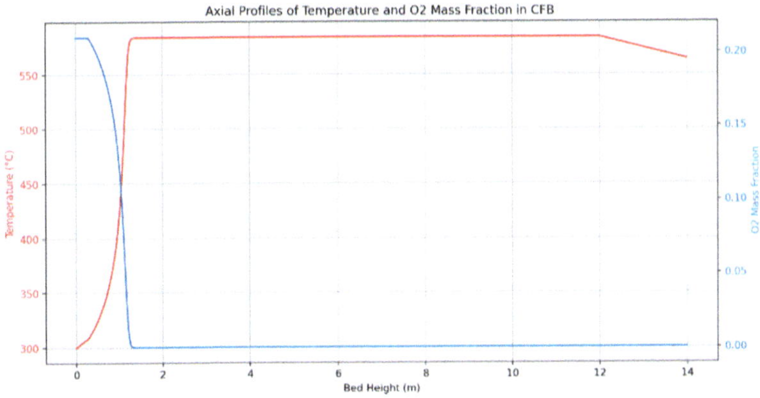

Fig. 1. Axial profile of temperature and oxygen mass fraction

2.3 Method and Model Simplification

The conservation equations for mass, momentum, species, and energy are discretized using the finite difference method on a computational grid of ~ 10^4–10^5 cells, with mesh independence verified. A preliminary Python-based solver, developed with NumPy and SciPy, is employed to perform zone wise simulations and test kinetic assumptions. This iteratively solved simplified momentum and species-energy balances with drag force coupling between gas and solids, providing convergence checks through residual monitoring and energy balance closure. The validated Python results are then used as initial and boundary inputs for detailed pressure based transient 3D CFD simulations in Ansys Fluent. PISO is selected for pressure-velocity coupling. The model assumes uniform sludge feeding to the BFB with focus on thermal zone development. The sludge contains 67% moisture, 68.9% volatile matter, 17.8% fixed carbon, and 13.8% ash as moisture free basis (MFB), fluidized by air containing 21% oxygen. The CFB is modeled as an entrained loop comprising a cyclone, return leg, and loop seal to maintain particle residence time. Gas turbulence is modeled using the k-ε closure, and radiation is incorporated via a gray gas approximation with calibrated sand absorption.

A KTGF based Euler-Euler multiphase framework is applied, where the solid sludge is considered as a granular continuum. Drying (~100 °C), devolatilization, and char oxidation through source terms dynamically coupled to gas phase oxygen and temperature. This approach enables specially and temporally resolved bed scale predictions, similar to Khiari et al. [4], but fully integrated in a multiphase CFD framework.

The model is solved in pressure based transient mode, representing regime shifts. Convection terms are discretized with second order upwind and time with second order implicit. Typical time step is 0.01–0.05 s with 20–40 iterations per time step. Residual targets for continuity, momentum, volume, species, and energy are within 10^{-4}-10^{-6}. The under-relaxation factors are 0.3–0.5 to ensure stable coupling of interphase drag, heat, and mass transfer. Regime shifts are run from 0–20 s for BFB drying and beyond 20 s for CFB rapid fluidization with secondary air, matching the UDF defined boundary ramps.

The Boundary Conditions Used in the Model Are as Shown in Table 4.

Table 4. Boundary conditions

Boundary	Type	Value/Specification
Sludge inlet ($C_{10}H_{16}.NO_3S_{0.2}$)	Mass flow inlet	0.14 – 0.28 kg/s 75% moisture
Primary air	Velocity inlet	21% O_2,1.9kg/kg wet sludge
Secondary air	Velocity inlet	21% O_2,50–70% 10–25 m/s,200–400 °C
Walls	No slip/adiabatic	-
Outlet	Pressure outlet	0 Pa (gauge), back flow composition
Radiation model	P1	-

3 Results and Discussions

The results are reported under PISO coupled transient operating conditions. The system advances from BFB dominated drying for 0–20 s into CFB combustion after 20 s with secondary air injection. This dynamic modeling reflects real operation where velocity and turbulence change over time. A low temperature region (<100 °C) in the bottom 10–15% of the bed corresponds to the drying zone, where incoming moist sludge is heated to evaporate moisture. Once all moisture is removed, the temperature begins to rise more sharply. In the central zone, around 6–8 m height from reactor base, the temperature capping is observed due to onset of volatile combustion driving the temperature to peak 750–800 °C. After this, temperature decreases slightly as char combustion is completed at experiencing heat losses in the upper free board.

The simulated temperature profiles align with typical observations in fluidized beds, which generally maintain uniform peak temperatures in the combustion zone [5]. The model accurately captures the effect of intense mixing on efficient heat distribution [1], and the temperature trends qualitatively agree with experimental BFB data. The rapid temperature rise in the volatile combustion zone reflects the balance between chemical heat release and convective losses, while radiative heat transfer between hot particles and walls becomes significant at peak temperatures. The dual-stage BFB-CFB design effectively supports lower temperatures for drying/pyrolysis and higher temperatures in the CFB due to strong heat recirculation, consistent with established fluidized bed heat transfer behavior [1].

3.1 Species Concentration

The simulated species concentration profile along the 14 m reactor height are presented in Fig. 2. Initially at the air inlet (Z = 0), the gas mixture contains approximately 21% oxygen and 79% nitrogen by volume. In the lower region of the bed (0–3 m), dominant by sludge drying, a significant increase of water vapor concentration is observed due to the rapid moisture evaporation from the sludge particle. This evaporation region is energy intensive leading to a localized suppression if temperature rise.

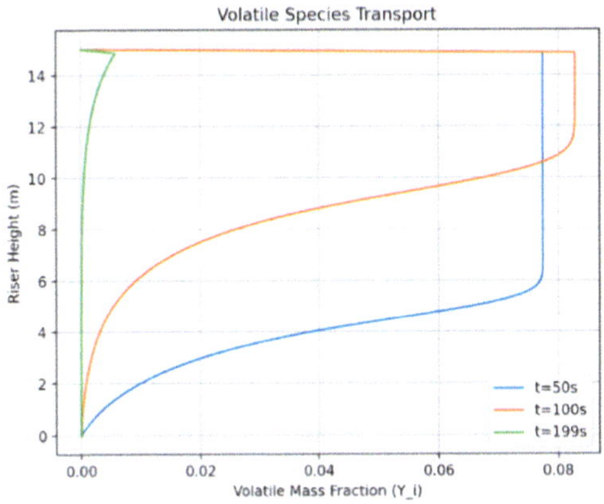

Fig. 2. Volatile species transport

Volatile release in the lower riser is driven by pyrolysis, with major species (CO, CH_4, H_2) quickly generated near the dense bed. By t = 199 s, volatiles are nearly depleted, showing effective devolatilization and combustion. Temperature peaks at the riser base, with secondary air promoting mid-riser combustion and sustaining temperatures > 850 °C. These high-temperature, oxygen-rich conditions ensure efficient oxidation of unburnt hydrocarbons and minimizing pyrogenic dioxin formation. Short residence time in the 250–400 °C window and controlled cooling prevent de-novo synthesis. Species trends align with fluidized bed literature, confirming combustion entirety and effective emission control.

3.2 Pressure Distribution

Figure 3 shows the axial pressure profile in the bed and riser. The pressure dropping nearly linear from the bottom to top of the bed. This liner trend is consistent with the hydrostatic weight of the suspended solids. In a fully fluidized bed, the pressure drop equals the weight of the particles per cross section plus a little viscous drop [1]. In this study, a bed height of 1 m and average solid density 2400 kg/m^3 would produce nearly 20 kPa pressure drop. The smooth line indicates that the drag model produces a uniform pressure gradient as expected in dense fluidization. Small fluctuations in pressure are often observed in experiments due to bubble formation. The predicted pressure drop is validated with measured values for a similar pilot-scale BFB [1]. In their study, the shyamlal-o brien drag law produced accurate pressure profile [1].This model uses a comparable drag closure, hence the reason the pressure trend is consistent.

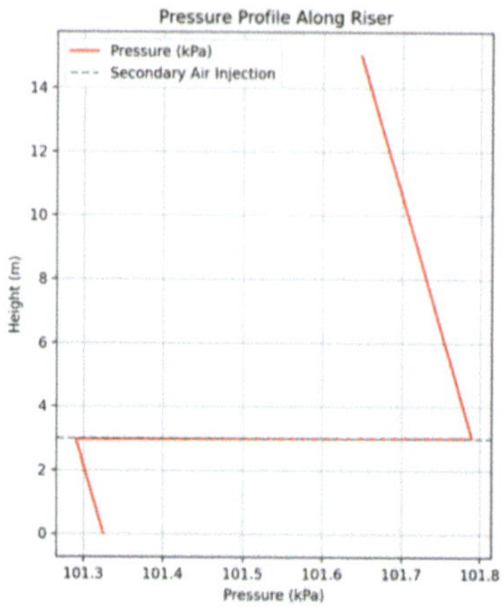

Fig. 3. Pressure profile along the CFB

The nearly constant gradient indicates the bed is in the bubbling fluidization regime. If flow rates were higher, one would expect a different profile shape. The uniform pressure increase also implies that flow regime is well developed and there is no significant channeling occurs.

3.3 Validation

The 2D outputs are compared with literature and pilot data to verify the model behavior. The overall pressure drop and fluidization behavior were checked against the 350 kW BFB data of Belhadj et al [1]. They report that the euler-euler CFD predicts the bed's minimum fluidization velocity and pressure drop vary closely within a few percent of experimental data. In this simulation the predicted pressure gradient of 20 kPa/m^2 and minimum velocity, agree similarly well with typical values for sand beds. This builds confidence in the drag and momentum sub-models. While full-scale temperature profiles exist, pilot studies show sludge particle temperatures of 500–700 °C. Peak temperature in this study is 850 °C, which is reasonable for ash-rich sludge. The model captures moisture removal, rapid heating, and nearly complete carbon conversion by the CFB exit. Predicted emissions (O_2: 1–2%, CO_2: 10–15%) set with industrial data, confirming combustion completeness and matching reported > 90% carbon conversion in incinerators [4] with rapid fluidization.

This study captures core combustion chemistry and flow behavior consistent with fluidization theory and benchmarks while excludes detailed species like NO_x due to scope limits. The model reliably predicts key hydrodynamics and thermochemical trends, supporting its credibility despite limited pilot-scale validation [1]. Although with the availability of limited sludge incineration data, comparison with BFB pilot data of Belhadj et al. [1] shows agreement with ~ 5% for pressure drop, and sludge incineration temperature align with Khiari et al. [4] for drying/pyrolysis.

4 Conclusions

A detailed Euler-Euler multiphase model of sludge incineration in a dual-stage fluidized bed was developed, capturing five key thermal zones from drying to char oxidation. The model integrates moisture evaporation, multi-step pyrolysis kinetics, granular stress (KTGF) [5], and full heat transfer mechanisms. Results show realistic drying, rapid temperature rise during volatile combustion, and consistent pressure and species profiles aligned with pilot data [1]. This confirms the model's ability to reproduce known fluidized bed combustion behavior. The model provides a strong foundation for future work, including bed optimization, pollutant prediction, and scale-up to industrial applications. This study focused on core combustion chemistry of C and H species. Pollutant emissions such as CO, NOx, SOx, Poly Aromatic Hydrocarbons (PAHs), Volatile Organic Compounds (VOCs) are not yet molded. These will be incorporated in future work using staged air and UDF based reburning sub-models.

References

1. Belhadj, E., Chilton, S., Nimmo, W., Roth, H., Pourkashanian, M.: Numerical simulation and experimental validation of the hydrodynamics in a 350 kW bubbling fluidized bed combustor. Int. J. Ener. Environ. Eng. **7**(1), 27–35 (2016). https://doi.org/10.1007/s40095-015-0199-4
2. Kamal, M.: Potential research works on fluidized bed combustors. Int. J. Eng. Appl. Sci. **2**(1), 160–166 (2025). https://doi.org/10.21608/ijeasou.2025.352787.1043
3. Syed-Hassan, S.S.A., Wang, Y., Hu, S., Su, S., Xiang, J.: Thermochemical processing of sewage sludge to energy and fuel: Fundamentals, challenges and considerations. Renew. Sustain. Energy Rev. **80**, 888–913 (2017). https://doi.org/10.1016/j.rser.2017.05.262
4. Khiari, B., Marias, F., Vaxelaire, J., Zagrouba, F.: Incineration of a small particle of wet sewage sludge: A numerical comparison between two states of the surrounding atmosphere. J. Hazard. Mater. **147**(3), 871–882 (2007). https://doi.org/10.1016/j.jhazmat.2007.01.109
5. Nemoda, S., Mladenovic, M., Paprika, M., Eric, A., Grubor, B.: Three phase Eulerian-granular model applied on numerical simulation of non-conventional liquid fuels combustion in a bubbling fluidized bed. Therm. Sci. **20**(1), 133–149 (2016). https://doi.org/10.2298/tsci151025196n
6. Magnussen, B.F., Hjertager, B.H.: On mathematical models of turbulent combustion with special emphasis on soot formation and combustion. Symposium (International) on Combustion **16**(1), 719–729 (1977)
7. Abdelaziz, O.Y., Gadalla, M.A., Ashour, F.H.: Sewage sludge combustion and co-firing for energy recovery: Kinetics and modelling. Fuel **210**, 331–340 (2017)

Numerical Investigation of $MA_3Bi_2I_9$-Based Lead-Free Perovskite Solar Cells: Impact of ETL Selection and Interface Band Alignment

W. G. A. Pabasara[1,2(✉)] and G. A. Sewvandi[1]

[1] Department of Materials Science and Engineering, Faculty of Engineering, University of Moratuwa, Moratuwa, Sri Lanka
ayomi@fot.ruh.ac.lk

[2] Department of Engineering Technology, Faculty of Technology, University of Ruhuna, Kamburupitiya, Sri Lanka

Abstract. This study presents a comprehensive numerical investigation of lead-free perovskite solar cells (PSCs) employing methylammonium bismuth iodide ($MA_3Bi_2I_9$) as the absorber material in an FTO/ETL/$MA_3Bi_2I_9$/Cu_2O/Ni device architecture. Using SCAPS-1D simulation software, the photovoltaic performance of three electron transport layers (ETLs), TiO_2, SnO_2, and ZnO was analysed under identical absorber thickness (1.4 μm) and defect density (1 × 1012 cm¯3). The resulting power conversion efficiencies (PCEs) were 16.16% for TiO_2, 14.82% for SnO_2, and 16.23% for ZnO, with all devices achieving high open-circuit voltages ($V_{oc} \approx 1.44$ V). All three ETLs yield a CBO of + 0.26 eV and a VBO of –0.95 eV, both within the literature-reported optimal thresholds (CBO < 0.30 eV and VBO < 0.2 eV) for achieving high efficiency. This alignment enables efficient hole extraction and creates an ideal spike-and-cliff band structure that reduces interfacial recombination while maintaining effective carrier flow. Therefore, the overall band alignment in all configurations supports favorable operation, and the remaining performance disparities stem from charge transport and interface quality. This work promotes $MA_3Bi_2I_9$ as a stable, non-toxic absorber and validates Cu_2O/Ni contacts as effective, low-cost alternatives to spiro-OMeTAD/Au.

Keywords: Lead-Free Perovskite · $MA_3Bi_2I_9$ · SCAPS · ETL · Band Alignment

1 Introduction

Photovoltaic (PV) technologies play a vital role by converting sunlight directly into electricity through the photovoltaic effect. Over the past decade, momentous research efforts have been dedicated to developing novel PV materials that offer higher efficiency, enhanced stability, and lower fabrication costs [1, 2]. Among these novel materials, perovskites have marked noteworthy progress due to their excellent optical and electronic properties, simple and cost-effective fabrication processes, and compatibility with flexible substrates [3–5].

M. Siriwardhana et al. (Eds.): IERS 2025, PTT, pp. 322–331, 2026.
https://doi.org/10.1007/978-981-95-7984-6_28

Despite impressive performance achieved by lead-based perovskites such as methylammonium lead halides ($CH_3NH_3PbX_3$), their environmental and human toxicity as well as stability concerns, have forced researchers to find stable lead-free alternatives [6–8]. Methylammonium bismuth iodide ($MA_3Bi_2I_9$) has gained significant attention as a promising lead-free perovskite absorber material due to its non-toxicity, favourable bandgap and enhanced environmental stability [9, 10]. However, the practical application of $MA_3Bi_2I_9$-based perovskite solar cells (PSCs) still needs to overcome significant performance challenges, mainly related to charge carrier recombination and inefficient carrier extraction at interfaces [11].

In this context, engineering appropriate electron transport layers (ETLs) and hole transport layers (HTLs) is crucial to enhance charge extraction, reduce recombination, and optimise overall device performance. ETL plays a significant role in enabling efficient electron transport while suppressing hole recombination at the absorber interface, thus directly affecting the open-circuit voltage (V_{oc}), short-circuit current density (J_{sc}), fill factor (FF), and power conversion efficiency (PCE) of the PSC [12, 13]. In this work, we systematically investigate and optimise the impact of three widely used ETLs; TiO_2, SnO_2, and ZnO in the FTO/ETL/$MA_3Bi_2I_9$/Cu_2O/Ni device architecture through inclusive numerical simulations using SCAPS-1D software.

By varying key parameters such as absorber layer thickness and defect densities, we elucidate their precise influence on photovoltaic performance and stability. The SCAPS-1D simulator enables detailed analysis by solving fundamental semiconductor equations. This approach allows us to identify optimal configurations to realise highly efficient, stable, and environmentally friendly PSC devices, thus advancing $MA_3Bi_2I_9$ as a viable candidate for sustainable photovoltaic applications.

2 Design and Modeling of Device

2.1 Device Architecture

Figure 1 schematically represents the structure of the simulated PSC. This device employs $MA_3Bi_2I_9$ as the primary absorber layer, integrated within an n–i–p architecture. This device architecture comprises a fluorine doped tin oxide (FTO) front contact, ETL, an intrinsic $MA_3Bi_2I_9$ absorber layer, Cu_2O as the HTL, and Ni as the back-metal contact.

The n–i–p configuration provides an essential benefit over conventional p–n junction devices by positioning a broad depletion region within the intrinsic absorber layer. This configuration considerably enhances the device's long-wavelength absorption, as deeper photon penetration enables the effective generation and separation of electron-hole pairs within the extensive intrinsic region. Consequently, the broadened depletion region substantially advances the quantum efficiency of PSC by facilitating efficient carrier separation and reducing recombination losses. In this study, the impact of different ETLs was systematically investigated, including TiO_2, SnO_2, and ZnO, to examine their respective impact on the photovoltaic performance of the $MA_3Bi_2I_9$-based PSCs. These materials are commonly used as ETLs due to their excellent thermal stability, suitable conduction band positions, high electron mobility, and compatibility with low-temperature, scalable fabrication processes [14–16]. Compared to organic ETLs, these

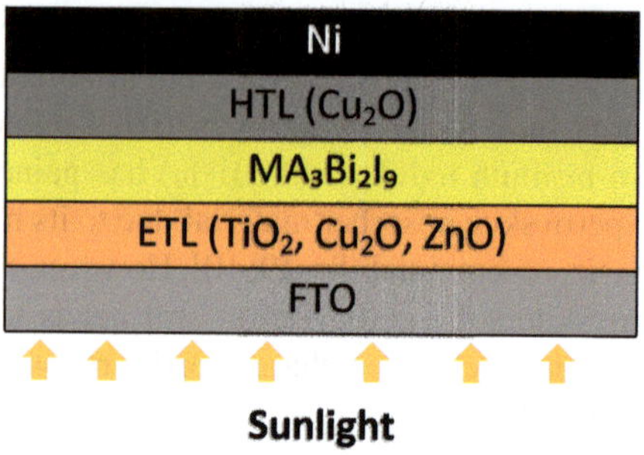

Fig. 1. Device architecture the proposed device

inorganic oxides serve superior environmental stability, moisture resistance, and chemical robustness, making them suitable for improving the durability of PSCs. Similarly, Cu_2O is employed as the HTL material owing to its high hole mobility, favourable band alignment with $MA_3Bi_2I_9$, and excellent stability under ambient conditions, providing a non-hygroscopic and cost-effective substitute to unstable organic HTLs like spiro-OMeTAD [17]. Each ETL material was rigorously analysed using SCAPS-1D simulation software. The simulations were conducted under ambient conditions of standard temperature of 300 K, frequency at 1 MHz, and illumination under the AM1.5G solar spectrum. The input parameters used for SCAPS 1D simulation are depicted in Tables 1 and 2.

Table 1. Input parameters used for the SCAPS 1D simulations [9, 18]

Parameter	FTO	TiO_2	SnO_2	ZnO	$MA_3Bi_2I_9$	Cu_2O
Thickness (μm)	0.4	0.05	0.05	0.05	1.4	0.05
Band gap (eV)	3.5	3.26	3.5	3.2	2.29	2.2
Electron affinity (eV)	4 [9]	4 [9]	4 [18]	4 [18]	4.26 [9]	3.4 [18]
Dielectric permeability relative	9	10	9	9	100	7.5
VB effective density of state (states/cm^3/eV)	2.2×10^{18}	2.2×10^{18} [18]	2.2×10^{17} [18]	3.7×10^{18} [18]	2×10^{18}	2×10^{19}
CB effective density of state (states/cm^3/eV)	1.8×10^{18}	1.8×10^{18} [18]	2.2×10^{16} [18]	1.8×10^{18} [18]	1.8×10^{19}	1×10^{19}
Hole mobility (cm^2/Vs)	20	100	100	100	4	200
Electron mobility (cm^2/Vs)	10	25	25	25	2	8600

(*continued*)

Table 1. (*continued*)

Parameter	FTO	TiO_2	SnO_2	ZnO	$MA_3Bi_2I_9$	Cu_2O
Shallow uniform donor density (cm^{-3})	1×10^{19}	1×10^{19}	1×10^{17}	1×10^{18}	1×10^{18}	0
Shallow uniform donor density (cm^{-3})	0	0	0	0	0	1×10^{18}
Defect density	1×10^{19}	1×10^{15}	1×10^{15}	1×10^{15}	1×10^{12}	1×10^{12}

Table 2. Interface parameters used in this perovskite solar cell

Interface	Defect type	Capture cross section area for electron and holes	Energetic distribution	Reference for defect energy levels, E_t	Interface defect density [cm^{-2}]
ETL/ $MA_3Bi_2I_9$	Neutral	1×10^{-19} 1×10^{-19}	Single	Above the VB maximum	1×10^{10}
$MA_3Bi_2I_9$/HTL	Neutral	1×10^{-19} 1×10^{-19}	Single	Above the VB maximum	1×10^{10}

3 Results and Discussion

3.1 Impact of Valance Band Offset (VBO), Conduction Band Offset (CBO) on the Performance of the Device

ETL and HTL play a dual role in PSC: they shuttle photogenerated carriers from the perovskite absorber to the corresponding electrodes and at the same time, block the opposite carriers, thereby suppressing recombination at the ETL/absorber and ab-sorber/HTL interfaces.

When sunlight excites the perovskite film, electrons and holes are created, separat-ed, and guided to their respective electrodes. The efficiency of this charge separation process largely depends on the energy mismatches, which are called as conduction-band offset (CBO) and valence-band offset (VBO) at the ETL/absorber and the absorber/HTL interfaces, respectively (shown in Eq. 1 and Eq. 2) [19].

$$CBO = X_{Absorber} - X_{ETL} \tag{1}$$

where $X_{Absorber}$ and X_{ETL} are the electron affinities of the absorber and ETL, respectively. Depending on the sign and magnitude of the CBO, three interfacial "barrier" types arise. There are three types of barriers: cliff-like, nearly flat, and spike-like. Cliff like barrier occurs when CBO is negative, indicating that the conduction band minimum (CBM) of ETL is lower than that of absorber layer [18]. A flat barrier happens when the CBO is zero, which signifies no energy difference and, therefore, no barrier for charge transfer. In contrast, a spike-like barrier with a positive CBO, is observed when the CBM of the ETL is higher than that of the absorber layer.

$$VBO = X_{HTL} - X_{Absorber} + E_{g},_{HTL} - E_{g},_{Absorber} \quad (2)$$

where E_g is the bandgap energy of the layers. VBO is given by Eq. 2, which also has three types, same as CBO. Cliff-like barrier is characterised by a negative VBO, which occurs when the valence maximum (VBM) of the HTL is greater than that of the perovskite absorber layer. A flat barrier signifies a negligible or zero band offset between the HTL and the absorber layer. Conversely, a spike-like barrier indicates a positive VBO, resulting from a lower VBM of the HTL than that of the absorber layer [20].

Table 3. VBO, CBO and barrier types at the interfaces

Absorber layer	ETL	CBO	VBO	Barrier type
MA3Bi2I9	TiO2	0.26	-0.95	spike (ETL/absorber) and cliff (absorber/HTL)
	SnO2	0.26	-0.95	spike (ETL/absorber) and cliff(absorber/HTL)
	ZnO	0.26	-0.95	spike (ETL/absorber) and cliff(absorber/HTL)

The studies have proved that the CBO below 0.30 eV and VBO below 0.2 eV were optimum to deliver higher efficiencies [21]. All three ETLs give a CBO of + 0.26 eV, safely below the 0.30 eV upper limit (Table 3). The VBO is –0.95 eV, which positions the HTL valence band above that of the absorber layer, meaning holes face essentially zero uphill barriers. Combined, these CBO and VBO offsets make the ideal spike and cliff pattern, respectively, that reduces interfacial recombination while preserving carrier flow. Consequently, careful tuning of CBO and VBO are very crucial parameters for designing highly efficient perovskite solar cells.

4 Band Diagrams

Figure 2(a) – Fig. 2(c) display the equilibrium band diagrams of the $MA_3Bi_2I_9$-based ar-chitectures employing TiO_2, SnO_2 and ZnO as ETLs, while Cu_2O serves as the HTL in all three devices. In all cases, CBO at the ETL/absorber side is fixed by the ETL electron affinity ($\chi = 4.00$ eV), giving a modest + 0.26 eV spike with respect to the $MA_3Bi_2I_9$ CBO. Moreover, this value lies below the 0.30 eV, "safe limit" and efficiently blocks electron back-injection without hindering forward transport. At the back interface, the VBO of –0.95 eV places the Cu_2O valence band above that of the absorber layer, so photogenerated holes travel energetically downhill into the HTL while electrons are repelled, reducing the interfacial recombination. As a result of that, the overall band alignment in Fig. 2 illustrates the desirable spike-and-cliff pattern for all three ETLs, emphasising that any remaining performance differences will stem from transport properties or interface chemistry rather than from band-edge energetics.

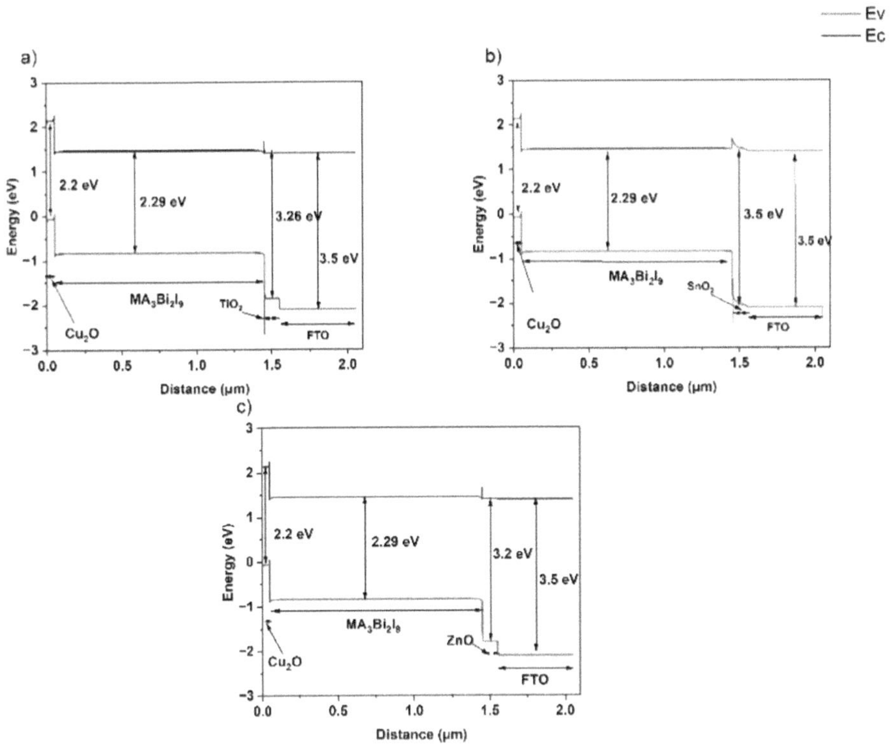

Fig. 2. Band diagrams of (a) TiO_2 (b) SnO_2 and (c) ZnO ETL based devices

5 Optimisation of the Performance of the Devices

The device performances were systematically optimised by carefully analysing and tuning the main parameters that significantly affect the performance of the PSC. $MA_3Bi_2I_9$ serves as the primary absorber layer in this device architecture, and its properties strongly influence the overall performance of the PSC. Specifically, the thickness of the perovskite absorber layer critically impacts the absorption of incident photons and charge carrier generation, with optimal thickness ensuring effective absorption without excessive recombination losses. Furthermore, defect density substantially influences charge carrier lifetime and recombination processes. A lower defect density minimises non-radiative recombination, improving charge extraction efficiency and device performance. Therefore, the thickness of the perovskite layer varied between 1 μm to 2 μm while defect density varied between 10^{10} cm^{-3} to 10^{16} cm^{-3} to investigate the performance variation and determine the optimum performance. The resulting optimised parameters, along with their corresponding photovoltaic performances, are summarised in Table 4.

Table 4. Optimised performances and parameters of the proposed devices

Configuration	Thickness (μm)	Defect Density (cm^{-3})	PCE (%)	V_{oc} (V)	J_{sc} (cm^{-3})	FF (%)
FTO/TiO_2/ $MA_3Bi_2I_9$/Cu_2O/Ni	1.4	1 × 1012	16.16	1.44	13.7	82.16
FTO/SnO_2/ $MA_3Bi_2I_9$/Cu_2O/Ni	1.4	1 × 10^{12}	14.82	1.43	12.58	82.26
FTO/ZnO/ $MA_3Bi_2I_9$/Cu_2O/Ni	1.4	1 × 10^{12}	16.23	1.44	13.75	82.16

As shown in Fig. 3, the I–V curve and the result table demonstrate the photovoltaic performance of $MA_3Bi_2I_9$-based PSCs with different ETLs (TiO_2, SnO_2, ZnO). From the IV curve, the ZnO-based PSC shows the highest current output, which aligns with the highest recorded J_{sc} of 13.75 mA/cm2. Similarly, the SnO_2-based PSC shows the lowest J_{sc} (12.58 mA/cm2), as per the given results. Although all devices have nearly identical V_{oc} (≈1.44 V), the PCE differs due to variations in J_{sc}, with ZnO yielding the highest PCE of 16.23% and SnO_2 the lowest (14.82%). The FF remains consistent across all devices, suggesting that the key performance difference is due to photocurrent. These results reveal that ZnO enables better charge extraction and collection, resulting in superior device performance.

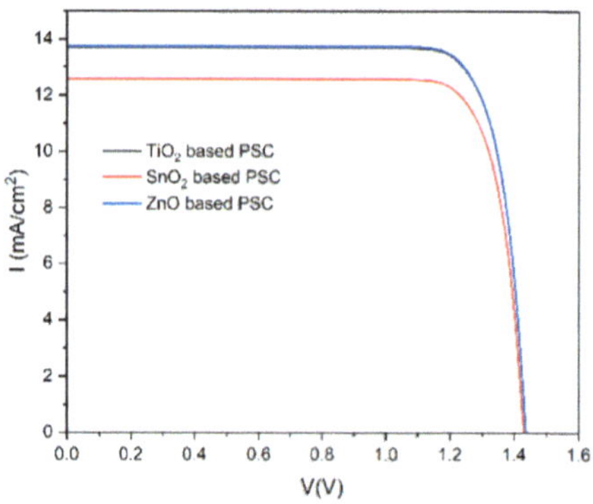

Fig. 3. IV Characteristics of optimised devices

Overall band alignment in Fig. 2 illustrates a combination of spike-like (at ETL/perovskite) and cliff-like (at HTL/perovskite) features across the three proposed devices. It suggests that the performance differences among the three devices are more likely attributed to differences in charge transport properties, interface defect densities, or interfacial chemistry rather than fundamental band-edge energetics. This is further supported by the nearly identical V_{oc} observed in all devices, indicating that barrier heights are not the limiting factor. Interfacial defects such as halide vacancies, oxygen vacancies, and Cu–I interphases can form under bias and illumination, causing increased recombination and resistance that degrade the IV characteristics over time. Therefore, optimising interfacial quality and ETL electronic properties is crucial for improving device efficiency (Table 5).

Table 5. Comparison of device performance with existing studies

Configuration	PCE (%)	V_{OC} (V)	J_{SC} (cm^{-3})	FF (%)
FTO/TiO2/$MA_3Bi_2I_9$/spiro-OMeTAD/Au [11]	14.07	1.44	8.23	88.09
FTO/TiO_2/ $MA_3Bi_2I_9$/spiro-OMeTAD/Au [9]	11.82	1.34	18.06	57.33
FTO/ZnO/ $MA_3Bi_2I_9$/Cu_2O/Ni (This study)	16.23	1.44	13.75	82.16
FTO/TiO_2/ $MA_3Bi_2I_9$/Cu_2O/Ni (This study)	16.16	1.44	13.7	82.16

The results show that the $MA_3Bi_2I_9$-based PSCs developed in the current study outperform previously reported simulated devices that employed spiro-OMeTAD as the HTL and Au as the back contact. While the spiro-OMeTAD-based devices demonstrated a higher FF (88.09%), their overall PCEs were lower due to reduced J_{sc} and stability concerns. In contrast, the TiO_2 and ZnO-based PSCs incorporating Cu_2O/Ni as HTL and back contact electrode achieved PCEs of 16.16% and 16.23%, respectively, with a consistent V_{oc} of 1.44 V. These results reveal the effective band alignment and reduced interfacial recombination enabled by Cu_2O, a low-cost and stable HTL. Although the FF in the current study is slightly lower (82.16%), the higher J_{sc} values and balanced device characteristics enhanced overall performance. Therefore, the proposed architecture offers a promising pathway for developing stable and efficient lead-free PSCs.

6 Conclusions

This This study demonstrates the potential of lead-free $MA_3Bi_2I_9$-based perovskite solar cells using a cost-effective FTO/ETL/$MA_3Bi_2I_9$/Cu_2O/Ni architecture. SCAPS-1D simulations revealed that ZnO, TiO_2, and SnO_2 ETLs yield efficiencies of 16.23%, 16.16%, and 14.82%, respectively, with consistently high V_{oc} values (~1.44 V). All configurations exhibited conduction band offsets (CBO) of + 0.26 eV and valence band offsets (VBO) of –0.95 eV, within the optimal range reported in literature for minimising interfacial recombination and ensuring carrier extraction. This ideal spike alignment at ETL/$MA_3Bi_2I_9$ and cliff alignment at $MA_3Bi_2I_9$/Cu_2O confirm that band-edge energetics are not the limiting factor, and performance variations can primarily be attributed to charge transport and interface quality. Moreover, the use of non-toxic $MA_3Bi_2I_9$ and low-cost Cu_2O/Ni contacts offers a promising, sustainable alternative to conventional spiro-OMeTAD/Au-based PSCs. Overall, the findings accentuate a feasible pathway toward efficient, stable, and environmentally friendly PSCs through targeted material selection and interface engineering.

Acknowledgements. The authors sincerely acknowledge Prof. Marc Burgelman and his team at the University of Gent, Belgium, for their invaluable effort in creating, maintaining and freely distributing the SCAPS-1D software, which was instrumental in performing the solar cell simulations in this research.

References

1. Dambhare, M.V., Butey, B., Moharil, S.V.: Solar photovoltaic technology: A review of different types of solar cells and its future trends. J. Physics: Conference Series **1913**(1), (2021)
2. Kabir, E., Kumar, P., Kumar, S., Adelodun, A.A., Kim, K.H.: Solar energy: Potential and future prospects. Renew. Sustain. Energy Rev. **82**, 894–900 (2018)
3. Kim, J.Y., Lee, J.W., Jung, H.S., Shin, H., Park, N.G.: High-Efficiency Perovskite Solar Cells. Chem. Rev. **120**(15), 7867–7918 (2020)
4. Kajal, P., Ghosh, K., Powar, S.: Manufacturing techniques of Perovskite solar cells BT - applications of solar energy. In: Tyagi, H., Agarwal, A.K., Chakraborty, P.R., Powar, S., (eds.) Springer Singapore, Singapore, pp. 341–364 (2018)
5. Pabasara, W.G.A., Sewvandi, G.A., Numerical Simulation and Optimisation of Stable $CH_3NH_3PbI_3$-based 2D/3D Mixed Dimensional Perovskite Solar Cell, Complex System Re-search Centre and Mathematical Institute of the Serbian Academy of Sciences and Arts (2025)
6. Zhang, Q., Hao, F., Li, J., Zhou, Y., Wei, Y., Lin, H.: Perovskite solar cells: must lead be replaced – and can it be done? Sci. Technol. Adv. Mater. **19**(1), 425–442 (2018)
7. Magdalin, A.E., et al.: Development of lead-free perovskite solar cells: Opportunities, challenges, and future technologies. Results in Engineering **20**, 101438 (2023)
8. Pabasara, W.G.A., Akmal, U.K.D.M., Wickramaarachchi, W.A.A.S., Swvandi, G.A.: Numerical simulation of lead-free bismuth chalcohalide based Perovskite solar cells. In: 2024 Moratuwa Engineering Research Conference (MERCon), pp. 436–441 (2024)
9. Jayawardane, S.T., et al.: Simulation-based performance analysis of lead-free bismuth perovskite solar cells: a comparative study of $Cs_3Bi_2I_9$ and $(CH_3NH_3)_3Bi_2I_9$-based perovskite solar cells. Adv. Theory and Simulat. **7**(7), 2400206 (2024)
10. Ganose, A.M., Butler, K.T., Walsh, A., Scanlon, D.O.: Relativistic electronic structure and band alignment of BiSI and BiSeI: Candidate photovoltaic materials. J. Mat. Chem. A **4**(6), 2060–2068 (2016)
11. Ahmad, K., Khan, M.Q., Alsulmi, A., Kim, H.: Numerical simulation and experimental study of methyl ammonium bismuth iodide absorber layer based lead-free perovskite solar cells. Chem. Eur. J. **29**(35), e202300513 (2023)
12. Bulowski, W., et al.: Optimisation of the ETL titanium dioxide layer for inorganic perovskite solar cells. J. Mater. Sci. **59**(17), 7283–7298 (2024)
13. Pandey, V., Gupta, A.K., Shriwastav, M.: Numerical simulation and optimisation of lead-based perovskite solar cell with inorganic HTL using SCAPS-1D. J. Opt. **53**(3), 2038–2046 (2024)
14. Li, J., et al.: Enhanced crystallinity of low-temperature solution-processed SnO_2 for highly reproducible planar perovskite solar cells. Chemsuschem **11**(17), 2898–2903 (2018)
15. Fatima, Q., Haidry, A.A., Zhang, H., El Jery, A., Aldrdery, M.: A critical review on advancement and challenges in using TiO_2 as electron transport layer for perovskite solar cell. Mat. Today Sustain. **27**, 100857 (2024)
16. Wu, X., et al.: ZnO electron transporting layer engineering realised over 20% efficiency and over 1.28 V open-circuit voltage. EcoMat **4**(4), e12192 (2022)

17. Nejand, B.A., Ahmadi, V., Gharibzadeh, S., Shahverdi, H.R.: Cuprous oxide as a potential low-cost hole-transport material for stable perovskite solar cells. Chemsuschem **9**(3), 302–313 (2016)
18. Ishraq, M.H., Kabir, M.R., Tarekuzzaman, M., Rahman, M.F., Rasheduzzaman, M., Hasan, M.Z.: ETL and HTL engineering in CH3NH3PbBr3 Perovskite for stable and efficient performance Photovoltaic devices applications using SCAPS- 1D. Adv. Theor. Simulations 7(9), 2400360 (2024)
19. Ijaz, S., et al.: A numerical approach to optimise the performance of HTL-free carbon electrode-based perovskite solar cells using organic ETLs. Heliyon **10**(7), 29091 (2024)
20. Utsho, K.I.F., et al.: Optimising $Cs_2CuBiBr_6$ double halide perovskite for solar applications: the role of electron transport layers in SCAPS-1D simulations. RSC Adv. **15**(3), 2184–2204 (2025)
21. Yogesh Kumar, S.K.C., Perez, T.D., Mathews, N.R., Mathew, X.: Band offset studies in MAPbI3 perovskite solar cells using X-ray photoelectron spectroscopy. Optical Materials **92**,425–431 (2019)

Numerical Simulation and Analytical Calculation of Heat Transfer in a Roof Assembly with Coir-Fiber Reinforced Natural Rubber Composite Insulation

W. S. A. Sudarshana[1(✉)], L. K. T. Srimal[2], and R. Gallage[2]

[1] Institute of Technology, University of Moratuwa, Homagama, Sri Lanka
asankas@itum.mrt.ac.lk
[2] University of Ruhuna, Galle, Sri Lanka

Abstract. Building energy usage must be reduced because of the growing demand for energy worldwide. In order to reduce, heat gain through the roof, researchers are investigating insulating materials for roofing assembly. Using analytical techniques and Computational Fluid Dynamics (CFD) simulation, this study examines the thermal performance of coir-fiber reinforced natural rubber latex composites in a pitched roof assembly. Scaled-down test huts were used to perform experimental measurements outdoors in the first section of the investigation. CFD simulations were then used to confirm the findings. In the second section, heat flux transmission decrease and thermal conductivity (k-value) for various insulating schemes were assessed using the validated CFD model. The results demonstrated that coir fiber reinforced natural rubber composites are effective insulation materials appropriate for passive construction applications due to their low thermal conductivity (≈ 0.07–0.08 W/m·K).

Keywords: Coir Fiber Composite · Natural Rubber Insulation · Sustainable Building Materials · Eco-Friendly Insulation

1 Introduction

1.1 Background of Study

Because of the rising energy demand, climate change, and growing sustainability agenda, energy efficiency and indoor thermal comfort are today two primary goals considered in the design of buildings [1]. Buildings use about one-third of total energy worldwide and almost 40% in commercial buildings is used for HVAC (Heating, Ventilation, Air Conditioning) systems [2]. As nations like Sri Lanka in the tropics aim to provide indoor comfort without reliance on mechanical cooling, indoor thermal comfort, economic and environmental factors remain drivers in critical building design scenarios [3]. Roof structures, especially in single-story buildings or buildings with roof floors, are one of the main avenues for solar heat gain and can account for roughly 35% of thermal transmittance into interior spaces [4]. Thermal insulation systems beneath roofing layers are a

M. Siriwardhana et al. (Eds.): IERS 2025, PTT, pp. 332–340, 2026.
https://doi.org/10.1007/978-981-95-7984-6_29

way to stop this [5]. There are several insulation options, but as we will discuss this chapter, natural fiber composites are a viable choice due to their low-cost, biodegradability, and light impact on the environment [6]. Coconut coir fiber is plentiful in Sri Lanka, has suitable thermal insulating properties such as having low thermal conductivity and acceptable resistance to moisture [7]. When reinforcement studied in natural rubber matrices with coir, coir-based composites provide better mechanical stability and are ideal for construction purpose roof insulation [8]. This research examines scope thermal performance through numerical simulation and an analytical model of heat transfer processes in roof assemblies.

1.2 Sri Lankan Standard (SLS) and Building Codes

The Sri Lanka Standards Institution's (SLSI) standards and the Urban Development Authority's (UDA) rules govern building thermal performance in Sri Lanka. Documents like SLS 962:1988 (Thermal Insulation of Buildings – General Principles) and SLS 1292:2003 (Specification for Energy-Efficient Buildings) offer pertinent foundations even though there is no national energy code that is strictly enforced. Furthermore, the Green Building Guidelines, which emphasize passive de-sign elements like ventilation, shading, and thermal insulation, are promoted by the Sri Lanka Sustainable Energy Authority. These frameworks support the use of locally accessible and environmentally friendly materials, including insulation made of coir. Further to the above, the Sustainable Energy Authority of Sri Lanka actively promotes the Green Building Guidelines which focus on the use of passive design features such as thermal insulation, shading and ventilation. These regulatory frameworks promote the use of environmentally-friendly and locally-sourced materials in building construction. This highlights the relevance and potential of coir-based insulation in meeting the requirements of national standards.

1.3 Previous Work

The analysis of the thermal properties related to natural fiber composites have been done worldwide and locally by many previous studies. Chamath et al. [5–8] have added greatly to the limitations of thermal conductivity of coir fiber using experimental, analytical and numerical methods and demonstrated the use of coir fibers in composite insulators with low thermal conductivity and sufficient mechanical properties [7]. Prior research focused on the different fabrication methods of the composites (hand lay-up, compression molding, VARTM, etc.), and then from those comparisons, it was determined that VARTM is one of the most efficient and consistent methods in creating the composites for thermal applications [9]. Figure 1 illustrates the equipment used in vacuum assisted resin transfer molding (VARTM) process.

Although progress has been made on the previous studies in thermal performance, and understanding of the thermal properties of materials, there remains a gap in research that combines numerical simulations, analytical modelling and experimental tests specifically for use in roof applications of coir-fiber reinforced natural rubber composites. This study sought to fill that gap with the development of a sustainable, cost-effective and non-destructive insulation component for the roof which focused on the climatic conditions experienced in Sri Lanka.

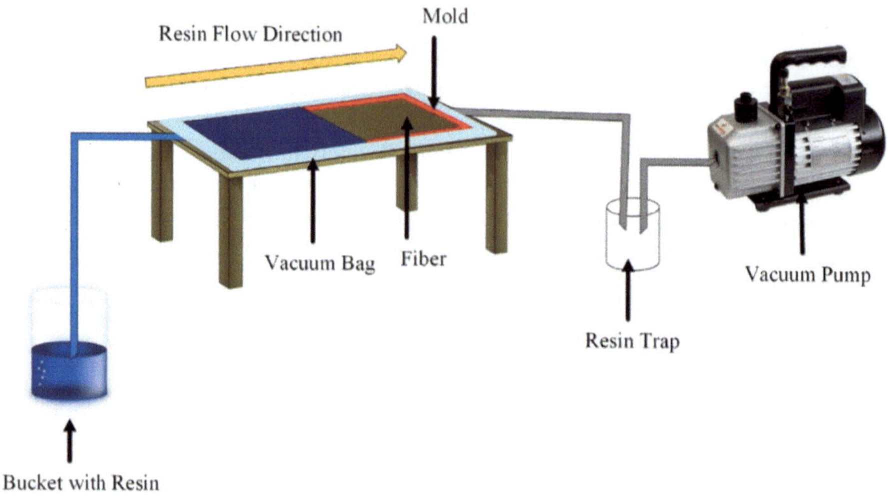

Fig. 1. Vacuum Assisted Resin Transfer Molding (VARTM) Technique

2 Experimental Measurements

Testing the thermal performance of coir-fiber reinforced natural rubber composite as roof insulation, involved testing in both simulated and actual conditions. Measurements are essential to validate numerical simulations and analytical heat transfer models, as well as to evaluate thermal conductivity and performance of the material as a reasonable roof application.

2.1 Data Acquisition Monitoring System

An operational data acquisition monitoring system was created and deployed comprising of scaled-down test huts, designed to mimic real residential roof assemblies. These huts were located outdoors to maintain natural environmental exposure to realistic photovoltaic materials for solar exposure, the variation of weather ambient temperature, as well as the wind's effect on the roof surface. A test hut was constructed using coir fiber–reinforced natural rubber composite mats roofing insulation, which were fabricated based on the author's previously obtained results [9]. Following Fig. 2 shows the 3D model of the scaled-down test hut with dimensions. Figure 3 illustrates the experimental construction of scaled-down test huts.

The Major Components of the Monitoring System Included
Thermocouples. Type-K thermocouples were installed at several depths within the roofing layers (above the insulation, within the composite layer, and at the ceiling) to measure temperature gradients across the insulation.

Heat Flux Sensors. Located on both sides of the composite insulation to record the rate of heat flow through the roof assembly.

Data Logger. An automated data logger with wireless capability that recorded temperature and heat flux data at designated intervals (5–10 min generally).

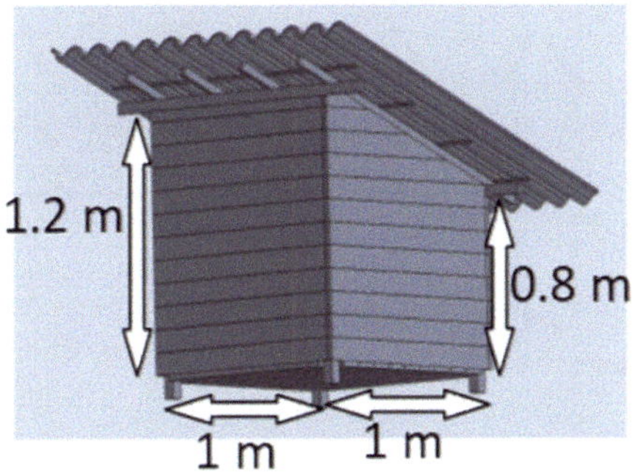

Fig. 2. Scaled-Down Test Hut

Fig. 3. Experimental Constriction of Scaled-Down Test Huts

Weather Station. Installed nearby to the test huts to record ambient temperature, humidity, solar irradiance, wind speed, and other inputs relevant to correlating thermal performance with environmental factors. This apparatus provided real-time monitoring of roof assemblies and complete thermal profiles of the assemblies over time, including during peak solar hours and night cooling. Following Fig. 4 shows location of sensors in the test hut.

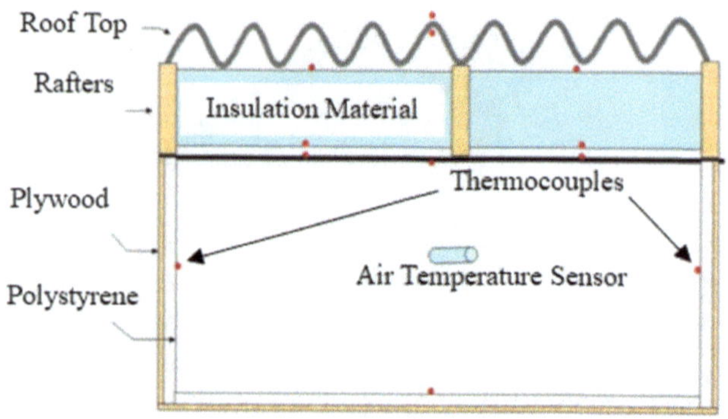

Fig. 4. Location of Sensors in the Scaled Down Test Hut

2.2 Calculation of Thermal Conductivity (K)

Thermal conductivity (k), an important parameter that indicates a material's ability to conduct heat, was calculated via a set of steady state and transient heat transfer approaches. The theory is based on one dimensional heat conduction equation (given in Eq. 1):

$$Q = -KA\frac{dT}{dx} \tag{1}$$

where, Q = Heat flux, K = Thermal conductivity, A = Cross section area, T = Temperature, $\frac{dT}{dx}$ = Temperature gradient

To improve accuracy, the trials were executed at different time of day and averaged in order to limit the effect of transient conditions. The data was also compared to known values found in the literature reviewed, and verified through: Guarded Hot Plate Method (where available in lab setup), Comparative Method - comparing against standard insulation materials (polyurethane foam, etc.), Numerical Validation - Numerical simulation using Computational Fluid Dynamics (CFD) tools like ANSYS to represent heat transfer across a similar boundary condition. Thermal Conductivity values calculated for coir-fiber/natural rubber composites consistently fell in the range of **0.07 – 0.08 W/m·K,** depending on the fiber-matrix ratio and orientation, which is certainly within acceptable limits for effective thermal insulation in residential buildings [9].

3 CFD Model

Computational Fluid Dynamics (CFD) modeling provides a dependable method of assessing heat transfer through roof assemblies made from coir-fiber reinforced natural rubber composites. Using a numerical simulation allows one to visualize temperature gradients, movement of air, and the performance of insulation material under realistic conditions. The use of CFD modeling can supplement the analytical and experimental results attained providing a full picture of thermal behavior in insulated roof structures.

3.1 Roof Configuration

The modeled roof assembly is meant to represent the construction of the experimental test huts. The assembly includes the following layers: Exterior roof covering (such as asbestos), Coir fiber–natural rubber composite insulation layer, Air gap (if applicable depending on the ventilation strategy), Interior ceiling board (such as gypsum or plywood). Dimensions of the model were scaled to be equal to the dimensions of the experimental test huts in order to maintain consistency with the physical and virtual testing environments. The composite layer was defined based on the best fiber-matrix combination selected from the literature review and earlier material testing [3].

3.2 Boundary Conditions

According to the climate data gathered from the Department of Meteorology, Sri Lanka, approximate outdoor environmental conditions, boundary conditions were set as follows: Ambient air temperature: Arranged for typical daytime peak (example, 32 °C for Sri Lanka). Solar radiation: Modeled as heat flux that was applied to the top surface of the roofing material; the value used (example, 800–1000 W/m2) was based on local Solar Irradiance data. Convection: Natural convection was applied to the air domain with surface convection coefficient defined for the interior and exterior surfaces. Initial internal temperature: At the standard indoor condition (example, 25 °C). Wall and floor boundaries: to be adiabatic or insulated; directed to keep lateral heat loss minimal to focus the heat flow only to the direction of vertical heat flow of the roof. The simulation was completed using ANSYS Fluent using steady-state heat transfer models. Mesh refinement to enable the model heated the composite layer specifically to adequately resolved thermal gradients.

4 Results and Discussion

This section shows a comparative review of results from CFD simulations that were also measured in open testing. It demonstrates the coir-fiber reinforced natural rubber composite insulation will effectively limit heat penetration and improve the human thermal comfort in roof structures.

4.1 CFD Results Validation Against Experimental Measurements

To confirm the reliability of the simulation, the CFD model was validated with the experimental temperature and heat flux data collected from the test hut set up. The temperature values at the various layers (roof surface, composite insulation, and room ceiling) indicated excellent correlation between the simulation and experiment. The average error between the CFD predicted internal surface temperature and experimental data was well within acceptable accuracy limits of ± 1.2 °C for passive thermal modeling. This confirmed The average error between the CFD predicted internal surface temperature and experimental data was well within acceptable accuracy limits of ± 1.2 °C for passive thermal modeling. This confirmed the appropriateness of the thermal properties and boundary conditions used in the simulation can be depended upon. Figure 5 shows the variation of indoor temperature over time in a test hut insulated with coir fiber reinforced natural rubber.

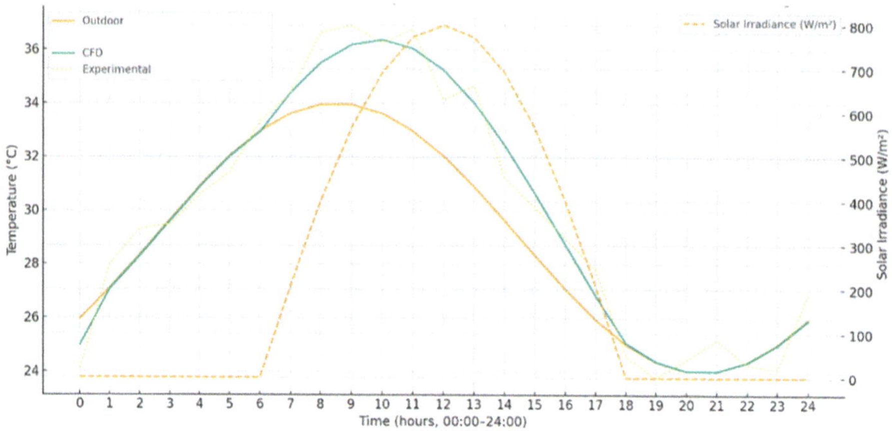

Fig. 5. The Variation of Indoor Temperature Over Time in a Test Hut

4.2 Heat Flux Penetration through the Pitch Roof Assembly

The magnitude of heat transfer through the roof directly affects thermal comfort. When a CFD analysis was performed, heat flux vectors were mapped across the roof cross-section to illustrate how far the heat penetrated through each material section. The coir fiber–natural rubber composite layer greatly limited the amount of heat transmitted across the exterior roof surface to the interior ceiling surface. A 36% reduction in peak heat flux for the composite insulation installation, on average, was achieved compared to a control roof without insulation.

4.3 Predicted Thermal Conductivity Value from CFD Model

The thermal conductivity (k) predicted from the CFD simulations correlated well with experimental or analytical predictions. Using steady-state simulations and Fourier's Law the predicted thermal conductivity of the coir fiber composite was:

$$K_{CFD} = 0.07W/mK$$

This value supports theoretical and experimental data suggesting that the thermal conductivity can range between 0.07–0.08 W/m·K depending on material density and fiber orientation. The alignment from these three techniques helps confirm the use of coir-fiber based composites for real-world roofing applications.

5 CFD Graphical Results

CFD simulation investigates the effects of roof slope on thermal insulation performance with two roof angle configurations of 20° and 30°, which are representative of roof configurations frequently observed in traditional Sri Lankan home design. Cut plots of temperature distribution were developed to highlight the thermal gradient through the roof configuration.

5.1 CFD Temperature Cut Plot of Roof Assemblies for a Roof Pitch of 20°

The CFD simulation of a roof with a 20° pitch demonstrated moderate thermal stratification within the insulation layer. The temperature cut plot demonstrates: Higher surface temperatures on the roofing material (~45–50 °C during peak hours), A gradual temperature drops through the composite insulation layer, a relatively stable indoor ceiling temperature (~28–30 °C) establishing the composite has insulation properties. The temperature changes for the 20° model demonstrates a slightly steeper thermal gradient in comparison to the 30° model (discussed below), due to more direct solar exposure from the shallower pitch.

5.2 CFD Temperature Cut Plot of Roof Assemblies for a Roof Pitch of 30°

The following simulation also illustrated a more favorable heat distribution for the 30° roof pitch: The higher slope decreased direct solar load during midday, Surface temperatures were a bit less than the 20° model (~42–47 °C), The interior ceiling temperature was better controlled in temperature (~27–29 °C). Again the insulation layer was effective; which the steeper pitch supplemented again the insulation layer was effective; which the steeper pitch supplemented the roof assemblies heat rejection performance.

6 Conclusions

This research focused on designing and developing coir-fiber reinforced natural rubber composite insulation specifically for pitch roofs, with the goal of enhancing energy efficiency and thermal comfort in residential buildings in tropical climates, with an emphasis on Sri Lanka. The thermal performance of this sustainable insulation material was studied via experimental measurements, analytical calculations, and CFD simulations. The findings confirmed that: The coir fiber-natural rubber composite has a low thermal conductivity ($\approx$ 0.07–0.08 W/m·K), which makes it an efficient insulating material appropriate for passive building use. The experimental test huts allowed for the insulation's real-world performance evaluation and quantified substantial reductions in internal ceiling temperature and heat flux in natural conditions. The CFD models, calibrated to reflect the actual performance, provided detailed insights into temperature distribution and heat flow and were demonstrated to accurately predict performance across varying roof pitches. Higher pitched roofs (e.g., 30°), performed better thermally due to a lower level of direct solar exposure, complimenting the coir based insulation layer's efficiency. This study provides a viable, sustainable, and local solution for enhancing thermal performance in domestic roofs, that aligns with Sri Lankan building standards and sustainability aspirations, whilst, also demonstrating how coir, an indigenous product usually recognized as waste because of the agricultural application, can be converted into a high-performance building product using composite engineering. One limitation of this study is its scalability to real roof structures, while a detailed cost analysis of coir fiber–reinforced natural rubber composites is identified as a potential area for future research.

References

1. Chamath, L.G., Srimal, L.K.T., Sewvandi, G.A.: Assessment of transverse thermal conductivity of coir fiber using experimental, analytical, and numerical methods. J. National Sci. Foundat. Sri Lanka (2023)
2. González-Torres, M., Pérez-Lombard, L., Coronel, J.F., Maestre, I.R., Yan, D.: A review on buildings energy information: Trends, end-uses, fuels and drivers. Energy Rep. **8**, 626–637 (2022)
3. Sudarshana, W.S.A., Srimal, L.K.T., Gallage, R.: Use of Coir Fiber-Reinforced Natural Rubber Composites for Roofing Insulation: A Review. Proceedings of 3rd International Research Conference, Research Unit, Institute of Technology, pp. 62–66. University of Moratuwa (2024)
4. Sudarshana, W.S.A., Gallage, R., Srimal, L.K.T.: Numerical Simulation and VARTM Process for Manufacturing of Coir Mats. Proceedings of the SLIIT International Conference on Engineering and Technology (2024)
5. Chamath, L.G., Kankanamge, L., Srimal, L.K.T., Gallage, R.: Design and Fabrication of a System to Measure Thermal Conductivity of Compressible Materials
6. Chamath, L.G., Srimal, L.K.T., Sewvandi, G.A.: Evaluating the Thermal Conductivity of Three-phase Insulation Composite using Analytical and Numerical Methods. In: Moratuwa Engineering Research Conference (MERCon), pp. 1–6 (2022)
7. Chamath, L.G., Srimal, L.K.T., Sewvandi, G.A., Gallage, R., Epaarachchi, J.: Optimizing the alkaline concentration for coir fiber treatment and estimation of lifetime. Sri Lanka J. National Sci. Foundat. Sri Lanka **52**(4), 515–525 (2024)
8. Chamath, L.G., Srimal, L.K.T., Sewvandi, G.A.: Effect of alkaline concentration on the surface properties of coir fibers. In: Proceeding of the National Engineering Research Symposium (NERS), pp. 7–12 (2020)
9. Sudarshana, W.S.A., Srimal, L.K.T., Gallage, R.: Production of Coir Fiber-Reinforced Natural Rubber Composite Using Vacuum-Assisted Resin Transfer Molding (VARTM) Process for Roofing Insulation Applications in Buildings. Engineer: J. Ins. Eng. Sri Lanka **58**(2), 1–14 (2025)

Urban Air Quality Mapping Using GIS: A Case Study of Gampaha District, Sri Lanka

K. P. D. D. Jayasekara[1(✉)] and H. I. D. Liyanage[2]

[1] National Engineering Research and Development Center of Sri Lanka, Jaela, Sri Lanka
dulinda@nerdc.lk

[2] National Research Council of Sri Lanka, Vidya Mawatha, Colombo 7, Sri Lanka

Abstract. Sri Lanka's rapid urbanization, industrialization, and increasing vehicular emissions have raised significant concerns about deteriorating air quality, mostly in densely populated regions such as Colombo, Gampaha, and Kandy. This study aimed to assess the spatial variability of urban air pollutants using a combination of ground-based measured data and Geographic Information System (GIS) mapping. The research examined distributions of particulate matter (PM10), nitrogen dioxide (NO_2), sulfur dioxide (SO_2), carbon monoxide (CO), and ozone (O_3) concentrations over the past 4 years. This study identified pollution hotspots and analyzed the distribution of air quality across the Gampaha district in Sri Lanka by using available atmospheric data in the Energy and Environ-mental Services Department of the National Engineering Research and Devel-opment Center of Sri Lanka. The findings highlighted the urgent need for tar-geted air quality management policies and suggested working frameworks for citizens within. This research highlights the importance of localized environ-mental conditions in shaping adaptive responses to urban air pollution in Sri Lanka.

Keywords: GIS · Air pollution · Air quality

1 Introduction

1.1 Urban Air Quality and Its Significance

Urban air quality refers to the condition of the air within cities and metropolitan regions, measured in terms of the concentration of pollutants like particulate matter (PM_{10} and $PM_{2.5}$), nitrogen oxide (NO_2), sulfur dioxide (SO_2), ozone (O_3), and carbon monoxide (CO). It's formed by a complex relationship of factors including industrial activities, vehicular emissions, population density, urban geography, climate, and even seasonal wind patterns.

Air quality holds critical significance due to its direct impact on public health. Prolonged exposure to pollutants like particulate matter, nitrogen oxides, and ground-level ozone contributes to respiratory and cardiovascular diseases, especially in vulnerable populations. Beyond health, poor air quality also affects ecosystems, accelerates infrastructure degradation, and imposes economic burdens through lost productivity

M. Siriwardhana et al. (Eds.): IERS 2025, PTT, pp. 341–351, 2026.
https://doi.org/10.1007/978-981-95-7984-6_30

and healthcare costs. Moreover, certain pollutants like aerosols act as climate forcers, linking air quality management to broader climate change mitigation efforts. In the context of urban planning, reliable air quality data informs policies on industries, transport, land use, and green infrastructure, making it a basis for designing sustainable, livable cities. Therefore, addressing urban air quality is not only a health related but also a vital connection of environmental and flexibility in rapidly urbanizing societies.

1.2 Air Quality Mapping as a Tool for Special Analysis

Air quality mapping is an important tool in spatial analysis that enables researchers and urban planners to visualize, interpret, and respond to pollution patterns across geographic spaces. By explaining raw environmental data, such as concentrations of PM or NO_2, into spatially referenced maps, this technique exposes the distribution of air pollutants across urban landscapes. It allows for the identification of pollution hotspots, fluctuations, and socio-environmental inequalities in exposure levels. Through Geographic Information Systems (GIS), satellite remote sensing, and increasingly, IoT-enabled sensor networks, air quality mapping provides a dynamic framework for analyzing how factors like land use, traffic density, and meteorological conditions influence air pollution. This spatial perspective is crucial not only for environmental monitoring but also for guiding targeted involvement, optimizing urban design, and informing policy decisions aimed at safeguarding public health and environmental equity.

1.3 Research Objective

General Objective:
This study aims to map and analyze the spatial and temporal distribution of major air pollutants—PM_{10}, NO_2, SO_2, CO, and O_3—across the Gampaha District by integrating available data with Geographic Information System (GIS) techniques.

Specific Objectives:

1. Identify pollution hotspots within the district and examine their relationship with urban infrastructure and land-use patterns, thereby providing a locally grounded perspective on emission sources.
2. Develop a replicable framework for integrating environmental data with spatial analysis, enabling improved monitoring and assessment of urban air quality.
3. Generate insights that support adaptive urban planning and targeted policy interventions, contributing to the broader discourse on environmental monitoring, urban sustainability, and climate-resilient development in developing cities.

1.4 Study Area

This study is centered on the Gampaha District, located in the Western Province of Sri Lanka. Gampaha is one of the country's most rapidly urbanizing districts, characterized by a mix of densely populated residential zones, expanding industrial estates, and significant transportation corridors. Its proximity to Bandaranaike International Airport and several major expressways contributes to sharp vehicular emissions, while ongoing urban extension and mixed land use make it an ideal candidate for analyzing the spatial dynamics of air pollution. The district also experiences in highest number and largest Export Processing Zones (EPZ) such as Katunayake EPZ and Biyagama EPZ. By focusing on Gampaha, this study captures the connection of suburban growth, infrastructural expansion, and environmental stress, offering an understanding into the air quality challenges faced by Sri Lanka's evolving urban landscapes. Table 1 shows some census and statistics data of the Gampaha District [1].

Table 1. Statistics data of the Gampaha District

Area	1387 km^2 (38% Western)
Population	2,373,000 (2023)
DS division	13
GN divisions	1177
Location in Globe	6°54′5" and 7°20′00″ North Latitude and 79°4′75" and 80°13′00" East longitude

2 Literature Review

Air pollution, particularly from fine particulate matter (PM10), has become a critical concern in urban environments due to its adverse effects on human health and environmental quality. Traditional monitoring networks are missing the spatial resolution needed to capture localized pollution patterns. In response, researchers have increasingly turned to Geographic Information Systems (GIS) as a powerful tool for visualizing, analyzing, and modeling the spatial distribution of air pollutants.

A developing Asian city, the Colombo city's air quality is rapidly deteriorating. One study [2], shows that, between 1998 and 2016, PM10 concentrations recorded at the Colombo Fort Air Quality Monitoring Station (AQMS) generally fluctuated between 55 and 100 μg/m^3, with a stable range of 72–82 μg/m^3 and these values are consistently exceeded Sri Lanka's national annual air quality standard of 50 μg/m^3. Not only in Asia, but also in Europe, the recent studies show that air pollution levels are leading to a decreasing in the average life expectancy [3]. Therefore, improving urban air quality in urban areas is a subject among researchers. Air pollution is recognized as a major public health hazard, leading causes of illness such as cancer, heart disease, and obesity. It significantly reduces both life expectancy and quality of life. Between 2008 and 2013, global levels of ambient air pollution rose by 8%, according to the World Health Organization. Alarmingly, about 43% of deaths from terminal lung disease are attributed to air pollution, which contributes to approximately 1.8 million deaths annually due to lung conditions and cancer [4]. Air pollution in Sri Lanka causes significant health impacts. An estimated 5,200 deaths annually are attributed to air pollution. At the Lady Ridgeway Children's Hospital, 60% of child patients seek treatment for respiratory illnesses [5]. Research conducted in 2008 found a strong correlation ($r = 0.717$) between PM10 levels and respiratory conditions such as bronchitis and emphysema in children. Approximately 20% of asthma cases at the hospital may be linked to elevated PM10 exposure [6]. Sri Lanka, with a population of around 21.5 million, experienced an estimated 7,260 deaths in 2019 due to ambient PM2.5 pollution [7]. A Sri Lankan study [8], measured PM2.5 and black carbon (BC) levels alongside GPS tracking, reporting a mean PM2.5 concentration of 148 $\mu g/m^3$. This value reflects the average exposure experienced by passengers traveling through the city in non-air-conditioned vehicles. Assessing the health impacts and economic costs of air pollution is vital for environmental policy making. For example, in 2012, Mumbai, with a population of 12.4 million, suffered an estimated USD 8 billion in health associated damages due to ambient air pollution. These assessments offer a quantifiable basis for rationalizing pollution control strategies and prioritizing interventions where the public health burden is greatest [9].

It is necessary to collect data to get an idea about the pollution level. One of the data collection methods is using satellites. Satellite derived NO_2 data for the years 2005, 2008, 2010, and 2012 were obtained from TEMIS [Tropospheric Emission Monitor-ing Internet Service], and the 'WHIPS' package developed by the University of Wisconsin was used to extract and grid these data over Sri Lanka, enabling spatial analysis of tropospheric nitrogen dioxide distribution during that period. Satellite imagery reveals a steady rise in NO_2 concentrations across Sri Lanka since 2008, both in magnitude and geographical spread. Emerging pollution hotspots are visible island wide, likely tied to post-civil war industrial growth. By 2012, notable in-creases are observed in the southern and northwestern regions, while the Colombo district shows a significant spatial expansion of NO_2 pollution [10]. Satellite images gives a spatial analysis. Researchers can uncover the spatial relationship between long-term exposure to air pollution and the prevalence of chronic illnesses such as bronchitis, asthma, and cancer by using GIS. This analytical approach allows for the visualization of pollution patterns over time and their correlation with health data, offering a powerful basis for targeted public health and environmental policy decisions [11]. Air quality in developing countries has deteriorated progressively. Rapid industrialization, urbanization, and factors such as vehicle emissions, use fuels, poorly maintained infrastructure, and weak environmental regulations are main causes. To visual representation of pollution trends and hotspots, researchers working on ArcGIS 10.8 to map recorded air quality parameters [12]. Air quality mapping is a worthy technique to aware the public. Awareness of air pollution can save the lives. 1,000 out of fewer than two million people are dying annually from air pollution in North Macedonia. This is why raising public awareness about particulate matter pollution is essential [13]. Air pollution contributes to the premature deaths of two million Indians every year [14].

The absence of instantaneous air quality monitoring systems significantly delays authorities from assessing current atmospheric conditions and responding promptly to pollution events. As urbanization accelerates, integrating air quality surveillance into smart city frameworks has become essential. Sri Lanka's existing ambient air quality monitoring network includes TSI BlueSky sensors operated by the University of Peradeniya and approximately 14 additional sensors managed by the National Building

Research Organization (NBRO). These instruments form the support of the country's efforts to monitor air pollution levels, though coverage and instantaneous accessibility may still be limited [15]. Air pollution contributed to approximately 6.7 million deaths globally in 2019, according to the World Health Organization. This overwhelming damage highlights that air pollution is not just an environmental issue, but one of the most urgent public health crises of our time. It affects people of all ages, particularly the elderly, children, and those with ill health conditions, and is linked to heart disease, stroke, respiratory illnesses, and cancers [16]. Fine particulate matter (PM2.5) and certain air toxics can penetrate the brain through blood, causing neuroinflammation, oxidative stress, and neuronal damage. Studies [17], suggest that air pollution exposure is associated with cognitive impairment, neuro-developmental disorders in children, and an increased risk of neurodegenerative diseases such as Alzheimer's and Parkinson's. Remote sensing technologies play a crucial role in air pollution monitoring by using advanced sensors and imaging systems to detect, quantify, and map atmospheric pollutants. These tools can be mounted on satellites, aircraft, drones, and ground-based platforms, offering broad spatial coverage and high temporal resolution. This makes them especially valuable for capturing real-time or long-term trends in air quality, detecting pollution hotspots, and supporting environmental policy and urban planning.

3 Methodology

3.1 Study Area

The research was conducted in the Gampaha District, located in the Western Province of Sri Lanka. The district consists of densely populated urban centers, semi-urban settlements, rural areas, and zones with industrial activities, providing diverse land-use characteristics relevant to air quality assessment.

3.2 Data Collection

Air quality data were obtained from the Energy & Environmental Services Department (E&ESD), which monitored approximately 35 sampling locations across the district. These locations were strategically selected to represent different environmental contexts, including:

- **Urban areas** – high population density and vehicular activity.
- **Rural areas** – lower population density with limited transport emissions.
- **Industrial zones** – potential emission sources from manufacturing and related activities.

At each sampling site, key air pollutants were monitored, including particulate matter (PM_{10}), nitrogen dioxide (NO_2), sulfur dioxide (SO_2), carbon monoxide (CO), and ozone (O_3). Each location was geotagged with latitude and longitude coordinates for spatial integration.

3.3 Data Preprocessing

PM_{10}, SO_2, NO_2, O_3 and CO are measured using standard measurement methods mentioned in the Gazette of the Democratic Socialist Republic of Sri Lanka, No. 1562/22, 2008 [18]. PM_{10}; Hi-volume sampling, Sulphur Dioxide (SO_2); Pararosanilines, Nitrogen Dioxide (NO_2); Colorimetric Saltzman, Carbon Monoxide (CO); non-dispersive infrared spectroscopy and Ozone (O_3); Ultraviolet photometric were the methods.

The raw air quality datasets underwent several preprocessing steps before spatial analysis:

- **Data cleaning** – removal of incomplete or inconsistent entries.
- **Standardization** – aligning pollutant concentration units to maintain comparability.
- **Temporal aggregation** – organizing data into daily, monthly, and annual averages to examine both short-term variations and long-term trends.

3.4 GIS-Based Spatial Analysis

Geographic Information System (GIS) techniques were applied to integrate, visualize, and analyze the air quality data. The following steps were carried out:

1. **Mapping of Monitoring Locations** – All sampling points were plotted on the district map using their geospatial coordinates.
2. **Interpolation** – Spatial interpolation methods such as Inverse Distance Weighting (IDW) and Kriging were applied to estimate pollutant concentrations between monitored points. This generated continuous surface maps showing pollutant distribution.
3. **Hotspot Identification** – Concentration maps for each pollutant were analyzed to locate areas with consistently high pollution levels.

3.5 Policy Relevance and Framework Development

Finally, the findings were used to establish a replicable GIS-based framework for urban air quality mapping. The outputs were designed to support adaptive planning, regulatory interventions, and future monitoring in rapidly urbanizing districts.

4 Findings and Discussion

Spanning four years (2021–2024) of careful observations and monitoring under consistently clear weather conditions, this study revealed a complex picture of air quality dynamics across Gampaha District's different landscapes. The study of pollutant concentrations over this period reflects both the pressures of urban growth and the delicate involvement of policy and environmental response. The colour coding of the Fig. 1 is set according to the Central Environmental Authority (CEA) standard limits [18] and the World Health Organization (WHO) Guidelines [19]. The Green area is showing the WHO Guidelines recommended area while the yellow area demonstrates the CEA limits. The Red area emphasis the hazardous area which means the areas having pollutants more than CEA recommends to live in.

Particulate Matter (PM_{10}): These microscopic particles, often invisible to the naked eye, are primarily generated from vehicular emissions, construction dust, industrial activities, and open burning. Over the four-year observation space, PM_{10} levels not only reflected the rhythm of daily life but also responded sensitively to wider infrastructural and policy changes. Figure 1 showed the spatial distribution of the PM_{10} in Gampaha District. Industrial zones regularly recorded the highest PM concentrations, with values frequently surpassing CEA's recommended limits. Residential and green-buffered areas, by contrast, consistently maintained lower averages, but surpassed with WHO's recommended guideline values emphasizing the mitigating role of vegetation and urban zoning.

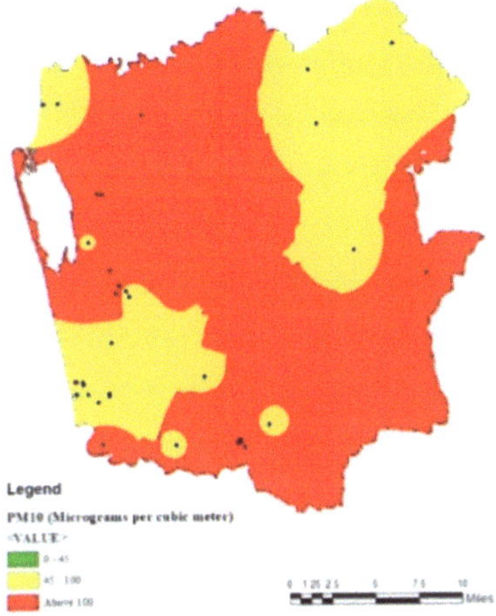

Fig. 1. Particulate Matter (PM_{10}) distribution through Gampaha District

Sulphur Dioxide (SO_2): Fig. 2 showed the spatial distribution of the Sulphur Dioxide (SO_2) in Gampaha District. Among the pollutants tracked, Sulphur dioxide (SO_2) provided a calming image of localized combustion sources, especially from fossil fuel use in industrial processes and certain categories of vehicular traffic. Although SO_2 levels generally remained within CEA air quality thresholds throughout much of the monitoring period, intermittent rushes may occur.

Nitrogen Dioxide (NO_2): A reactive gas primarily emitted from motor vehicle exhaust and combustion processes proved to be a dynamic and informative indicator of urban activity across the study period. Its spatial distribution shows that the Gampaha District is covered with a negligible area of safety zone according to the WHO guidelines for air quality. This means the NO_2 pollution takes place all over the district and it's spreading every corner. Figure 3 illustrates the visual representation of NO_2 distribution through Gampaha district.

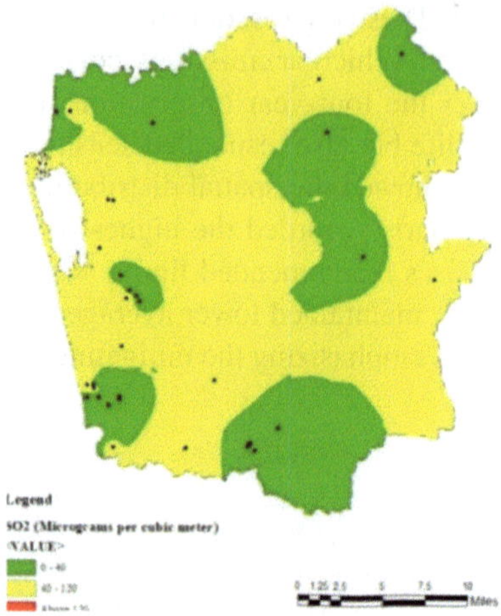

Fig. 2. Sulphur Dioxide (SO_2) distribution through Gampaha District

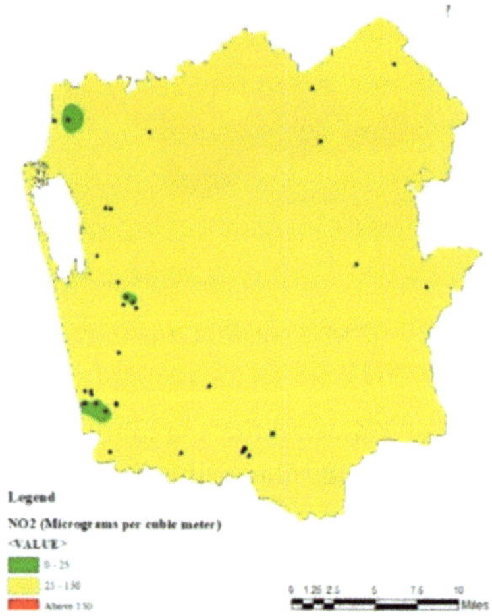

Fig. 3. Nitrogen Dioxide (NO_2) distribution through Gampaha District

Carbon Monoxide (CO): A colorless and odorless gas largely emitted through incomplete combustion, featured an even distribution in all over the district. Zones characterized by high vehicular density or industrial areas are hotspots when considering the values of the WHO guidelines for air quality. Like NO_2, the CO also spreads evenly through the district, showing the circulation of poor air quality in the Gampaha district. Figure 4 displays the distribution of CO across the Gampaha district, emphasizing the low quality of life.

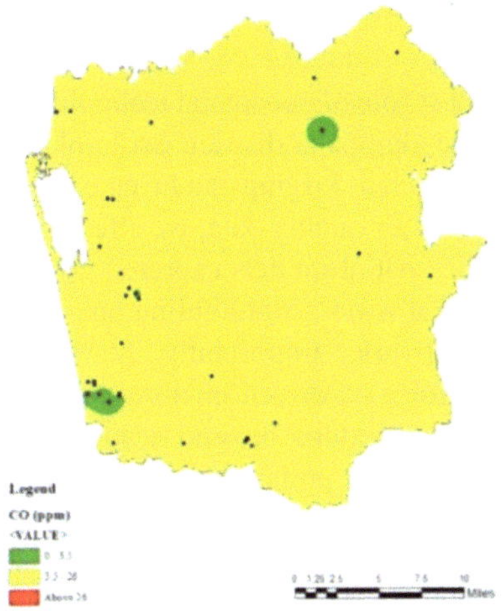

Fig. 4. Carbon Monoxide (CO) distribution through Gampaha District

Ozone (O_3): Unlike primary pollutants directly emitted from sources, ground-level ozone is a secondary pollutant formed through photochemical reactions involving nitrogen oxides (NO_X) and volatile organic compounds (VOCs) under sunlight. In Gampaha district, clear weather conditions, especially during the dry, sun concentrated months, ozone levels rise. O_3 levels rarely breached critical thresholds during the study. Their presence served as an indirect image of combustion footprint within the district's evolving industrial landscape. Figure 5 indicates the O_3 distribution throughout the district, narrating a safe air quality with respect to O_3.

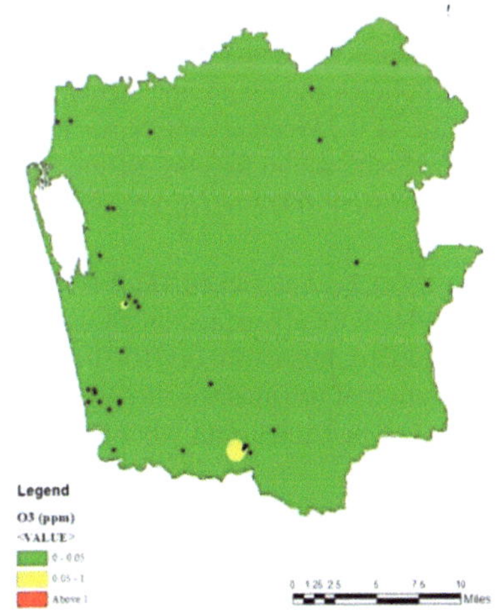

Fig. 5. Ozone (O_3) distribution through Gampaha District

5 Conclusions

This study highlights the robust role of geospatial technologies in interpreting com-plex air quality patterns across the Gampaha district, Sri Lanka. By connecting GIS-based spatial interpolation methods such as kriging and inverse distance weighting (IDW), and integrating them with measured air quality data, we effectively mapped PM10 and other pollutant concentrations. The spatial models captured critical hotspots, particularly in urban landscapes and industrial zones, highlighting areas that often go undetected by limited ground monitoring networks. Importantly, the research contributes to a growing recognition of the need for data-driven environ-mental governance. The insights gained from our spatial analysis extend beyond academic and industrial monitoring. This research is an essential tool for policy-makers, urban planners, and public health stakeholders. With the air pollution increasingly linked to respiratory and cardiovascular diseases, particularly among vulnerable populations, localized data is imperative for making decisions. Our findings can inform urban zoning strategies and public awareness campaigns by improving both environmental quality and health outcomes. Moreover, this study opens pathways for future interdisciplinary research. The integration of satellite-derived remote sensing data, measured and available data, and socio-economic indicators such as population density, housing quality, and access to healthcare would deepen our understanding of exposure gaps and resilience across communities. In conclusion, this work confirms that spatial analysis, when paired with robust air quality monitoring systems, is not just a technical exercise; it is a vital instrument for environmental justice and sustainable development. As Sri Lanka continues to urbanize, this kind of research will shape a cleaner and healthier future.

References

1. Department of Census and Statistics, Gampaha District: Grana Niladhari Divisions Statistics – 2020, Ministry of Economic Policies and Plan Implementation
2. Rathnayake, L.R.S.D., Sakura, G.B., Weerasekara, N.A.: Review of Outdoor Air Pollution in Sri Lanka Compared to the South Asian Region. Nature Environ. Pollut. Technol. Int. Q. Sci. J. **22**(2), 609–621 (2023). ISSN: 0972-6268, https://doi.org/10.46488/NEPT.2023.v22i02.006
3. Badach, J., Voordeckers, D., Nyka, L., Van Acker, M.: A framework for Air Quality Management Zones - Useful GIS-based tool for urban planning: Case studies in Antwerp and Gdansk. Build. Environ. **174**, 106743 (2020). https://doi.org/10.1016/j.buildenv.2020.106743
4. Samad, A., Vogt, U.: Investigation of urban air quality by performing mobile measurements using a bicycle (MOBAIR). Urban Climate **33**, 100650 (2020). https://doi.org/10.1016/j.uclim.2020.100650
5. Ruzaik, F.: Air pollution caused devastating impacts on children's health in urban area of Sri Lanka. Int. J. Sci. Res. Pub. **10**(12), 409–421 (2020). https://doi.org/10.29322/IJSRP.10.12.2020.p10845
6. Thishan Dharshana, K.G., Coowanitwong, N.: Ambient PM10 and respiratory illnesses in Colombo City, Sri Lanka. J. Environ. Sci. Health Part A **43**, 1064–1070 (2008). https://doi.org/10.1080/10934520802060035
7. Dhammapala, R., Basnayake, A., Premasiri, S., Chathuranga, L., Mera, K.: PM2.5 in Sri Lanka: Trend Analysis, Low-cost Sensor Correlations and Spatial Distribution, Aerosol and

Air Quality Research. Taiwan Association for Aerosol Research **22**(5), 2020. ISSN: 1680-8584, https://doi.org/10.4209/aaqr.210266

8. Nandasena, S.: Air Pollution exposure levels to pedestrians and commuters in different locations in city of Colombo. National Institute of Health Sciences, Seventh National Symposium on Air Quality Management in Sri Lanka August 16-17 (2018)
9. Kumar, A., Gupta, I., Brandt, J., Kumar, R., Dikshit, A.K., Patil, R.S.: Air quality mapping using GIS and economic evaluation of health impact for Mumbai City. India Journal of the Air & Waste Management Association **66**(5), 470–481 (2016). https://doi.org/10.1080/10962247.2016.1143887
10. Liyanage, P.L.S., et al.: Investigation of air Pollution in Sri Lanka using satellite observations. SAITM Research Symposium on Engineering Advancements. Colombo, Sri Lanka (2014)
11. Pandeya, M., Singha, V., Vaishyaa, R.C., Shuklab, A.K.: Analysis & Application of GIS Based Air Quality Monitoring- State of Art. Int. J. Eng. Res. Technol. (IJERT) **2**(12), 3788–3796 (2013). ISSN: 2278-0181
12. Kingsley, I.E., Oghenerume, A.C., Jonah, O.T., Ejike, U.R.: Application of GIS Technique for Mapping of Day Time Air Quality Index in Nnamdi Azikiwe University, Awka Campus. Int. J. Res. (IJR) Nigeria **11**(2) (2024). e-ISSN: 2348-6848 p-ISSN: 2348-795X
13. Jovanovska, E.M., Batz, V., Lameski, P., Zdravevski, E., Herzog, M.A., Trajkovik, V.: Methods for Urban Air Pollution Measurement and Forecasting: Challenges, Opportunities and Solutions, Atmosphere **14**(9), 1441 (2023). https://doi.org/10.3390/atmos14091441
14. Chavan, A.J., Daspute, A.B.: Assessment of Ambient Air Quality Status by using GIS Technology. J. Emerg. Technol. Innov. Res. (JETIR) **9**(9), f229–f236 (2022). ISSN-2349-5162
15. Santhiyago, P.: Air Quality Management in Sri Lanka, Air Resource Management Unit, Central Environmental Authority
16. Safarov, R., et al.: Spatial Analysis of Air Pollutants in an Industrial City Using GIS-Based Techniques: A case Study of Pavlodar, Kazakhastan, Sustainability (2024)
17. Tikader, M., Mukhopadhyay, D., Dabhadker, K.: Remote Sensing and GIS, Air Pollution Mitigation: A Bibliometric Review of Chhattisgarh, India, Int. J. Environ. Climate Change **14**(4), 174–193 (2024). https://doi.org/10.9734/IJECC/2024/v14i44106
18. The Gazette of the Democratic Socialist Republic of Sri Lanka No.1562/22 – FRIDAY (2008)
19. World Health Organization: https://iris.who.int/bitstream/handle/10665/345334/9789240034433-eng.pdf. Last accessed on 25 June 2025

Author Index

A
Abeysinghe, A. M. M. H. B. 47
Abeyweera, P. S. 3
Abeyweera, S. K. 3
Adhipaththu, W. A. M. P. L. 47
Amarasinghe, A. D. U. S. 300
Anuruddha, T. A. S. 36
Ariyarathna, S. M. W. T. P. K. 225
Athuraliya, J. P. D. S. 121
Atthanayake, I. U. 250, 311
Attygalle, D. 161

B
Bandara, J. M. S. J. 235

C
Chandrasekara, Vihanga 196
Charuni, W. K. P. G. N. 131

D
de Alwis, Ajith 290
De Silva, K. T. K. M. 131, 140
Dewanthi, J. M. H. B. 140
Dharmawickrama, S. P. H. E. 225

F
Fernando, A. J. 186
Fernando, K. J. P. 73
Fernando, P. K. A. P. 208
Fernando, T. N. 186
Fernando, W. P. D. 280

G
Gallage, R. 332
Gamage, R. R. 150
Gangadevi, R. 113
Gokul, B. C. Mohana 113

J
Jayasekara, K. P. D. D. 341
Jayasinghe, S. P. R. 121
Jayaweera, W. M. I. S. 161

K
Kaluarachchi, Lalani 290
Kannangara, C. H. 280
Karunanayake, M. S. H. 186
Karunasena, H. C. P. 150
Kawmudi, W. N. 250
Kulatunga, S. L. 73
Kumara, P. N. M. 73
Kumara, U. G. A. S. 140
Kuok, K. K. 264

L
Lakmal, H. K. I. S. 14, 26, 73, 103
Lakshmikanthan, Chitra 171
Liyanage, H. I. D. 341

M
Maduranga, M. W. P. 73
Mahanama, K. R. R. 311
Mahela, H. K. A. 300
Marasinghe, S. D. 208

N
Narayana, M. 300
Navodya, W. D. 26
Nikhil, A. 113
Nirmal, W. C. 73
Nissanka, Indrajith D. 290

P
Pabasara, W. G. A. 322
Pathiraja, P. W. A. W. 140
Pathmasiri, T. K. K. S. 140
Peiris, S. T. K. 14
Peiris, T. Chandana 121

M. Siriwardhana et al. (Eds.): IERS 2025, PTT, pp. 353–354, 2026.
https://doi.org/10.1007/978-981-95-7984-6

Pramuditha, N. P. A. K. G. 186

R
Raja, S. Senthil 113
Rajakaruna, Thumula 103
Ranasinghe, N. Y. 47
Rankothge, R. G. P. D. 150
Rathnapriya, K. G. S. M. 150
Rathnayaka, R. M. H. K. 208
Rathnayake, L. N. 121
Ratnasinghe, P. A. 264

S
Samaranayake, G. D. K. 36
Samararathne, U. M. 250
Samarawickrama, J. G. 36
Sameera, U. V. H. 47
Sampath, W. A. D. 150
Sarmilan, S. 300
Senanayake, S. M. V. P. D. 186
Senarath, J. B. 280
Senarathna, S. M. S. S. 36
Senevirathne, I. H. D. 150
Sewvandi, G. A. 322
Shanuka, Himasha 58
Silva, P. T. S. 208
Silva, S. A. P. S. 300
Siriwardana, Geethal 89
Siriwardhana, K. A. C. G. 311
Snehitand, Kalluru 113
Somasundara, D. H. G. S. R. 311
Srimal, L. K. T. 332
Subasinghe, S. M. U. S. T. J. A. 225
Sudarshana, W. S. A. 332
Sugathapala, A. G. T. 235

T
Tharaka, A. L. K. 140
Thilakarathne, B. L. Sanjaya 47
Thilakarathne, Roshan 58, 196
Thilakshan, T. 235

V
Venkatesan, Rattan Kumar 171

W
Weerakkody, W. A. D. D. Ransika 89
Weerasinghe, Y. S. P. 36
Wimalarathna, D. R. S. K. 36

MIX
Papier aus verantwortungsvollen Quellen
Paper from responsible sources
FSC® C105338

If you have any concerns about our products,
you can contact us on
ProductSafety@springernature.com

In case Publisher is established outside the EU,
the EU authorized representative is:
Springer Nature Customer Service Center GmbH
Europaplatz 3, 69115 Heidelberg, Germany

Printed by Libri Plureos GmbH
in Hamburg, Germany